2D Materials: Chemistry and Applications

(Part 2)

Edited by

Vinay Deep Punetha

*Centre of Excellence for Research, P.P. Savani University,
Surat-394125, Gujarat, India*

2D Materials: Chemistry and Applications (Part 2)

Editor: Vinay Deep Punetha

ISBN (Online): 978-981-5305-24-1

ISBN (Print): 978-981-5305-25-8

ISBN (Paperback): 978-981-5305-26-5

General:

1. Any dispute or claim arising out of or in connection with this License Agreement or the Work (including non-contractual disputes or claims) will be governed by and construed in accordance with the laws of the U.A.E. as applied in the Emirate of Dubai. Each party agrees that the courts of the Emirate of Dubai shall have exclusive jurisdiction to settle any dispute or claim arising out of or in connection with this License Agreement or the Work (including non-contractual disputes or claims).
2. Your rights under this License Agreement will automatically terminate without notice and without the need for a court order if at any point you breach any terms of this License Agreement. In no event will any delay or failure by Bentham Science Publishers in enforcing your compliance with this License Agreement constitute a waiver of any of its rights.
3. You acknowledge that you have read this License Agreement, and agree to be bound by its terms and conditions. To the extent that any other terms and conditions presented on any website of Bentham Science Publishers conflict with, or are inconsistent with, the terms and conditions set out in this License Agreement, you acknowledge that the terms and conditions set out in this License Agreement shall prevail.

Bentham Science Publishers Pte. Ltd.
80 Robinson Road #02-00
Singapore 068898
Singapore
Email: subscriptions@benthamscience.net

BENTHAM SCIENCE

CONTENTS

PREFACE

This second part of the book, "2D Materials: Chemistry and Applications," aims to provide comprehensive coverage of recent developments in various other 2D materials, extending its discussion beyond graphene to 2D Boron Nitride, Germanene, Silicene, Stanene, Mxene, and Transition Metal Chalcogenides. Recognizing the importance of these materials, detailed discussions have been conducted on topics such as their fundamental structure, surface chemistry, physiochemical, and optoelectronic properties in the backdrop of their applications in diverse fields ranging from electronics and energy storage to advanced composites and catalysis.

The first chapter of the second part of this book discusses one of the most promising fields where graphene can revolutionize the current state of affairs pertaining to advanced energy storage systems. The chapter examines the structural properties of graphene that facilitate enhanced capacitance and quick ion exchange and explores how rapid charging and discharging cycles lead to higher power densities. The book further explores the critical evaluation of the toxicity associated with the graphene family of materials and current strategies for remediation. The exploration of the toxicity of graphene materials and strategies for remediation is crucial as it addresses public health concerns and environmental safety, ensuring the responsible development and application of these advanced materials.

The discussion on graphene is summarised with sustainable and cost-effective practices and recent significant advancements in its production. This chapter highlights how cost-effective production of graphene remains elusive and the environmental concerns associated with its manufacturing processes that continue to pose significant challenges. It elaborates on the need for cost-effective production as it ensures more economically viable products for both manufacturers and consumers. This is particularly important in scaling up new technologies like graphene, where lowering costs can lead to broader adoption and integration into various industries.

Expanding the scope of this exploration, subsequent discussions introduce and elaborate on other members of the 2D material family, such as boron nitride and germanene. These materials are discussed in terms of their synthesis, functionalization, and wide-ranging applications, from energy systems to electronic devices. Moreover, a concise discussion on the relatively newer 2D materials such as Silicene, Stanene, Mxenes, and Transition Metal Chalcogenides enriches the discourse with the inclusion of an in-depth discussion on their unique electronic properties that make them promising candidates for applications in Nano electronics and photonics.

The comprehensive coverage of two-dimensional materials beyond graphene makes it an essential guide for researchers, engineers, and policymakers involved in the development of next-generation technologies and for ensuring the responsible integration of these materials into future products and applications.

Vinay Deep Punetha
Centre of Excellence for Research
P.P. Savani University, Surat-394125
Gujarat, India

List of Contributors

Abbas Zaarifi Department of Physics, Yasouj University, Yasouj, 75918-74934, Iran

Anton Kuzmin Scientific Laboratory "Advanced Composite Materials and Technologies,", Plekhanov Russian University of Economics, Stremyanny Ln, 36, 117997, Moscow, Russia
Department of Mechanization of Agricultural Products Processing, National Research Mordovia State, University, Bolshevistskaya st., 68, 430005, Saransk, Republic of Mordovia, Russia

Cinzia Casiraghi School of Chemistry, Manchester University, Manchester M13 9PL, United Kingdom

Ghayas Uddin Siddiqui Department of Chemical Engineering, Jeju National University, 63243, Jeju, South Korea

Gaurav Nath Department of Materials and Geosciences, Technische Universität Darmstadt, Darmstadt, Germany

Golnaz Taghavi Pourian Azar The Functional Materials and Chemistry Research Group, Research Centre for Manufacturing and Materials (CMM), Coventry University, Coventry, CV1 5FB, United Kingdom

Ghiasi Limanjoobi Seyedeh Hanieh Centre for Manufacturing and Materials (CMM) Coventry University, CV1 5FB, Coventry, United Kingdom

Isha Kumari Department of Chemistry, Bhagini Nivedita College, University of Delhi, New Delhi, India

Junaid Ali OptoElectronics Research Laboratory(OERL), Department of Physics, COMSATS University Islamabad, Pakistan

Khaled Pervez School of Chemistry, Manchester University, Manchester M13 9PL, United Kingdom

Neha Faridi Defence Institute of Bio-Energy Research, DRDO, Haldwani, Uttarakhand-263139, India

Mamoona Hayat OptoElectronics Research Laboratory(OERL), Department of Physics, COMSATS University Islamabad, Pakistan
Department of Chemistry, COMSATS University Islamabad, Pakistan

Mayank Punetha Centre of Excellence for Research, P.P. Savani University, Surat-394125, India

Pawan Singh Dhapola Center for Solar Cells and Renewable Energy, Department of Physics, Sharda University, Greater Noida, India

Pramod K. Singh Center for Solar Cells and Renewable Energy, Department of Physics, Sharda University, Greater Noida, India

Poulomi Sengupta1 Department of Chemistry, School of Science, Indrashil University, Rajpur, District Mehsana, Gujarat, India

Rajesh Kumar Department of Chemistry, S.S.J. University, Campus Almora-263601, Uttarakhand, India

Rakshit Pathak P.P Savani University, Surat, India

Sushant Kumar Center for Solar Cells and Renewable Energy, Department of Physics, Sharda University, Greater Noida, India

Siddhant B. Patel Chemical and Biochemical Engineering, School of Engineering, Indrashil University, Rajpur, District Mehsana, Gujarat, India

Saira Arif Department of Chemistry, COMSATS University Islamabad, Pakistan

Sadafara Pillai Departmet of Chemistry, School of Science, P.P. Savani University, Surat-394125, Gujarat, India

Shalini Bhatt Centre of Excellence for Research, P.P. Savani University, Surat-394125, Gujarat, India

Vinay Deep Punetha Centre of Excellence for Research, P.P. Savani University, Surat-394125, Gujarat, India

CHAPTER 1

Advanced Graphene-Based Supercapacitors for Energy Storage Applications

Isha Kumari[1,†], Sushant Kumar[2,†], Pawan Singh Dhapola[2,*] and Pramod K. Singh[2,*]

[1] *Department of Chemistry, Bhagini Nivedita College, University of Delhi, New Delhi, 110043, India*

[2] *Center for Solar Cells and Renewable Energy, Department of Physics, Sharda University, Greater Noida, India*

Abstract: Graphene-based supercapacitors (SC) are rising as the most efficient and smart energy storage systems. Nonpareil physiochemical properties of graphene offer immense potential for their use in developing next-generation energy storage and portable devices. Since the rise of graphene, this material has been seen as the best alternative to activated carbon in SC applications. Being a 2D material, its high surface area enables it to store electrostatic charge even after high cycling. Since the first graphene-based SC was fabricated in 2008, this material has been explored beyond the boundaries of pristine graphene. The recent invention paved the way for ultrafast charging devices with excellent efficiency. However, the widespread use of these devices in daily life seems far-fetched, but recent results in graphene-based architectures are fetching these possibilities to life. In the last decade, various revamped and manipulated graphene derivatives have also been investigated and found to have great potential in SC applications. These derivatives have shown tremendous specific capacitance with enhanced cyclability. Graphene derivatives can even exhibit capacitance retention of almost 100% after 20,000 cycles. This book chapter discusses the current state of affairs in various graphene-based SC devices, such as crumpled graphene, graphene-metal oxide composites, graphene-based aerogels, graphene nanoparticle systems, graphene-based fibers, graphene/carbon-based hybrid composites for their potential application in the fabrication of efficient energy devices. This comprehensive study aims to analyze current trends and the opportunities and challenges offered by graphene and its derivatives in the development of next-generation SCs.

Keywords: DSSC, EDLC, Graphene, Supercapacitor, Solar-cell, Storage.

---

[*] **Corresponding authors Pawan Singh Dhapola and Pramod K. Singh:** Center for Solar Cells and Renewable Energy, Department of Physics, Sharda University, Greater Noida, India; E-mails:pramodkumar.singh@sharda.ac.in and pawan.dhapola@sharda.ac.in
[†] These authors contributed equally to this work.

Vinay Deep Punetha (Ed.)

INTRODUCTION

Graphene has revolutionized the field of supercapacitors and graphene-based supercapacitors are believed to revamp the industry within five to ten years [1]. Moreover, the extensive research on the use of graphene derivatives in SC fabrication has opened the possibility of eventually utilizing these architectures in many different applications [2]. Various methods to synthesize graphene for SCs are explored by the research community, which include widespread epitaxial growth (EG) and chemical vapor deposition (CVD) method, mechanical/chemical exfoliation of sp^2 hybridized graphite bulks, synthesis in a microwave plasma reactor, arc-discharge and chemical reduction of GO [3 - 5]. A complete discussion on the methods of synthesis of graphene is beyond the scope of this chapter.

The conception of the 'idea of SC' originates from the poor performance of batteries, especially in their time to charge. However, batteries are well-celebrated devices for their high energy density [6]. Capacitors, on the other hand, exhibit ultrafast charging; unfortunately, their capacity to store charges is limited. In addition, capacitors can store energy only for a short time. An SC is a mesmerizing device that contains the best things of both batteries and capacitors. It shows ultrafast charging and the tendency to retain more charges with a more extended period of charge retention [7]. The first categorization of SCs comes from the mode of functioning of its components and architecture.

The SCs are categorized into the following three types, viz. electrical double-layer capacitors (EDLCs), pseudocapacitors (PCs), and hybrid SCs (HSCs) [8]. The EDLCs are superior high-power density devices with capacitances of several thousand farads. The high power density of these materials is due to the energy-fast adsorption and desorption also known as electrosorption, which generally occur on porous electrodes [9 - 12]. The fundamental architecture of EDLCs is very similar to that of lithium-ion batteries with two electrodes, an electrolyte, and a separator. Mostly porous materials are encouraged for the electrosorption of charged ions as electrodes, while electrolytes act as an active source of charged particles. On the other hand, PCs differ from EDLCs in the architecture of electrodes, which plays a crucial role in the mechanical differences in the energy storage process. In PCs, the electrode consists of active materials that exhibit simultaneous oxidation and reduction [13]. Energy storage is achieved *via* multiple methods, such as electrosorption, redox reactions, and intercalation of charges. External potential induces rapid and reversible redox changes on the electrode, and this facilitates fast migrations of charges between the electrode and electrolyte. The HSCs are fabricated using one electrode for each of the EDLCs and PC to assemble the best properties of both of them into one. The hybrid storage system exhibits high power of EDLCs and high energy of PCs [14]. Fig. (**1**) illustrates the architectural differences in EDLCs, pseudocapacitors, and HSCs.

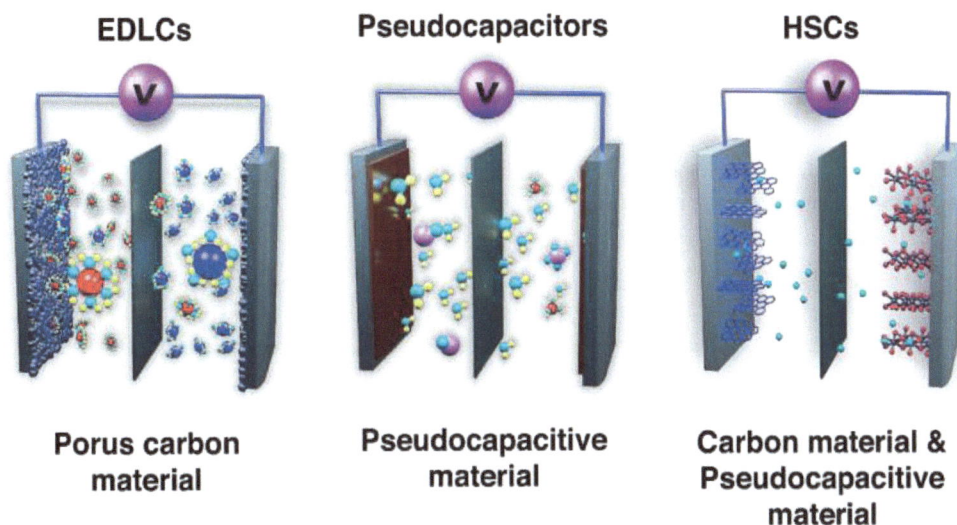

Fig. (1). Illustration showing mechanistic differences in EDLSs, pseudocapacitors and HSCs.

The stature of graphene in developing next-generation SCs is on the top owing to its excellent properties, such as the unique consistency of atoms in creating a honeycomb lattice with sp^2 hybridized carbon atoms [15]. In addition, properties such as electrical conductivity, high specific surface area, excellent mechanical performance, and very high theoretical capacitance bolster its claim to deliver the best available materials. The 2D framework of graphene and its derivatives allows rapid charge migration along the 2D plane and reduces the struggle in ionic diffusion. Though pristine graphene offers some challenges in fabricating electrodes, the extended pi-cloud prevents superior dispersion by stimulating π- π interactions between the layers [16]. These inter-layer interactions affect the specific gravimetric capacitance adversely. Various graphene-based derivatives are used to manage the reduction in specific gravimetric capacitance, which includes employing porous graphene sheets, 3D graphene architectures, graphene aerogels, and many more. The 3D graphene derivatives show pores of 2.1 to 3.4 nm and high SSA (295 m²/g). It was observed that interconnected pores significantly reduce the diffusion distances from the ion sources to the electrodes. In addition, the stacked graphene sheets greatly enhance charge storage in the SCs [4, 7, 13]. In addition, the strategy to dope graphene surfaces with suitable functionalities has shown promising results in enhancing the performance of the electrodes. In addition, defective and wrinkled graphene sheets have also been synthesized and used to develop efficient electrodes. Another major challenge in using pristine graphene comes from the low packing density. In pristine structures with a high surface area, very low volumetric capacitance is observed along with low energy density due to their poor packing density. Graphene has also been used as an assisting material in bringing out the best from the other materials that can be used to fabricate the electrodes. The primary Pseudo-capacitive materials, such as oxides and hydroxides of 'd' block elements and various conducting polymers, suffer from some limitations or drawbacks, which can be rectified using graphene. One of the significant drawbacks of these materials is inferior electrical conductivity and incident volume change. These disadvantages prevent the fabrication of efficient electrodes by reducing the power density and desired sustenance after repeated cycles [15]. The problems described above can be addressed using graphene-based composite materials. The reinforced graphene in composite materials offers conducting networks and facilitates the redox reactions of 'd' block oxides/hydroxides and conducting polymers [16, 17]. In addition, the layers of graphene assist in the superior dispersion of 'd' block oxide/hydroxide and their nanoparticles and play the role of the conductive matrix. The conductive matrix enhances conduction and ameliorates electrode performance [18 - 20]. This book chapter thoroughly discusses the significant recent development of graphene-based materials and their electrochemical performance. The review will offer a thorough comprehension of the importance attributed to materials associated with graphene.

CRUMPLED GRAPHENE FOR SUPERCAPACITORS

Regarding supercapacitor performance parameters, graphene has been deemed the optimal choice for electrode material due to its theoretically predicted surface area. However, the presence of irreversible stacking interactions among individual graphene sheets leads to a reduction in specific surface area compared to its theoretical values [21]. 2-D graphene sheets are converted into the 3-D crumpled graphene structure to resolve this limitation. It is a new carbon nanostructure drawing notice because of its 3-D open system and its permanence in an aqueous solution [22]. The mechanical properties of graphene, *i.e.*, number of layers, Young's modulus, interfacial energy, *etc.*, govern its deformation, and the deformation in shape enhances some properties of it, *i.e.*, transmittance, chemical potential, wettability, conductivity and expansion for energy storage [23]. Theoretically, different strains, pressure differences, uniaxial tension, and temperature are responsible for corrugations. Experimentally, the deformations are because of varying reduction techniques, changes in temperature, fast evaporation of aerosol elements (containing graphene), and allocating 2-dimensional graphene on a required substrate. Different corrugations' formation depends on the amount of strain and the synthesis technique used. The reasons responsible for the appearance of wrinkles are defects [24, 25], functional groups [26], and substrate-induced corrugation, while ripples are the intrinsic tendency of graphene, which is owed to thermal fluctuations. Crumpled graphene shows distinctive characteristics like elevated conductivity, high specific surface area, and permanence towards graphitization which gives it a superior capacitance and therefore locates an application in energy storage devices, specially supercapacitors [27].

Any material must have a high surface area to use as an electrode material in supercapacitors and achieve large specific capacitance because a high surface area means high accommodation of charges and ions. By inserting inorganic/ organic interlayer pillars or spacers between sheets or utilizing functional groups of chemical stitching such as Boron, Sulphur, and Nitrogen, the surface area and physical/ chemical characteristics of graphene can be modified [28, 29]. To support it, Wang *et al.* utilized a hydrothermal procedure to insert a carbon nanotube in the graphene sheets as an interlayer spacer. He takes different ratios of CNT and graphene for the same. He gets the highest value of capacitance, 318 F/g, at the ratio of 1:1. This value is consistent with the theoretical value of crumpled graphene [28]. Also, Tang *et al.* (2012) chemically stitched methacrylate groups for the structural corrugations and increment of the spacing interlayer to the graphene sheets, which tends to take almost the rectangular shape of the CV graph, showing good super capacitive performance [29]. Li *et al.* (2018) optimized the content of CNT to use it to increase the inter-layer spacing of graphene by introducing it into the graphene sheets. They obtained 206 F/g specific capacitance in EMIMBF4 (1-ethyl-3-methylimidazolium tetrafuoroborate) electrolyte [30]. Yu Y *et al.* (2014) used diethylene glycol as an interlayer spacer for the increment in the surface area of the rGO. They obtained a specific capacitance value of 237 F/g with 2000 cyclic stability [31]. Some of the significant results have been summarised in Table **1**.

Table 1. Different crumpling processes used in recent studies and their specific capacitance concerning the surface area.

S/no.	Crumpling method	Surface area (m2/g)	Specific Capacitance (F/g)	Ref.
1.	Dopamine-assisted nitrogen doping of graphene.	401	128	[32]
2.	Hydrazine-mediated aerosol spray pyrolysis for water reduction.	255	150	[33]
3.	Utilization of freezing technique.	433	212	[34]
4.	Thermal expansion-induced nitrogen-doped graphene formation.	237	270	[35]

(Table 1) cont.....

S/no.	Crumpling method	Surface area (m2/g)	Specific Capacitance (F/g)	Ref.
5.	Carbonization of 1,10-Phenanthroline monohydrochloride monohydrate (Phen) and Melamine (MA) for graphene synthesis.	1150	284	[36]
6.	In situ ZnO implantation method.	253	300	[37]
7.	Thermal exfoliation for the synthesis of Ni2P/GS composite.	722	2240	[38]

GRAPHENE-METAL OXIDE COMPOSITES

On the mechanism of charge storing phenomenon, three types of supercapacitors are there, starting with electrochemical double layer capacitor (EDLC), pseudocapacitor, and hybrid supercapacitor. From them, the pseudocapacitors mechanism is based on the pseudo reaction in electrolytes, and the efficiency of the cell is due to the redox reaction in electrolytes. This storage is done by reduction-oxidation reactions, electrosorption, and intercalation processes known as pseudocapacitance [39, 40]. The most common materials for electrodes used in pseudocapacitors are metal oxides/transition metals and conducting polymers which make pseudocapacitors to get more energy densities than electrochemical double-layer supercapacitors [40]. But in terms of cycle life and power density, pseudocapacitors are inferior to EDLCs [41]. And to enrich these properties of pseudocapacitors, graphene can be a good candidate [41]. Therefore metal/transition metal oxides can be decorated with graphene oxide, where graphene acts as a sustained matrix to build up electroactive species in the nano dimension, which tends to a greater surface area and modified electrochemical performances [42].In metal oxide-graphene composites, graphene can act as a functional component for restraining metal oxide or as s substrate. Hence these composites will raise their storage reaction in the field of energy conversions [43, 44]. There are different methods used to decorate graphene with metal oxide, *i.e.,* atomic layer deposition method, sol-gel method, solution mixing method, solvothermal method, self-assembly method, microwave method, electrochemical deposition method, reduction method, encapsulation method, layer-to-layer assembly method, thermal production method, co-precipitation procedure, photochemical synthesis, *etc.* Recently many studies have been performed on graphene composites with different metal oxides.

Beka *et al.* [45] produced a structure of nickel-cobalt sulfide in a 3D core-shell by using hydrothermal steps on chemical vapor deposition to grow graphene for supercapacitor applications where $NiCo_2S_4$ acts as a core and carbon nanosheets act as a shell. Excellent capacitive performance is shown by the structure of the core/shell with the support of graphene having a high surface area. The electrode using this graphene/NCS/CNS composites exhibits outstanding cycling stability of 93% with 5000 cycles and very high areal capacitance of 15.6 F/cm^2 at a $10mA/cm^2$ density of current.

Zhai *et al.* [46] synthesized manganese dioxide-decorated graphene nanoparticles using electrostatic adsorption. Graphene particles in a water medium possess negative charges, but manganese dioxide needs positive charges for adsorption. They used a micro-emulsion process to eliminate this problem by integrating manganese dioxide in hexadecyltrimethylammonium bromide. Using these two novel approaches, the large aromatic molecules integrated graphene sheets possess specific capacitance hiked by 40% and 250%, respectively.

Hassan *et al.* [47] developed a one-pot synthesis method to synthesize a ruthenium-based hybrid composite. This synthesis includes the formation of rGO from graphene oxide and $Ru^{3+}(RuCl_3)$ in Ru nanoparticles using a single-step method without a reductant. The prepared composite's supercapacitive properties are studied with 1 M of $NaNo_3$ neutral electrolyte in a three-electrode setup. And it gives a specific capacitance of 270 F/g with energy and power density of 15Wh/kg and 76.4 kW/Kg, respectively, and cyclic stability over 5000 cycles at 24A/g.

Some more metal oxide/graphene composites synthesized by other researchers are given in Table. **2**, containing their different super-capacitive properties.

Table 2. Studies on metal oxide/graphene composites with their super capacitive performances.

S. No.	Metal Oxide/Graphene Composites.	Specific Capacitance (F/g)	Power Density (kW/kg)	Energy Density (Wh/Kg)
1.	MnO_2/Gr [48]	30	5	30.4
2.	$Ni_3(PO_4)_2$/Gr [49]	125	0.5	49.2
3.	Co-Ni/Gr [50]	156	2.5	23.9
4.	Au/Gr [51]	174	0.2	29.4
5.	Ru/Gr [52]	270	76.4	15
6.	Ag/Gr [53]	472	1.5	41

GRAPHENE-BASED AEROGELS

Graphene-Based Aerogels' restacking issue limits its extraordinary properties as an electrode material. This re-stacking happens because of π-π interactions and Van der Waals forces among graphene layers which the formation of graphene aerogels can solve, and the layer can be chemically bonded together [54]. Like solid, liquid, gas, and plasma, aerogels can be considered a new state of matter. It can be defined as the lightest materials with large surface areas, high porosity, and low density. Recently aerogels have brought attention to their applications in supercapacitors as an electrode material. They possess good physicochemical and mechanical properties, tunable surface areas, and pore sizes. As the characteristics mentioned above of hybrid aerogels can be tuned, it is believed that innovative ranges of aerogels can be produced in the future.

Aerogels can be classified into two classes which are single-component and composite aerogels. Single-component aerogels consist of organic, oxide, carbon, and chalcogenide, while hybrid aerogels are further classified into micro, nano, multi-component, and gradient aerogels [55]. Because of the 3D structures of graphene aerogels, it has an elevated possibility, especially in energy storage devices technology. Some latest studies on applying graphene aerogels in supercapacitors are given below—Chen, T.T. *et al.* [56] synthesized a 3D graphene aerogel using a two-step hydrothermal method. The obtained graphene aerogel was measured in three electrode systems with 6M KOH as an electrolyte which shows the highest capacity of 410 F/g at 0.1 A/g. Moreover, a solid-state supercapacitor is also fabricated, showing no capacitance loss after 5000 cycles of cycling at a current density of 5A/g.

Song, Z. *et al.* [57] uses hydrothermal process to synthesize a composite of Fe_2O_3 and 3D graphene aerogel (Fe_2O_3/GA). Fe_2O_3 particles were nut shelled homogenously in graphene aerogel for the same. The electrochemical performances of the prepared composite were examined using different techniques, *i.e.*, Cyclic voltammetry, Galvanostatic charge-discharge (GCD) with a three-electrode system in an electrolyte 0.5 M Na_2SO_4 environment. The specific capacitance shows 81.3 F/g at 1A/g current density, in the potential operating range of -0.8V to 0.8V. Liu, Y. *et al.* [58] synthesized triple composites by combining rod-like $MnCO_3$ and MnO_2 hybrid nanostructure to particle form by graphene oxide and then fabricated aerogel-based asymmetric supercapacitors. The prepared triple composite MnO_2/$MnCO_3$/rGOaerogels possess high mechanical strength and electrical conductivity and showed an energy density of 17.8 Wh/kg at 400 W/kg power density in a potential range of 0-1.6V. Aken, K.L.V., *et al.* [59] obtained single-wall carbon nanotube aerogels by drying them at the critical point. The electronic and ionic conductivity of prepared aerogels is improved and shows good electrochemical performances with high charge-discharge stability over 10000 cycles. Other than the above-mentioned studies, some other studies on aerogels with different materials and methods are provided in Table **3**.

Table 3. Studies on graphene aerogels with different materials and methods.

S. No.	Material and Method	Specific Capacitance	Cyclic Retention	Refs.
1.	MoS2/graphene using in-situ thermal decomposition-reduction method [60].	862.5 mAh/g at 0.1 A/g	109.6%	[60]
2.	Graphene/polyaniline using Hydrothermal [61].	520.3 F/g at 1A/g	----	[61]
3.	N-doped holey graphene using Two-step hydrothermal treatment [62].	318.3 F/g at 0.5 A/g	98.4%	[62]
4.	Boron-doped graphene using the Hydrothermal method [63]B	308.5 F/g at 1 A/g	92%	[63]
5.	Glucose/graphene using hydrothermal reduction and CO_2 activation method [64].	305.5 at 1 A/g	98.5%	[64]
6.	MoS2/chemically using modified graphene Hydrothermal route [65].	268 F/g at 1 A/g	93%	[65]

Along with the advantages mentioned above of graphene aerogels, it has some disadvantages, *i.e.*, costly production, time-consuming synthesis process, and fragile and brittle nature. They still need some improvement to be used in today's world. But for high energy storage, we definitely need aerogels as they have a very high surface area, even more, significant than graphene oxide.

GRAPHENE/CARBON BASED HYBRID COMPOSITES

Combining graphene with conducting polymers has allowed supercapacitor electrodes to be made without binder, improving conductivity and eliminating the step needed to form the electrodes separately. In pseudocapacitor applications, Polypyrrole (PPy), polythiophene (PT), and Polyaniline (PANI) are more frequently used conducting polymers. It is possible to increase the energy storage capacity of nanocomposites through in-situ or dispersion polymerization, which provides films that can be directly used as electrodes for supercapacitors, eliminating the need for additional steps. Another technique for creating composites of conducting polymer and graphene is electro-polymerization. This method allows a polymer matrix to expand using a single polymerization phase more quickly. Different techniques, such as chronoamperometry, cyclic voltammetry, and chronopotentiometry, can achieve electro-polymerization. When electro-polymerization is carried out using a three-electrode setup, counter electrode, and reference electrode are used. Contrasted with G-paper (147 F/g), the high value of specific capacitance (233 F/g) was attained using PANI-flexible GO's composite paper with excellent strength [66].

Using the PANI and GO interfacial interactions, nanofibers of the PANI/GO composite were created, providing a power density of 80 W/Kg and an energy density of 7.1 Wh/Kg. After 1000 cycles, the retention in capacitance was approximately 80.6%, which was significantly greater than that of pristine PANI (25%) and went (73%) [67]. Non-stacked PPy/Go nanocomposites were made with good GO dispersion in the PPy matrix, and at a scan rate of 100 m/s, they displayed an increase in the capacitance of about 92 F/g in comparison to pure PPy [68]. Composites of PPy and reduced GO are produced when pyrrole is photopolymerized in the presence of GO. Following 1000 cycles, retention increased composite conductivity (610 S/m) above pure PPy (0.012 S/m), and a specific capacitance of 376 F/g was noted, with roughly 84% of capacitance [69]. Carbon nanotubes, activated carbon, and carbon dots were employed to make a hybrid composite in addition to graphene. For instance, the interfacial coupling of liquid crystalline graphene oxide led to the combined electrical, electrochemical, and mechanical properties of MWCNT and PEDOT: PSS, which were used as electrodes for energy storage and flexible multifunctional 3D architecture. A specific capacitance of 318 F/g was measured at a scan rate of 5 mV/s and a current density of 1 A/g in an electrolyte of 1 M H_2SO_4 [70]. The Ni-Al layer double hydroxide (1 OH), MWCNT, and rGO sheet composite electrode material for the supercapacitor was produced utilizing a simple one-step ethanol solvothermal technique, provides a specific capacitance value of 1869 F/g at the current density of 1mA/cm2 [71]. Various graphene-based metal oxides can enhance the aggregation and restacking of graphene bands emitted due to the van der Wall interaction. Recent research has demonstrated the viability of using hydrothermal and solvothermal processes to create high-quality graphene metal oxide nanocomposites. The hydrothermal method was used to develop MnO-based composites. According to reports, MnO-based composites favorably affect the supercapacitors' charge-discharge stability, making them suitable for a range of applications [72].

CONCLUSION

Recent findings have unveiled the potential of graphene as a substitute for batteries, by employing supercapacitors in a broad spectrum of applications. Numerous efficient and cost-effective methodologies have been devised for the synthesis of porous nanomaterials. Notably, commercial laser systems have demonstrated rapid and single-step approaches to construct highly porous and conductive three-dimensional networks comprising two-dimensional graphene sheets. Furthermore, advancements in 3D printing and cutting-edge laser cutting tools have expanded the scalability prospects with remarkable precision. Diverse forms of graphene, including crumpled graphene, graphene composites, and aerogels, exhibit remarkable capabilities for the fabrication of supercapacitors with superior energy density. While progressing towards thin-film batteries, enhancing power density remains a challenge in the years to come. Moreover, emerging synergies can be harnessed, as studies indicate the potential benefits of employing laser processing for both graphene materials and pseudocapacitive components. This integrated approach allows, for instance, the introduction of oxygen vacancies in metal oxides or the photochemical transformation of conductive polymers, facilitating their utilization alongside graphene materials produced using novel laser 3D printers. Nevertheless, certain obstacles must be addressed for the widespread adoption of graphene-based supercapacitors. Firstly, large-scale manufacturing of high-quality and uniform porous graphene materials is imperative. Optimization is still necessary in various aspects, such as ensuring intimate contact and adherence of pseudocapacitive components to the graphene network, selecting ideal pseudocapacitive materials, and exploring potential synergies with the electrolyte to establish an optimized voltage range and cycling stability. Overcoming the challenges of graphene sheet agglomeration and restacking is a concern prevalent in all-carbon electric double-layer capacitors, which can be mitigated through intercalation of active nanoparticles within the graphene material. Should these expectations be met, graphene supercapacitors and micro-supercapacitors are poised to emerge as competitive alternatives or complements to conventional lithium-ion batteries and thin film batteries, playing a vital role in future wearable and portable electronic devices.

ACKNOWLEDGEMENT

We would like to acknowledge Ms. Himani Pant for her valuable editorial support in the preparation of this document.

REFERENCE

[1]　Liu, C.; Yu, Z.; Neff, D.; Zhamu, A.; Jang, B.Z. Graphene-based supercapacitor with an ultrahigh energy density. *Nano Lett.,* **2010**, *10*(12), 4863-4868.
[http://dx.doi.org/10.1021/nl102661q] [PMID: 21058713]

[2]　Li, L.; Song, B.; Maurer, L.; Lin, Z.; Lian, G.; Tuan, C.C.; Moon, K.S.; Wong, C.P. Molecular engineering of aromatic amine spacers for high-performance graphene-based supercapacitors. *Nano Energy,* **2016**, *21*, 276-294.
[http://dx.doi.org/10.1016/j.nanoen.2016.01.028]

[3]　Wu, Z.S.; Feng, X.; Cheng, H.M. Recent advances in graphene-based planar micro-supercapacitors for on-chip energy storage. *Natl. Sci. Rev.,* **2014**, *1*(2), 277-292.
[http://dx.doi.org/10.1093/nsr/nwt003]

[4]　Punetha, V.D.; Dhali, S.; Rana, A.; Karki, N.; Tiwari, H.; Negi, P.; Basak, S.; Sahoo, N.G. Recent advancements in green synthesis of nanoparticles for improvement of bioactivities: A review. *Curr. Pharm. Biotechnol.,* **2022**, *23*(7), 904-919.
[http://dx.doi.org/10.2174/1389201022666210812115233] [PMID: 34387160]

[5]　Sahoo, N.G.; Sandeep, M. A process of manufacturing. Indian Patent 352780, 2016.

[6]　Li, Z.; Song, B.; Wu, Z.; Lin, Z.; Yao, Y.; Moon, K.S.; Wong, C.P. 3D porous graphene with ultrahigh

surface area for microscale capacitive deionization. *Nano Energy,* **2015**, *11*, 711-718.
[http://dx.doi.org/10.1016/j.nanoen.2014.11.018]

[7]   Stoller, M.D.; Park, S.; Zhu, Y.; An, J.; Ruoff, R.S. Graphene-based ultracapacitors. *Nano Lett.,* **2008**, *8*(10), 3498-3502.
[http://dx.doi.org/10.1021/nl802558y] [PMID: 18788793]

[8]   Reina, A.; Jia, X.; Ho, J.; Nezich, D.; Son, H.; Bulovic, V.; Dresselhaus, M.S.; Kong, J. Large area, few-layer graphene films on arbitrary substrates by chemical vapor deposition. *Nano Lett.,* **2009**, *9*(1), 30-35.
[http://dx.doi.org/10.1021/nl801827v] [PMID: 19046078]

[9]   Kumar, S.; Singh, P.K.; Punetha, V.D.; Singh, A.; Strzałkowski, K.; Singh, D.; Yahya, M.Z.A.; Savilov, S.V.; Dhapola, P.S.; Singh, M.K. *In-situ* N/O-heteroatom enriched micro-/mesoporous activated carbon derived from natural waste honeycomb and paper wasp hive and its application in quasi-solid-state supercapacitor. *J. Energy Storage,* **2023**, *72*, 108722.
[http://dx.doi.org/10.1016/j.est.2023.108722]

[10]   Song, B.; Sizemore, C.; Li, L.; Huang, X.; Lin, Z.; Moon, K.; Wong, C-P. Triethanolamine functionalized graphene-based composites for high performance supercapacitors. *J. Mater. Chem. A Mater. Energy Sustain.,* **2015**, *3*(43), 21789-21796.
[http://dx.doi.org/10.1039/C5TA05674H]

[11]   Lu, X.; Li, L.; Song, B.; Moon, K.; Hu, N.; Liao, G.; Shi, T.; Wong, C. Mechanistic investigation of the graphene functionalization using p-phenylenediamine and its application for supercapacitors. *Nano Energy,* **2015**, *17*, 160-170.
[http://dx.doi.org/10.1016/j.nanoen.2015.08.011]

[12]   Li, L.; Zhang, J.; Peng, Z.; Li, Y.; Gao, C.; Ji, Y.; Ye, R.; Kim, N.D.; Zhong, Q.; Yang, Y.; Fei, H.; Ruan, G.; Tour, J.M. High-performance pseudocapacitive microsupercapacitors from laser-induced graphene. *Adv. Mater.,* **2016**, *28*(5), 838-845.
[http://dx.doi.org/10.1002/adma.201503333] [PMID: 26632264]

[13]   Wu, Z.K.; Lin, Z.; Li, L.; Song, B.; Tuan, C.C.; Li, Z.; Moon, K.; Bai, S.L.; Wong, C.P. Capacitance enhancement by electrochemically active benzene derivatives for graphene-based supercapacitors. *RSC Advances,* **2015**, *5*(102), 84113-84118.
[http://dx.doi.org/10.1039/C5RA18108A]

[14]   Chen, J.; Xu, J.; Zhou, S.; Zhao, N.; Wong, C.-P. Template-grown graphene/porous Fe 2 O 3 nanocomposite: A high-performance anode material for pseudocapacitors. *Nano Energy,* **2015**, *15*, 719-728.
[http://dx.doi.org/10.1016/j.nanoen.2015.05.021]

[15]   Wu, Z.S.; Parvez, K.; Feng, X.; Müllen, K. Graphene-based in-plane micro-supercapacitors with high power and energy densities. *Nat. Commun.,* **2013**, *4*(1), 2487.
[http://dx.doi.org/10.1038/ncomms3487] [PMID: 24042088]

[16]   Song, B.; Li, L.; Lin, Z.; Wu, Z.K.; Moon, K.; Wong, C.P. Water-dispersible graphene/polyaniline composites for flexible micro-supercapacitors with high energy densities. *Nano Energy,* **2015**, *16*, 470-478.
[http://dx.doi.org/10.1016/j.nanoen.2015.06.020]

[17]   Liu, W-W.; Feng, Y-Q.; Yan, X-B.; Chen, J-T.; Xue, Q-J. Superior micro-supercapacitors based on graphene quantum dots. *Adv. Funct. Mater.,* **2013**, *23*(33), 4111-4122.

[18]   Niu, Z.; Zhang, L.; Liu, L.; Zhu, B.; Dong, H.; Chen, X. All-solid-state flexible ultrathin micro-supercapacitors based on graphene. *Adv. Mater.,* **2013**, *25*(29), 4035-4042.
[http://dx.doi.org/10.1002/adma.201301332] [PMID: 23716279]

[19]   Cao, X.; Qi, D.; Yin, S.; Bu, J.; Li, F.; Goh, C.F.; Zhang, S.; Chen, X. Ambient fabrication of large-area graphene films *via* a synchronous reduction and assembly strategy. *Adv. Mater.,* **2013**, *25*(21), 2957-2962.

[http://dx.doi.org/10.1002/adma.201300586] [PMID: 23606536]

[20]   Mathew, E.E.; Balachandran, M. Crumpled and porous graphene for supercapacitor applications: A short review. *Carbon Lett.,* **2021**, *31*(4), 537-555.
[http://dx.doi.org/10.1007/s42823-021-00229-2]

[21]   Mao, S.; Wen, Z.; Kim, H.; Lu, G.; Hurley, P.; Chen, J. A general approach to one-pot fabrication of crumpled graphene-based nanohybrids for energy applications. *ACS Nano,* **2012**, *6*(8), 7505-7513.
[http://dx.doi.org/10.1021/nn302818j] [PMID: 22838735]

[22]   Deng, S.; Berry, V. Wrinkled, rippled and crumpled graphene: An overview of formation mechanism, electronic properties, and applications. *Mater. Today,* **2016**, *19*(4), 197-212.
[http://dx.doi.org/10.1016/j.mattod.2015.10.002]

[23]   Mohan, A.N.; Manoj, B.; Ramya, A.V. Probing the nature of defects of graphene like nano-carbon from amorphous materials by raman spectroscopy. *Asian J. Chem.,* **2016**, *28*(7), 1501-1504.
[http://dx.doi.org/10.14233/ajchem.2016.19739]

[24]   Wang, C.G.; Lan, L.; Liu, Y.P.; Tan, H.F. Defect-guided wrinkling in graphene. *Comput. Mater. Sci.,* **2013**, *77*, 250-253.
[http://dx.doi.org/10.1016/j.commatsci.2013.04.051]

[25]   Ramya, A.V.; Manoj, B.; Mohan, A.N. Extraction and characterization of wrinkled graphene nanolayers from commercial graphite. *Asian J. Chem.,* **2016**, *28*(5), 1031-1034.
[http://dx.doi.org/10.14233/ajchem.2016.19577]

[26]   El Rouby, W.M.A. Crumpled graphene: Preparation and applications. *RSC Advances,* **2015**, *5*(82), 66767-66796.
[http://dx.doi.org/10.1039/C5RA10289H]

[27]   Wang, Y.; Wu, Y.; Huang, Y.; Zhang, F.; Yang, X.; Ma, Y.; Chen, Y. Preventing graphene sheets from restacking for high-capacitance performance. *J. Phys. Chem. C,* **2011**, *115*(46), 23192-23197.
[http://dx.doi.org/10.1021/jp206444e]

[28]   Tang, L.A.L.; Lee, W.C.; Shi, H.; Wong, E.Y.L.; Sadovoy, A.; Gorelik, S.; Hobley, J.; Lim, C.T.; Loh, K.P. Highly wrinkled cross-linked graphene oxide membranes for biological and charge-storage applications. *Small,* **2012**, *8*(3), 423-431.
[http://dx.doi.org/10.1002/smll.201101690] [PMID: 22162356]

[29]   Li, J.; Tang, J.; Yuan, J.; Zhang, K.; Yu, X.; Sun, Y.; Zhang, H.; Qin, L-C. Porous carbon nanotube/graphene composites for high-performance supercapacitors. *Chem. Phys. Lett.,* **2018**, *693*, 60-65.
[http://dx.doi.org/10.1016/j.cplett.2017.12.052]

[30]   Yu, Y.; Sun, Y.; Cao, C.; Yang, S.; Liu, H.; Li, P.; Huang, P.; Song, W. Graphene-based composite supercapacitor electrodes with diethylene glycol as inter-layer spacer. *J. Mater. Chem. A Mater. Energy Sustain.,* **2014**, *2*(21), 7706-7710.
[http://dx.doi.org/10.1039/C4TA00905C]

[31]   Wang, J.; Ding, B.; Xu, Y.; Shen, L.; Dou, H.; Zhang, X. Crumpled Nitrogen-Doped Graphene for Supercapacitors with High Gravimetric and Volumetric Performances. *ACS Appl. Mater. Interfaces,* **2015**, *7*(40), 22284-22291.
[http://dx.doi.org/10.1021/acsami.5b05428] [PMID: 26399912]

[32]   Luo, J.; Jang, H.D.; Huang, J. Effect of sheet morphology on the scalability of graphene-based ultracapacitors. *ACS Nano,* **2013**, *7*(2), 1464-1471.
[http://dx.doi.org/10.1021/nn3052378] [PMID: 23350607]

[33]   Yan, J.; Xiao, Y.; Ning, G.; Wei, T.; Fan, Z. Facile and rapid synthesis of highly crumpled graphene sheets as high-performance electrodes for supercapacitors. *RSC Advances,* **2013**, *3*(8), 2566.
[http://dx.doi.org/10.1039/c2ra22685e]

[34]   Zou, Y.; Kinloch, I.A.; Dryfe, R.A.W. Nitrogen-doped and crumpled graphene sheets with improved

supercapacitance. *J. Mater. Chem. A Mater. Energy Sustain.,* **2014**, *2*(45), 19495-19499.
[http://dx.doi.org/10.1039/C4TA04076G]

[35]   Xu, X.; Yang, J.; Zhou, X.; Jiang, S.; Chen, W.; Liu, Z. Highly crumpled graphene-like material as compression-resistant electrode material for high energy-power density supercapacitor. *Chem. Eng. J.,* **2020**, *397*, 125525.
[http://dx.doi.org/10.1016/j.cej.2020.125525]

[36]   Zhu, J.; Dong, S.; Xu, Y.; Guo, H.; Lu, X.; Zhang, X. Oxygen-enriched crumpled graphene-based symmetric supercapacitor with high gravimetric and volumetric performances. *J. Electroanal. Chem.,* **2019**, *833*, 119-125.
[http://dx.doi.org/10.1016/j.jelechem.2018.11.032]

[37]   Cuihua, An. Effects of highly crumpled graphene nanosheets on the electrochemical performances of pseudocapacitor electrode materials. *Electrochimica Acta.,* **2014**, *133*, 180-187.
[http://dx.doi.org/10.1016/j.electacta.2014.04.056]

[38]   Dsoke, S.; Pfeifer, K.; Zhao, Z. The role of nanomaterials for supercapacitors and hybrid devices. In: *Frontiers of Nanoscience*; Elsevier, **2021**; Vol. 19, pp. 99-136.
[http://dx.doi.org/10.1016/B978-0-12-821434-3.00001-6]

[39]   Sarno, M. Nanotechnology in energy storage: The supercapacitors. In: *Studies in Surface Science and Catalysis*; Elsevier, **2020**; Vol. 179, pp. 431-458.
[http://dx.doi.org/10.1016/B978-0-444-64337-7.00022-7]

[40]   Li, Z.; Xu, Z.; Wang, H.; Ding, J.; Zahiri, B.; Holt, C.M.B.; Tan, X.; Mitlin, D. Colossal pseudocapacitance in a high functionality-high surface area carbon anode doubles the energy of an asymmetric supercapacitor. *Energy Environ. Sci.,* **2014**, *7*(5), 1708-1718.
[http://dx.doi.org/10.1039/C3EE43979H]

[41]   Liu, X.; Qi, X.; Zhang, Z.; Ren, L.; Liu, Y.; Meng, L.; Huang, K.; Zhong, J. One-step electrochemical deposition of nickel sulfide/graphene and its use for supercapacitors. *Ceram. Int.,* **2014**, *40*(6), 8189-8193.
[http://dx.doi.org/10.1016/j.ceramint.2014.01.015]

[42]   Yang, S.; Feng, X.; Ivanovici, S.; Müllen, K. Fabrication of graphene-encapsulated oxide nanoparticles: Towards high-performance anode materials for lithium storage. *Angew. Chem. Int. Ed.,* **2010**, *49*(45), 8408-8411.
[http://dx.doi.org/10.1002/anie.201003485] [PMID: 20836109]

[43]   Ng, Y.H.; Iwase, A.; Kudo, A.; Amal, R. Reducing graphene oxide on a visible-light BiVO$_4$ photocatalyst for an enhanced photoelectrochemical water splitting. *J. Phys. Chem. Lett.,* **2010**, *1*(17), 2607-2612.
[http://dx.doi.org/10.1021/jz100978u]

[44]   Beka, L.G.; Li, X.; Liu, W. Nickel cobalt sulfide core/shell structure on 3D graphene for supercapacitor application. *Sci. Rep.,* **2017**, *7*(1), 2105.
[http://dx.doi.org/10.1038/s41598-017-02309-8] [PMID: 28522809]

[45]   Zhai, The preparation of graphene decorated with manganese dioxide nanoparticles by electrostatic adsorption for use in supercapacitors. *Carbon,* **2012**, *50*(14), 5034-5043.
[http://dx.doi.org/10.1016/j.carbon.2012.06.033]

[46]   Hassan, H.K.; Atta, N.F.; Hamed, M.M.; Galal, A.; Jacob, T. Ruthenium nanoparticles-modified reduced graphene prepared by a green method for high-performance supercapacitor application in neutral electrolyte. *RSC Advances,* **2017**, *7*(19), 11286-11296.
[http://dx.doi.org/10.1039/C6RA27415C]

[47]   Yu, G.; Hu, L.; Vosgueritchian, M.; Wang, H.; Xie, X.; McDonough, J.R.; Cui, X.; Cui, Y.; Bao, Z. Solution-processed graphene/MnO$_2$ nanostructured textiles for high-performance electrochemical capacitors. *Nano Lett.,* **2011**, *11*(7), 2905-2911.
[http://dx.doi.org/10.1021/nl2013828] [PMID: 21667923]

[48] Mirghni, A.A.; Madito, M.J.; Oyedotun, K.O.; Masikhwa, T.M.; Ndiaye, N.M.; Ray, S.J.; Manyala, N. A high energy density asymmetric supercapacitor utilizing a nickel phosphate/graphene foam composite as the cathode and carbonized iron cations adsorbed onto polyaniline as the anode. *RSC Advances,* **2018**, *8*(21), 11608-11621.
[http://dx.doi.org/10.1039/C7RA12028A] [PMID: 35542801]

[49] Maaoui, H.; Singh, S.K.; Teodorescu, F.; Coffinier, Y.; Barras, A.; Chtourou, R.; Kurungot, S.; Szunerits, S.; Boukherroub, R. Copper oxide supported on three-dimensional ammonia-doped porous reduced graphene oxide prepared through electrophoretic deposition for non-enzymatic glucose sensing. *Electrochim. Acta,* **2017**, *224*, 346-354.
[http://dx.doi.org/10.1016/j.electacta.2016.12.078]

[50] Dhibar, S.; Das, C.K. Silver nanoparticles decorated polypyrrole/graphene nanocomposite: A potential candidate for next-generation supercapacitor electrode material. *J. Appl. Polym. Sci.,* **2017**, *134*(16), 44724.
[http://dx.doi.org/10.1002/app.44724]

[51] Jaikumar, A.; Santhanam, K.S.V.; Kandlikar, S.G.; Raya, I.B.P.; Raghupathi, P. Electrochemical deposition of copper on graphene with high heat transfer coefficient. *ECS Trans.,* **2015**, *66*(30), 55-64.
[http://dx.doi.org/10.1149/06630.0055ecst]

[52] Zhang, Q.; Zhang, Y.; Gao, Z.; Ma, H-L.; Wang, S.; Peng, J.; Li, J.; Zhai, M. A facile synthesis of platinum nanoparticle decorated graphene by one-step γ-ray induced reduction for high rate supercapacitors. *J. Mater. Chem. C Mater. Opt. Electron. Devices,* **2013**, *1*(2), 321-328.
[http://dx.doi.org/10.1039/C2TC00078D]

[53] Korkmaz, S.; Kariper, İ.A. Graphene and graphene oxide based aerogels: Synthesis, characteristics and supercapacitor applications. *J. Energy Storage,* **2020**, *27*, 101038.
[http://dx.doi.org/10.1016/j.est.2019.101038]

[54] Du, A.; Zhou, B.; Zhang, Z.; Shen, J. A Special material or a new state of matter: A review and reconsideration of the aerogel. *Materials,* **2013**, *6*(3), 941-968.
[http://dx.doi.org/10.3390/ma6030941] [PMID: 28809350]

[55] Chen, T.T.; Song, W.L.; Fan, L.Z. Engineering graphene aerogels with porous carbon of large surface area for flexible all-solid-state supercapacitors. *Electrochim. Acta,* **2015**, *165*, 92-97.
[http://dx.doi.org/10.1016/j.electacta.2015.02.008]

[56] Song, Z.; Liu, W.; Xiao, P.; Zhao, Z.; Liu, G.; Qiu, J. Nano-iron oxide (Fe$_2$O$_3$)/three-dimensional graphene aerogel composite as supercapacitor electrode materials with extremely wide working potential window. *Mater. Lett.,* **2015**, *145*, 44-47.
[http://dx.doi.org/10.1016/j.matlet.2015.01.040]

[57] Liu, Y.; He, D.; Wu, H.; Duan, J.; Zhang, Y. Hydrothermal self-assembly of manganese dioxide/manganese carbonate/reduced graphene oxide aerogel for asymmetric supercapacitors. *Electrochim. Acta,* **2015**, *164*, 154-162.
[http://dx.doi.org/10.1016/j.electacta.2015.01.223]

[58] Van Aken, K.L.; Pérez, C.R.; Oh, Y.; Beidaghi, M.; Joo Jeong, Y.; Islam, M.F.; Gogotsi, Y. High rate capacitive performance of single-walled carbon nanotube aerogels. *Nano Energy,* **2015**, *15*, 662-669.
[http://dx.doi.org/10.1016/j.nanoen.2015.05.028]

[59] Wang, S.; Wang, R.; Zhao, Q.; Ren, L.; Wen, J.; Chang, J.; Fang, X.; Hu, N.; Xu, C. Freeze-drying induced self-assembly approach for scalable constructing MoS$_2$/graphene hybrid aerogels for lithium-ion batteries. *J. Colloid Interface Sci.,* **2019**, *544*, 37-45.
[http://dx.doi.org/10.1016/j.jcis.2019.02.078] [PMID: 30825799]

[60] Yang, F.; Xu, M.; Bao, S.J.; Wei, H.; Chai, H. Self-assembled hierarchical graphene/polyaniline hybrid aerogels for electrochemical capacitive energy storage. *Electrochim. Acta,* **2014**, *137*, 381-387.
[http://dx.doi.org/10.1016/j.electacta.2014.06.017]

[61]   Xu, P. A high surface area N-doped holey graphene aerogel with low charge transfer resistance as high performance electrode of non-flammable thermostable supercapacitors. *Carbon,* **2019**, *149*, 452-461.
[http://dx.doi.org/10.1016/j.carbon.2019.04.070]

[62]   Li, J.; Li, X.; Xiong, D.; Wang, L.; Li, D. Enhanced capacitance of boron-doped graphene aerogels for aqueous symmetric supercapacitors. *Appl. Surf. Sci.,* **2019**, *475*, 285-293.
[http://dx.doi.org/10.1016/j.apsusc.2018.12.152]

[63]   Liu, K.K.; Jin, B.; Meng, L.Y. Glucose/graphene-based aerogels for gas adsorption and electric double layer capacitors. *Polymers,* **2018**, *11*(1), 40.
[http://dx.doi.org/10.3390/polym11010040] [PMID: 30960024]

[64]   Yang, M.; Jeong, J.M.; Huh, Y.S.; Choi, B.G. High-performance supercapacitor based on three-dimensional MoS2/graphene aerogel composites. *Compos. Sci. Technol.,* **2015**, *121*, 123-128.
[http://dx.doi.org/10.1016/j.compscitech.2015.11.004]

[65]   Wang, D.W.; Li, F.; Zhao, J.; Ren, W.; Chen, Z.G.; Tan, J.; Wu, Z.S.; Gentle, I.; Lu, G.Q.; Cheng, H.M. Fabrication of graphene/polyaniline composite paper *via in situ* anodic electropolymerization for high-performance flexible electrode. *ACS Nano,* **2009**, *3*(7), 1745-1752.
[http://dx.doi.org/10.1021/nn900297m] [PMID: 19489559]

[66]   Wu, Q.; Xu, Y.; Yao, Z.; Liu, A.; Shi, G. Supercapacitors based on flexible graphene/polyaniline nanofiber composite films. *ACS Nano,* **2010**, *4*(4), 1963-1970.
[http://dx.doi.org/10.1021/nn1000035] [PMID: 20355733]

[67]   Bora, C.; Dolui, S.K. Fabrication of polypyrrole/graphene oxide nanocomposites by liquid/liquid interfacial polymerization and evaluation of their optical, electrical and electrochemical properties. *Polymer,* **2012**, *53*(4), 923-932.
[http://dx.doi.org/10.1016/j.polymer.2011.12.054]

[68]   Pham, H.D.; Pham, V.H.; Oh, E.S.; Chung, J.S.; Kim, S. Synthesis of polypyrrole-reduced graphene oxide composites by *in-situ* photopolymerization and its application as a supercapacitor electrode. *Korean J. Chem. Eng.,* **2012**, *29*(1), 125-129.
[http://dx.doi.org/10.1007/s11814-011-0145-y]

[69]   Islam, M.M.; Aboutalebi, S.H.; Cardillo, D.; Liu, H.K.; Konstantinov, K.; Dou, S.X. Self-assembled multifunctional hybrids: Toward developing high-performance graphene-based architectures for energy storage devices. *ACS Cent. Sci.,* **2015**, *1*(4), 206-216.
[http://dx.doi.org/10.1021/acscentsci.5b00189] [PMID: 27162972]

[70]   Yang, W.; Gao, Z.; Wang, J.; Ma, J.; Zhang, M.; Liu, L. Solvothermal one-step synthesis of Ni-Al layered double hydroxide/carbon nanotube/reduced graphene oxide sheet ternary nanocomposite with ultrahigh capacitance for supercapacitors. *ACS Appl. Mater. Interfaces,* **2013**, *5*(12), 5443-5454.
[http://dx.doi.org/10.1021/am4003843] [PMID: 23647434]

[71]   Gund, G.S.; Dubal, D.P.; Patil, B.H.; Shinde, S.S.; Lokhande, C.D. Enhanced activity of chemically synthesized hybrid graphene oxide/Mn3O4 composite for high performance supercapacitors. *Electrochim. Acta,* **2013**, *92*, 205-215.
[http://dx.doi.org/10.1016/j.electacta.2012.12.120]

[72]   Punetha, V.D.; Rana, S.; Yoo, H.J.; Chaurasia, A.; McLeskey, J.T., Jr; Ramasamy, M.S.; Sahoo, N.G.; Cho, J.W. Functionalization of carbon nanomaterials for advanced polymer nanocomposites: A comparison study between CNT and graphene. *Prog. Polym. Sci.,* **2017**, *67*, 1-47.
[http://dx.doi.org/10.1016/j.progpolymsci.2016.12.010]

CHAPTER 2

Multifaceted Applications of Nanoparticle-Functionalized Graphene

Poulomi Sengupta[1,*] and **Siddhant B. Patel[2]**

[1] Department of Chemistry, School of Science, Indrashil University, Rajpur, District Mehsana, Gujarat, 382740, India

[2] Chemical and Biochemical Engineering, School of Engineering, Indrashil University. Rajpur, District Mehsana, Gujarat, 382740, India

Abstract: Recently, graphene sheets have attracted a huge awareness for their special optical, mechanical, magnetic, electronic, and thermal characteristics. This has been possible due to the thin yet robust two-dimensional structural arrangement. The special properties may further be enhanced by smart chemical modifications on the two-dimensional structure. Meanwhile, nanoparticles also have come up as an emerging platform for their size, shape, surface area, optoelectronic properties and flexibility in functionalization. To utilize the advantages of both worlds, the scientific community has combined graphene with metallic nanoparticles. This event has brought about extreme enhancements in the above properties. Both inorganic and organic nanoparticles have been attached to the graphene surface. However, the attachment of metallic nanoparticles has increased their applications in developing sensors and catalysts. In this literature review, we want to concentrate on synthesizing and functionalizing graphene with different metallic nanoparticles. At the same time, we would discuss their applications in various fields.

Keywords: Bioimaging, Clean water, Catalysts, Graphene, Graphene oxide, Nanoparticles, Nano-graphene hybrid, Photocatalysis, Reduced graphene oxide, Sensor, SERS.

INTRODUCTION

Graphene is a miraculous material formed by the two-dimensional combination of carbon atoms. Structurally it is analogous to well-known aromatic allotropes of elemental carbon, like graphite, carbon nano-tubes (CNT), and fullerene (Fig. **1**). Like its peers, graphene is also formed by alternatingly placed carbon-carbon single and double bonds in a ring of 6 carbon atoms. It has a honeycomb-like structure. The sole difference between graphene and other carbon allotropes

---

*[*] **Corresponding author Poulomi Sengupta:** Department of Chemistry, School of Science, Indrashil University, Rajpur, District Mehsana, Gujarat, 382740, India; E-mail: poulomisg@gmail.com*

Vinay Deep Punetha (Ed.)

(fullerene, CNT) is its stable two-dimensional structure and atomic height. It can grow as tall as a grown-up man. At the same time, its cross section is one single carbon atom providing light weight, transparency, and flexibility. The presence of alternating single and double bonds within the molecular structure confers very high strength to the material, significantly enhancing its mechanical properties. It has a very high breaking strength of almost 42 Nm^{-1} and a Young's modulus of around 1.0 TPa [1]. Surprisingly, it is 200 times more robust than steel of the same diameter; and although lighter in weight, it is harder than diamond. Graphene has zero band gap; because of the delocalized electrons graphene can efficiently conduct electricity and heat [2]. It exhibits an amazing ambipolar electric field effect with a high charge carrier mobility (up to 10,000 $cm^2V^{-1}s^{-1}$) at an atmospheric temperature and pressure. It enables ballistic electron transfer at a very high speed, only 300 times slower than the speed of light. The conjugated system also helps graphene absorb a wide range of wavelengths. Being single molecular in height, graphene has a huge surface area (>2,000 m^2/g) compared to graphite (10 m^2/g) and CNT (1,300 m^2/g). Graphene has been explored very extensively for this odd combination of strength with transparency, flexibility, and conductivity.

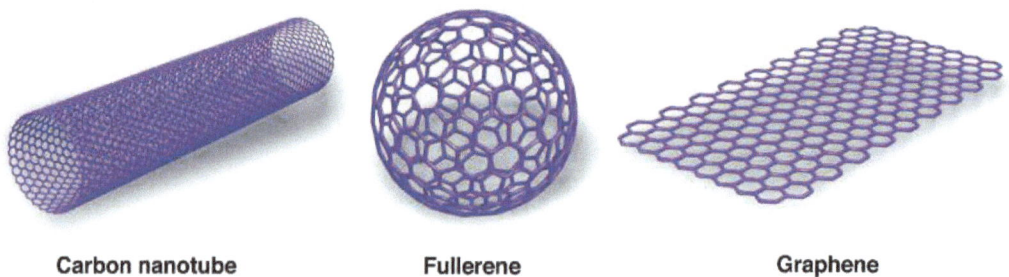

Carbon nanotube   **Fullerene**   **Graphene**

Fig. (1). Comparison: Carbon nanotube, fullerene, and graphene.

The modern world has witnessed many applications of graphene, including medical sensors, barcode materials, sports items, electronics, OLED (organic light emitting diode), energy storage, photovoltaic, superconductors, and environmental remediation. However, having a height of one atom, the loading capacity of graphene is low. rGO (reduced graphene oxide) has a few functional groups present in it; still the exposure to new functionalization is restricted. Hence this limitation restricts graphene from expanding its applications further. Moreover, graphene and its derivatives carry unsatisfied valency, surface defects, and oxygenated functional groups (Fig. **2**). Hence it is possible to expand its functionality by combining it with different objects.

Nanoparticles, on the other hand, have brought booming success in the past three decades. Nanoparticles having a greater surface-to-volume ratio can magically increase the surface area; moreover, the metallic nanoparticles possess unique optical [3], electronic, thermal [4], magnetic [5], and catalytic [6] properties, which are exhaustively explored in different research papers, patents, and products. Despite so many wonderful properties of nanomaterials, they are dimensionless. Hence to convert it into a tangible material a three-dimensional/two-dimensional support is essential.

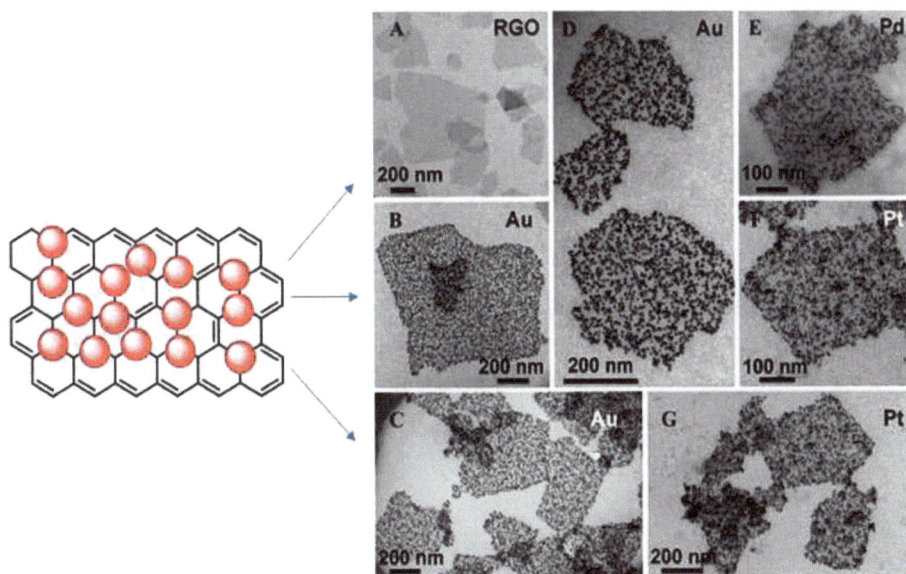

Fig. (2). Schematic and real representations [12] of nanoparticle-decorated graphene sheet. Permission obtained from the publisher.

The material scientists noticed the complementarity of nanoparticles and graphene; according to them, the combination can synergistically enhance both material's beneficial properties. Now attachment of nanoparticles on the graphene sheet could have been challenging. Quite gratifyingly, in rGO, the functional groups present may act as a special nucleation point for the formation of novel metal nanoparticles. Besides, the defects present in graphene may initiate the nucleation and accommodation of metallic nanoparticles. As a result, many nanoparticle-decorated graphene has been possible in recent years [7 - 9]. Recently, with the help of AI (artificial intelligence), nanocomposite designing of graphene has been possible [10]. The deposition of different metals has enriched different properties in graphene. Most importantly, the introduction of 3-dimensional nanoparticles has enhanced the surface area. Hence the main applications of the graphene sheet, like sensors and catalysis, have greatly benefitted. Nanoparticles have a versatile surface chemistry that allows easy conjugation of different payloads [11, 12]. The surface can also be passivated, leading to comfortable solvent dispersibility which expands the use of graphene in various mediums.

Graphene has been a focus of attention for numerous scientific sectors. The accidental discovery of single molecular graphene by a scotch tape in 2004 [13]. followed by a Nobel prize in 2010 bestowed it an overnight popularity. Its unique structure, interesting morphology, flexibility, and strength, have attracted scientists for years. However, the deposition of nanoparticles has expanded its applications to a tremendous extent. Hence, in this chapter, we plan to focus on the application parts, basic routine synthesis, and characterization techniques.

SYNTHESIS OF GRAPHENE

The conversion of graphene to different functional materials and exploration of their applications have been very much active since the facile synthesis of single-layered graphene. This event extensively enhanced the surface area; attachment with the nanoparticles further gave it a newer dimension. Typically, there are a few popular methods for successfully preparing graphene.

Mechanical Exfoliation (ME)

In this procedure, an adhesive tape is placed on the smooth surface of graphene, peeled off, and placed on another surface. As a result, a single-layered graphene is produced on the second surface [14]. The quality of graphene synthesized in this process is very pure. However, it is limited to a small-scale synthesis. The large-scale production of graphene following this simple method is not possible.

Liquid Phase Exfoliation

The liquid phase exfoliation method is easy to adapt in any industrial setting. In this exfoliation method, graphene is dipped in different suitable solvents and exfoliated in the presence of mild chemicals [15]. Ultrasound vibrations and ionic liquids are also used in a few cases to separate the single-atomic graphene layers easily. High yield and low operational cost are two key points for liquid phase exfoliation.

Chemical Vapor Deposition (CVD)

In this process, a thermocatalytic deposition of hydrocarbon on a metal surface produces graphene. Metals like Ni and Co are popularly used here. It is a relatively novel process for generating graphene with a vast surface area [16].

After successfully isolating single-atom graphene, graphene oxide is prepared by chemical oxidation. Numerous methods are available for the same. However, Hummer's method is the most accepted one where graphene oxide is prepared by the oxidation of potassium permanganate, sodium nitrate, and sulphuric acid [17]. Following oxidation, nanoparticles can be put on the graphene surface (Fig. **3**).

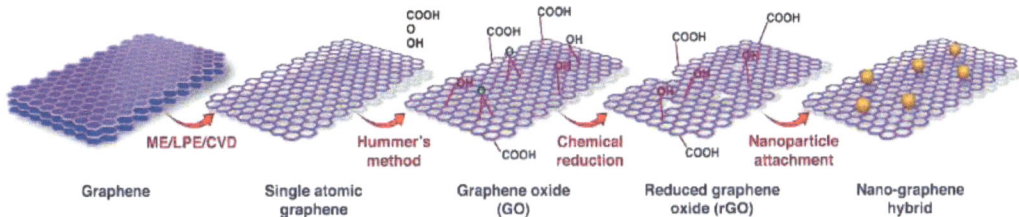

Fig. (3). The schematic representation for the preparation of nanoparticle-graphene hybrid.

Reduction Method

Metal nanoparticles, especially the noble metal nanoparticles can be synthesized following the reduction method. The significant advantages of the reduction method are: (a) It is a single-pot synthesis method, (b) It is highly systematic and comparatively simple to perform, and (c) It can be a bit challenging to control the size and morphology of the NPs. Nevertheless, microwave-facilitated reactions can address that. Some key references are found here [18, 19].

Hydrothermal Method

Metal oxide nanoparticles and quantum dots graphene hybrids can be prepared by the hydrothermal methods. The special features of this method are: (a) nanoparticles with high crystalline properties and optically narrow size distribution can be synthesized, and (b) high temperature and long reaction times can partially or completely reduce GO (graphene oxide) on its own [20, 21].

Electrochemical Method

In the electrochemical method, metal nanoparticles, mainly noble metal nanoparticles are synthesized. The special features are, that it is (a) A simple, straightforward, fast, and green technique, and (b) Low cost, easy to mimic, and automated. Moreover, it is highly stable and reproducible, and its (c) Density, size, and morphology can be controlled by pulsed current. The electrochemical method also allows for the simultaneous reduction of GO [22, 23].

Ex-situ Method

The key features of this method are (a) Nanoparticles are synthesized in advance; the precise control of the size, shape, and density of the nanoparticles are maintained and are used to form hybrids, (b) *Ex-situ* methods, for example, covalent or noncovalent interactions such as van der Waals interactions, hydrogen bonding, π–π stacking, or electrostatic interaction also stabilize nanoparticles on graphene's surface [24].

APPLICATIONS

Ever since the discovery of a single molecular graphene sheet from graphite, scientists from all the major fields have tried to exploit the structural advantages of graphene. Drug delivery, tissue engineering, bioimaging, biosensing – these fields are always challenging as their deliverables are hydrophobic molecules that aggregate *in vivo* with the loss of activity. Graphene, due to its high surface area, was thought to be an excellent choice for delivering therapeutic payloads. However, graphene is hydrophobic, rather GO or rGO would have been better candidates as delivery vehicles. For drug delivery, the relatively hydrophilic rGO was a better choice, as there were hydroxy groups for the chemical conjugation of therapeutics and π-bonds for non-bonding interactions with hydrophobic molecules. Hence rGO was explored extensively for gene, drug delivery, tissue engineering applications, catalytic properties, and biosensing. However, there are indications of cytotoxicity [24] and hematotoxicity [25, 26] arising out of graphene and GO. Many studies on the safety concerns of graphene analogues showed dose-dependent cytotoxicity [27 - 29] and *in vivo* toxicity.

Nanoparticles came out to be of great significance at this point. Nanoparticles, due to their comfortable surface chemistry, and functionalization capabilities, can be conjugated or grown *in situ* on graphene/GO surfaces. The lattice defects and functional groups present on rGO help in that. Nanoparticles have a few highly desirable and powerful optical, electronic, chemical, magnetic, and structural properties. These upon combination with graphene, become even stronger and can be applied in several biological applications like drug/gene delivery, sensing, targeting, photothermal therapy, and bioimaging (Fig. **4**).

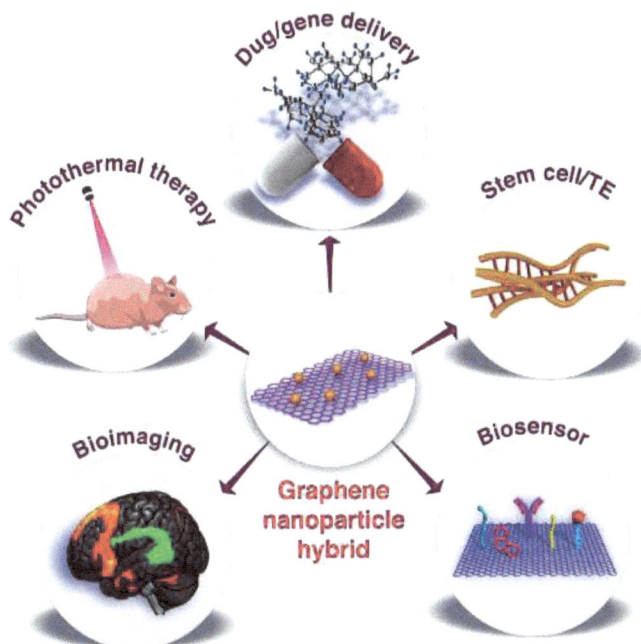

Fig. (4). Applications of nanoparticle-graphene hybrid materials.

Applications in Water Purification

For many countries in the world, access to fresh water is a challenge. Not that there is no resource for freshwater but with the advent of humanity, with heavily growing industries, the water is gradually getting polluted. There are two major types of pollutants. 1. Organic molecules, 2. Heavy metals. Dye and pharmaceutical industries are the primary reasons behind organic pollutants, whereas the electronics, polymer, and pesticide industries are the significant contributors of heavy metals in water. Undoubtedly, both pollutants are detrimental to human health. Heavy metals may bind with proteins, and nucleic acids, which may lead to gene modification. Water pollution due to heavy metals is a direct threat to human health. On the other hand, organic pollutants can interfere with different biological activities such as metabolism, digestion, *etc.*, which leads to several difficult-to-cure diseases. Cadmium (Cd), chromium (Cr), arsenic (As), copper (Cu), mercury (Hg), manganese (Mn), lead (Pb), *etc.*, although these metals are present in traces, still, they are harmful to flora and fauna. Different methods are available for removing heavy metals from the wastewater such as, ion exchange, adsorption, reverse osmosis, solvent extraction, precipitation and amalgamation, filtration by a membrane, photocatalytic treatment, coagulation, *etc.* Among these, the adsorption technique is more economical and popularly used. For adsorption, activated charcoal is an excellent option [30]. Nevertheless, there are problems associated with the adsorption technique; those are sludge formation, pretreatment and posttreatment of the membrane. Magnetic nanomaterials are good alternatives for adsorption and easy separation. However, functionalization and stability are significant issues [31]. Graphene and rGO having a very large surface area, can be used for metal ion adsorption. Magnetic iron-based nanomaterials deposited on graphene can be employed in organic molecule adsorption. Graphene separates the magnetic nanoparticles to avoid their aggregation problem. At the same time, iron oxide nanoparticles cover graphene and prevent the aggregation of graphene layers between themselves by π-π overlap. Moreover, organic pollutants can comfortably adsorb on graphene and its derivatives by π-stacking and non-bonding interactions like electrostatic and weak interactions. Some examples of magnetic nanoparticles decorated GO or rGO are explored for water purification.

In Table **1**, some representative examples of nano-graphene are listed which have been explored in water decontamination:

Table 1. Examples of nano-graphene have been explored in water decontamination.

S. No.	Composites	Removed Pollutant	Advantages	Refs.
1.	Fe_3O_4-graphene ₙₐₙₒₛₕₑₑₜ	Cationic dye	The presence of less oxygenated functional groups increases dye adsorption.	[32]
\2.	Fe_3O_4-graphene nanosheet@SiO_2	Methylene blue	Endothermic and spontaneous adsorption.	[33]
3.	Fe_3O_4 graphene composite	Organic pollutant	Purification by absorbing UV light.	[34]
4.	Amine- functionalized mesoporous Fe_3O_4 graphene	Heavy metals	Highly efficient	[35]
5.	Magnetic graphene oxide	Methadone	Highly efficient	[36]
6.	transition-metal ferrite nanophotocatalysts	Methylene blue	Photodegradation	[37]
7.	Fe_3O_4-graphene	Aniline, p-chloroaniline	Highly efficient	[38]
8.	Fe_3O_4-CNT-rGO	Hydrazine	Electrochemical sensing	[39]
9.	Graphene oxide/$MnFe_2O_4$	Pb(II)	Enhanced removal	[40]
10.	Fe_3O_4-GO	Cu(II), Co(II), fulvic acid, rhodamine B, acid blue, malachite green	Easy and efficient separation, magnetic separation.	[41 - 43]

Applications in Biological Systems

Biological systems comprise complex organs, cells, tissues, ducts, solid bones and so on. It is very difficult to deliver, track, inhibit, or promote any biological process due to its complicated structure. Hence the search for a better delivery vehicle, biosensors, and imaging agent is always on. Once the nanoparticle-graphene hybrid material was discovered, its application in drug delivery, biosensing and bioimaging was extensively explored. The main applications of biological applications of nanoparticle-graphene hybrid materials are discussed here:

Drug/gene Delivery

Most of the time, therapeutics like drugs are hydrophobic; hence it is essential to conjugate it with cargo. Graphene-nanoparticle hybrids have significantly expanded their surface area. Hence chemical conjugation with drugs is easier here. Along with the drugs, targeting groups can also be conjugated, which will deliver the cargo to the destination location in the body. Here in the following example (Fig. **5**), graphene carrying the targeting group (folic acid) and chemotherapeutic drug (doxorubicin) can selectively home into cancer cells.

Biosensor Application

Biosensors are chemicals that are used to detect and or analyze a certain amount of analyte in a mixture of chemicals or a family of analytes. Because of its electron density, conductivity, and catalytic properties, graphene molecules can work as biosensors. Metallic nanoparticles already have excellent optoelectronic properties. As a result, the sensing capabilities increased by the combination with graphene. Biosensors are relevant in numerous places where analyte measurement is essential, like food safety, environmental monitoring, defence, drug delivery, quantification of any health hazard, *etc.*

Generally, biosensors can be obtained by combining two basic elements: a receptor and a transducer. The receptor controls the interaction with different analytes, and the transducer converts the interaction with a measurable signal. The efficiency of biosensor nanoparticles depends on the limit of detection (LoD), sensitivity to the target, linear and dynamic ranges, reproducibility, and selectivity. Other important parameters are the sensor's response time, operational ease and cost, storage stability, and portability. Most importantly, the ideal sensor should be reusable. Graphene has several positive points to be used as a sensor: it has a high surface area, excellent electrical properties, good immobilization properties, very high mechanical strength, and advantageous optical properties. Based on these advantages, four types of hybrid nano-graphene-based sensor materials are in research; FRET (Forster resonance energy transfer), FET (field-effect transistor), electrochemical, and SERS (surface-enhanced Raman scattering). Various Au, Pt, SiO$_2$, QD (quantum dot), UCNP (up-conversion nanoparticles) - based nanoparticles have been vastly explored for biosensing applications. In the following example, AgNP-decorated nanoparticles are investigated as resistivity-based methane sensors. In Fig. (**6**), silver nanoparticle-embedded graphene sheets sensed methane gas at a low temperature.

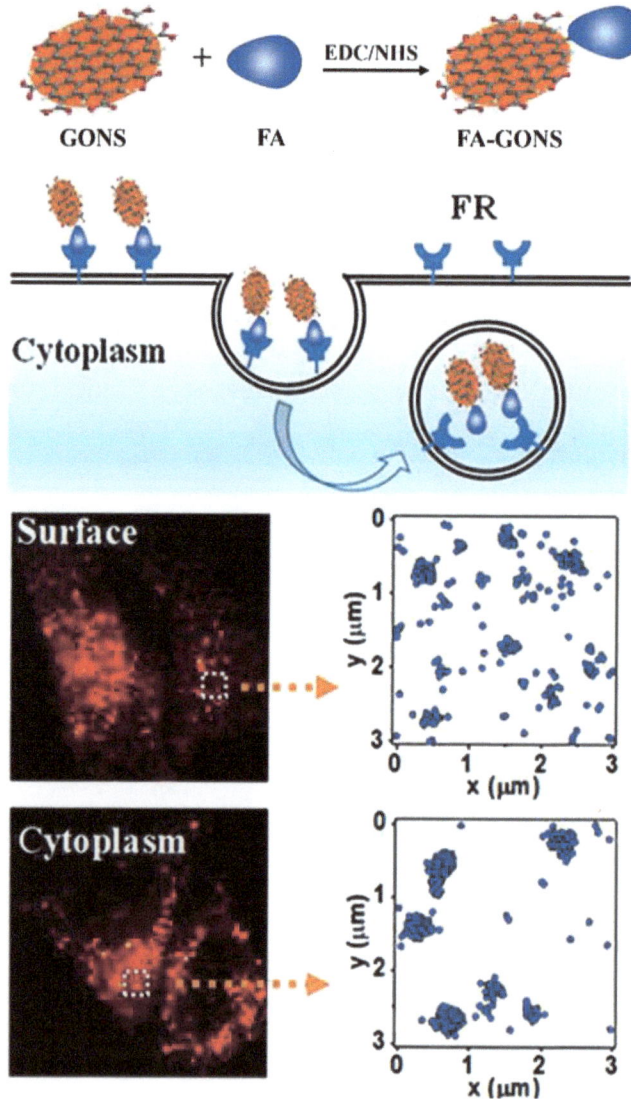

Fig. (5). Doxorubicin-loaded graphene nanostructures. Drug targeting was achieved by folic acid [44]. Image used with permission..

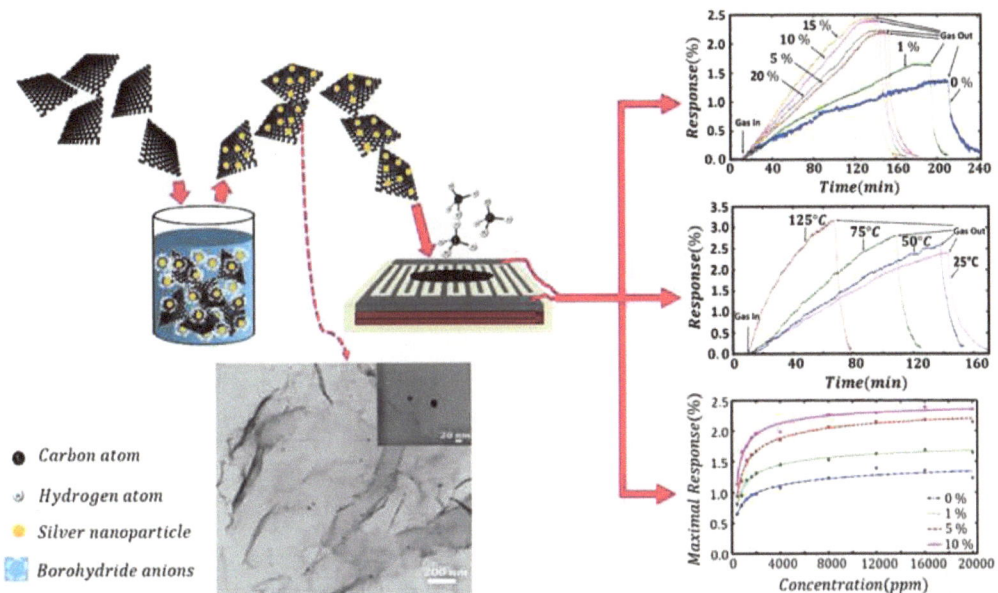

Fig. (6).   AgNP-decorated graphene nanohybrid for low temperature methane gas sensing [45]. Image obtained with permission from ACS..

Bioimaging

It is essential to image the internal organs to clearly understand the complex biological systems. In the traditional imaging technology involving organic dyes, the dyes may face several problems during live organ/cell imaging. The organic dyes often photo-bleach with time, they may interact with different chemicals already present inside the body, or get too dilute in the complex organ such that it is unable to create a proper image of the related organs. For this reason, a revolution in the imaging system is necessary. GO-enwrapped dye-loaded metal nanoparticles are one of the primitive examples of graphene-nano hybrid particles [46]. Later, graphene-based Fe_3O_4 particles were widely explored and reported to image biological systems by MRI (magnetic resonance imaging) [47 - 49].

In Table **2**, we have consolidated the recent reports on nanoparticle-graphene hybrid materials used in biological systems.

Table 2. Recent reports on nanoparticle-graphene hybrid materials application in biological systems.

S. No.	Composites	Application Type	Refs.
1.	Ag/ZnO-GO	Wound dressing	[50]
2.	MoS_2-GO	Antibacterial agent	[51]

(Table 2) cont.....

S. No.	Composites	Application Type	Refs.
3.	Fe-oxide-chitosan-GO	Antibacterial agent	[52]
4.	CuS-Dox-GO	Drug delivery	[53]
5.	Zirconia-NP-rGO	Cancer drug detection	[54]
6.	Cisplatin-Chitosan-GO	Intercellular drug accumulation	[55]
7.	QD-Graphene	Drug delivery and real-time monitoring	[56]
8.	Ag-GO	Thermoluminiscent	[57]
9.	ZnO-GO	Selective cytotoxicity to glioblastoma cells	[58]
10.	Ag-PEI-GO	Antibacterial and antibiofilm.	[59]
11.	AuNP-Graphene hydrogel	Detection of cell-released NO.	[60]
12.	AgNP-Graphene	Drug release monitored by SERS.	[61]
13.	PINIPAM-GO-Fe_3O_4	NIR light, magneto, and pH-responsive drug release.	[62]
14.	PEG-AgNP-GO	Long-term antibacterial activity.	[63]
15.	QD-GO	Fluorescent imaging, monitoring, drug delivery.	[64]
16.	Zirconium dioxide-rGO	Electrochemical assay of protein kinase activity.	[65]
17.	Fe_3O_4-GO	Drug delivery, tumor MRI imaging, targeted delivery	[66]
18.	Lanthanum hydroxide-GO	Preventing the evolution of antimicrobial resistance.	[67]
19.	Graphene/CuO_2 Nanoshuttles	Interrupting bacterial respiration.	[68]
20.	ZnO-GO	Inhibiting Cd-induced hepatotoxicity.	[69]
21.	rGO/ZnO_2-Ag	Bacterial growth Suppressor by O_2	[70]
		nanobubbles	
22.	Graphene-QD	Theranostic applications	[71]
23.	Fe_3O_4@rGO	Combating hepatocellular carcinoma by dual-drug	[72]
24.	GO-MgO	Antimicrobial, antioxidant, and anticancer activities.	[73]
25.	AgNPs-GO	Anticancer activity against specific cells.	[74]
26.	AuNP-GO	Covalent drug delivery system.	[75]
27.	TiO_2/mesoporous ZnO–GO	Cytotoxicity of Curcumin to Caco cells increased	[76]

Photocatalytic Applications

Graphene is a two-dimensional material composed of a single layer of carbon atoms arranged in a crisscross honeycomb-like lattice. Due to its excellent properties, such as high electrical conductivity, large surface area, and high mechanical strength, graphene has attracted a significant amount of attention in the world of photocatalysis. The application of graphene-based materials for photocatalytic reactions has made significant strides in recent years. Here are some of the recent developments in photocatalysis involving graphene:

Enhanced Photocatalytic Performance

Nanomaterials based on graphene have been utilized as potent and productive photocatalysts for the degradation of numerous organic and inorganic pollutants. Photocatalytic degradation using graphene involves the use of light to activate the graphene surface, leading to the initiation of electron-hole pairs that can drive the decomposition of organic and inorganic materials. The degradation of organic dyes, which are frequently found in wastewater from the textile, paper, and dye industries, is one example of a photocatalytic decomposition reaction using graphene. Fig. (**7**) depicts the typical steps involved in the photocatalytic degradation of organic dyes using graphene.

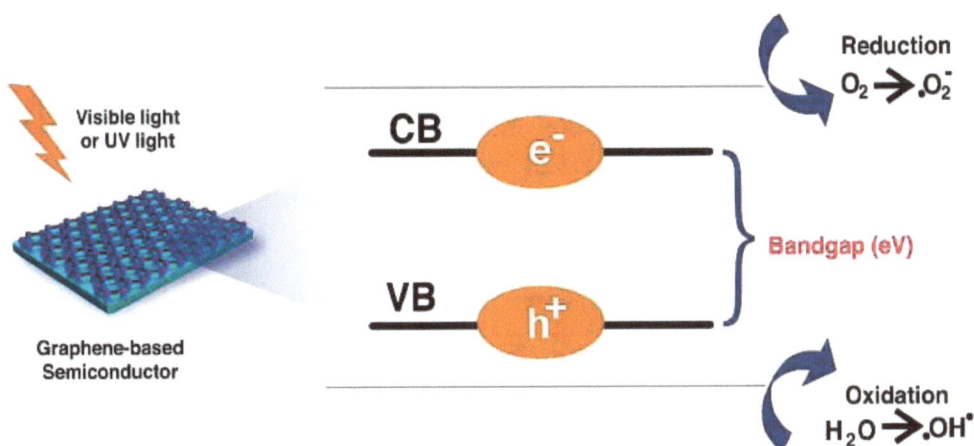

Fig. (7). General photocatalysis mechanism for graphene-based semiconductors..

The mechanism of photocatalysis can be explained in a few simple steps:

a. Graphene is exposed to light, typically in the ultraviolet (UV) or visible region.
b. The energy from the light is absorbed by the graphene, generating electron-hole pairs.
c. The electrons generated by the graphene are used to reduce oxygen molecules (O^2) to produce superoxide radicals ($\cdot O^{2-}$).
d. Subsequently, dye degradation is caused by the reaction of dyes with superoxide radicals.

Overall, this process offers a method for removing harmful organic dyes from wastewater utilizing renewable energy sources. Using a nanoparticle-graphene hybrid as a photocatalyst for organic degradation provides several benefits, including a high surface area, excellent conductivity, and stability under harsh reaction conditions.

The degradation of inorganic pollutants, such as heavy metals, is another example of a photocatalytic degradation reaction involving graphene. In this instance, the photocatalytic reaction involves graphene-based nanocomposites containing metal ions, such as copper or silver, that can act as co-catalysts to boost the photocatalytic activity of graphene. The photocatalytic degradation of heavy metals by shall be replaced with "that of typically involves the same steps as described previously, but the reaction pathway may involve distinct intermediate species and mechanisms.

Researchers have demonstrated that combining graphene with other photocatalytic materials, such as TiO_2, ZnO, and CdS, can significantly boost photocatalytic activity [77 - 79]. For instance, recently Yu *et al.* (2022) [80] synthesized reduced graphene oxide (rGO)/ titanium dioxide (TiO_2) nanocomposites and have observed improved degradation of Rhodamine B as compared to that achieved with the help of TiO_2 reported by Kiwaan *et al.* (2020) under UV light condition [81]. Table **3** lists some highly cited recent developments in graphene-nanocomposites.

Table 3. Recently reported research papers on the photocatalytic performance of graphene-based nanomaterials.

S. No.	Nanoparticle-Graphene Details	Substrate	Mechanism/special Features	Refs.
1.	rGO/TiO_2	Methylene Blue	TiO_2/rGO demonstrated 3.3 times degradation as compared to TiO_2.	[82, 83]
2.	rG)/TiO_2	Rhodamine B	Photocatalytic degradation was observed under the visible light range.	[84]
3.	GO/TiO_2	Rhodamine B	Ag and rGO increased charge separation and reduced photo-generated electron-hole pair recombination.	[85]
4.	BiOCl/CQDs/rGO	**Ciprofloxacin**	Due to enhanced light absorption and charge	[86]
			Separation.	
5.	CdS/rGO	Methylene blue	The negatively charged rGO/CdS attracted cationic dye molecules, which degraded faster.	[87]
6.	Bi_2WO_6/rGO	Rhodamine B	rGO sheets improved the separation of electron-hole pairs.	[88]

(Table 3) cont.....

S. No.	Nanoparticle-Graphene Details	Substrate	Mechanism/special Features	Refs.
7.	$Ag/rGO/Bi_2WO_6$	Rhodamine B	rGO's electron transfer capability considerably improves photogenerated electron-hole pair separation and transmission.	[89]
8.	ZnO/(rGO)	Phenol	Enhanced charge separation and decreased recombination of photogenerated electron-hole pairs.	[90]
9.	$g-C_3N_4/GQDs$	Antibiotics	The photocatalytic reaction rate was 3.46 times that of pure graphitic carbon nitride (g-C_3N_4).	[91]
10.	TiO_2nanotube/3- D graphene foam	Rhodamine B	Well photo induced electron transport leads to superior photocatalysis degradation. Improved absorption performance and	[92]
			photocurrent response.	
11.	Cu_2O/GO	Rhodamine B	Enhanced light absorption and charge separation.	[93]
12	WO_3/Graphene- based materials	MB, Rhodamine B, ciprofloxacin	Graphene's improved light absorption and charge separation.	[94 - 97]
13	ZnO/G	Methyl Orange	A higher degradation rate was achieved as compared to pure ZnO.	[98]
14.	$\alpha-Fe_2O_3$/graphene	Rhodamine B	Exhibited high (~98%) catalytic activity under visible light.	[99]

In general, graphene-based materials have demonstrated great promise for enhancing the photocatalytic activity of diverse photocatalysts. The improved charge separation, reduced recombination of photo-generated electron-hole pairs, enhanced light absorption, and efficient charge transfer and separation between graphene-based materials and photocatalysts are responsible for the enhanced photocatalytic activity.

Production of Hydrogen via Photocatalysis:

By splitting water molecules, graphene-based materials have also been utilised for photocatalytic hydrogen production. Multiple studies have demonstrated the effectiveness of graphene-based materials for this application.

Graphene photocatalytic water splitting reactions typically include similar steps as mentioned earlier:

a. Light, typically in the ultraviolet (UV) or visible range, is used to expose graphene.
b. Light energy is absorbed by graphene, resulting in the formation of electron-hole pairs.
c. The graphene electrons are used to reduce protons (H$^+$) to produce hydrogen gas (H$_2$).
d. The generated holes by graphene are used to oxidize water (H$_2$O) to produce oxygen gas (O$_2$).

From Fig. (**8**), it can be observed that in this case of water splitting, the dye molecule is replaced with water. Overall, this process enables the production of hydrogen gas, a clean and renewable fuel, using renewable energy sources. Nano-graphene composites as a photocatalyst for hydrogen production have several advantages, including a large surface area, high conductivity, and stability under harsh reaction conditions.

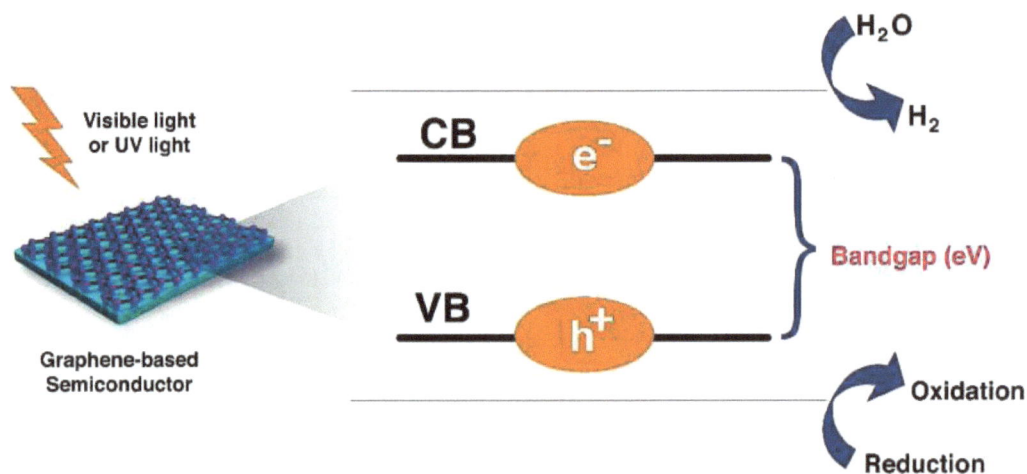

Fig. (8). General reaction mechanism for water splitting using graphene-based semiconductors.

As a result, various composites of graphene have been widely explored. Graphene/TiO$_2$ composites have been extensively studied for hydrogen production *via* photocatalysis. Hernández-Majalca. *et al*. (2019) synthesized a composite of graphene oxide/TiO$_2$ nanoparticles and discovered that it was significantly more effective than pure TiO$_2$ [98]. This was done to enhance the photocatalytic activity to produce hydrogen. The graphene oxide was responsible for the increased light absorption and the charge separation, both of which contributed to the improved photocatalytic activity. A few highly cited recent developments of graphene-nanocomposite have been enlisted in Table **4**.

Table 4. Recently reported research on photocatalytic hydrogen production with the help of graphene-based materials.

S. No.	Nanoparticle-graphene details	Process	Mechanism/special features	Refs.
1.	rGO/TiO$_2$	Water-splitting	Self-tuning optoelectronic properties were observed. H$_2$ production: 6500 μmolH$_2$/g	[100]
2.	GO/g-C$_3$N$_4$	Water Splitting (under visible light irradiation by using 10 vol% TEA solutions as a sacrificial reagent and 1.0 wt% Pt as a cocatalyst).	Comparatively very high (12 times) H$_2$ production was observed with GO/g-C$_3$N$_4$ as compared to g-C3N4.	[101]
3.	Metal or metal oxide/Graphene/g- C$_3$N$_4$	Water-splitting	Improved charge separation and decreased recombination of photo-generated electron-hole pairs which lead to superior H$_2$ production.	[102 - 105]
4.	CdS/GQDs	Water-splitting	The H$_2$ production was 2.7 times compared to pure CdS	[106]
5.	G/ZnO	Water-splitting	Higher H$_2$ production was observed as compared to pure ZnO.	[107]
6.	γ-Fe$_2$O$_3$/rGO/C$_3$N$_4$	Water-splitting	Hight hydrogen and oxygen evolution rate were achieved with the help of graphene-based materials.	[108]
7.	Metal/rGO/Metal Oxide	Water-splitting	Enhances charge separation and decreases recombination of photogenerated electron-hole pairs which enhances H$_2$ production.	[109, 110]

Graphene-based materials have demonstrated great promise for enhancing photocatalytic hydrogen production activity. The improved photocatalytic activity can be attributed to enhanced light absorption, efficient charge separation, decreased recombination of photo-generated electron-hole pairs, and enhanced charge transfer and separation between graphene-based materials and photocatalysts.

Fuel Cell Applications:

Because of its unique properties, such as its high surface area, excellent electrical conductivity, and mechanical strength, graphene has shown great promise for use in fuel cells. Recent research has focused on making catalysts, electrodes, membranes, and nanocomposites for fuel cells out of graphene to make them work better, last longer, and be cheaper.

Proton Exchange Membrane Fuel Cells (PEMFC):

Graphene-based materials could be used as a catalyst support in fuel cells. In this case, graphene serves as a substrate for depositing catalyst nanoparticles, which can improve the fuel cell's catalytic activity.

Graphene-based materials can also be used as a membrane material in fuel cells. Because of their high proton conductivity and stability, graphene oxide (GO) and reduced graphene oxide (rGO) have been investigated as potential PEM materials [111 - 113].

A graphene-based PEMFC operates in the same way as a conventional PEMFC (shown in Fig. **9**), with the difference being the material used for the PEM. Hydrogen is supplied to the anode of a graphene-based PEMFC, where it is split into protons and electrons on the surface of the graphene-supported catalyst. Protons migrate through the graphene-based PEM to the cathode, recombining with oxygen to form water, resulting in an electrical current in the external circuit. The electrons complete the circuit by flowing through the external circuit to the cathode.

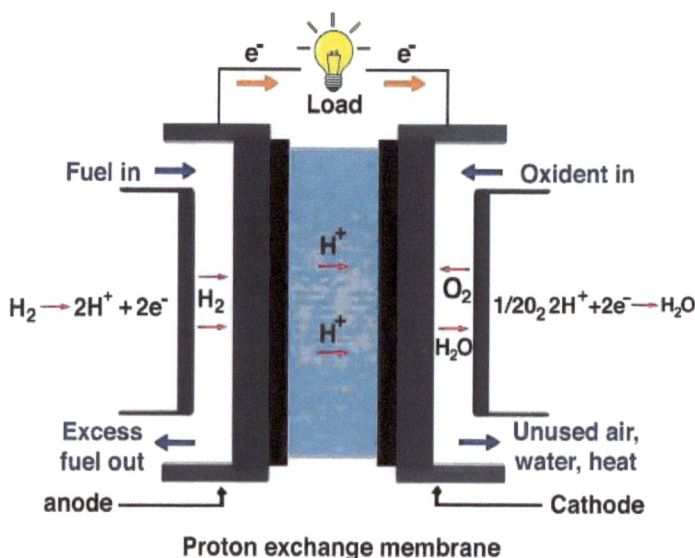

Fig. (9). Schematic diagram for the working principle of PEMFC.

Graphene-based materials have shown great promise for their use in fuel cells due to their unique properties. However, further research is needed to address the remaining challenges and fully exploit their potential. Because of their high power density and efficiency, graphene-based membranes have shown great promise for use in proton exchange membrane fuel cells (PEMFCs), which are widely used in transportation applications. Peng *et al.* (2016) developed a graphene-based composite membrane for PEMFCs by adding a graphene oxide layer to a commercial Nafion membrane (GO-Nafion-PEM) [114]. Compared to PEM, the composite membrane showed better proton conductivity and mechanical stability, making it a promising alternative for use in PEMFCs. Further, recent research on applying graphene-based material in PEMFC is listed in Table **5**. However, there are still obstacles to overcome, such as increasing the mechanical stability of graphene-based membranes and decreasing their susceptibility to water poisoning.

Table 5. Recently reported research on fuel cell application of graphene-based nanomaterials.

S. No.	Application	Nanoparticle- Graphene Details	Mechanism/special Features	Refs.
1.	PEMFC	Single-layer graphene (SLG)	After 100 h of galvanostatic discharging, the SLG on the anode has hydrogen crossings of only 1.75×10^{-4} mol s^{-1}, which was far lower than MEAs (8.16×10^{-4} mol s^{-1}).	[115]
2.	HT-PEMFC	Polybenzimidazole (PBI)/sulfonated graphene oxide	sGO filler increased composite membrane proton conductivity.	[116]
3.	DMFC	GQDs/MWCNTs Composite	GQDs/MWCNTs electrode-based DMFC exhibited the highest current signals for methanol oxidation as compared to GQDs/modified glassy carbon electrode.	[117]
4.	DMFC	rGO/Pt-Pd	Pt-Pd/rGO outperformed Pt/rGO and Pd/rGO in terms of electrochemical surface area, which leads to superior catalytic activity.	[118]
5.	ORR	Graphene doped with Fe/N/S and incorporating Fe_3O_4 nanoparticles	14.2 times higher mass activity achieved as compared to commercial Pt/C.	[119]

Direct Methanol Fuel Cell (DMFC)

Materials based on graphene have been investigated for their potential use as catalysts and catalytic supports in DMFCs, a specific type of fuel cell that works on methanol. The following is an explanation of the mechanism that is utilized by graphene-based DMFCs. The graphene-based catalyst or catalytic support is commonly positioned on the anode of a DMFC. After being adsorbed onto the surface of graphene, methanol is oxidized at the anode, which results in the production of carbon dioxide, water, protons, and electrons. An exterior circuit is responsible for producing electricity, and a proton exchange membrane is responsible for transporting protons to the cathode compartment, where they react with oxygen to make water. An external circuit produces the electricity. After that, the water is recirculated back to the anode, where it is used once more in the reaction that oxidizes the methanol, thus bringing the cycle to its conclusion.

Graphene-based materials can potentially improve the performance of DMFCs by hastening the oxidation of methanol that takes place at the anode. It has been established that graphene oxide, in particular, possesses better catalytic activity for the oxidation of methanol than traditional carbon black. In addition, nitrogen-doped graphene has been investigated for its potential use as a catalyst for the oxidation of methanol [120 - 122]. These studies have demonstrated that the material possesses increased levels of both activity and stability. DMFCs can have their performance and stability improved by using membranes or electrode materials that are based on graphene. These materials can be employed in the devices. It has been established that membranes built from graphene oxide have higher proton conductivity and selectivity than conventional Nafion membranes [123, 124]. Further, recent research reported on the application of graphene-based material in DMFC is listed in Table **5**.

Catalyst and Electrode

The high activity and stability of catalysts based on graphene have been established by several fuel cell reactions, including the ORR (oxygen reduction reaction), the (HER) hydrogen evolution reaction, and the (MOR) methanol oxidation reaction. Usually, for such applications, both strategies, single-atom doping, and heteroatom doping is carried out [125]. For instance, Liu *et al.* (2018) reported the development of a PtCu/GO ORR and EOR catalyst [126]. As compared to the activity of commercial Pt/C catalysts, the ORR activity of the catalyst was 5.3 times higher and EOR activity was 2.36 times higher. Raja *et al.*, 2022 developed heteroatom-doped RuO_2-P-rGO and studied HER and OER (oxygen evolution reaction) [127]. They found that RuO_2-P-rGO contributed to its enhanced stability in long-term HER and OER.

Graphene electrodes might be useful for making fuel cells that are flexible, lightweight, and portable. These characteristics make them a good candidate for use in wearable technology and as power sources in remote locations [128].

Research conducted over the course of the past few years has shown, on the whole, that materials based on graphene offer a great deal of potential for use in fuel cells in the capacities of catalysts, electrodes, membranes, and nanocomposites. When it comes to the commercialization of fuel cells for a variety of applications, additional research is required to optimize the performance, durability, and cost-effectiveness of the materials that make up these fuel cells.

Organocatalytic Applications

GO and rGO for their semiconductive and conductive nature can be used as catalyst supports for performing many organic/inorganic reactions. The availability of electrons on the GO and rGO surface can attract metal ions and stabilize them as nanoparticles. Moreover, the availability of electrons can attract organic molecules having excess electrons (π electrons from aromatic systems). These heterogeneous catalysts are highly stable under ambient conditions, inexpensive for large-scale preparation, and excellent in performance, and regeneration [129]. The graphene-based catalysts have been reported to carry out several special reactions (arylation reaction, A^3-coupling reaction, oxidation reaction, click reaction, reduction reaction, hydrogenation, Aldol condensation, Suzuki coupling, Suzuki-Miyaura coupling, Heck reaction, Friedel Crafts, Aza Michael reaction, ring opening of epoxides, ring-opening polymerization) which are very low yielding if conducted without or with a metal catalyst. To everybody's surprise, nano-GO and nano-rGO have worked wonderfully for the effective conversion of the above organic reactions. A selective list (Table **6**) has been included here.

Table 6. Graphene and graphene-nano composites in heterogeneous catalysis.

Serial No.	Composites	Detection Application	Advantage	Refs.
1.	GO-Pd, G-Pd	Suzuki coupling, Suzuki-Miyaura coupling, Heck reaction.	GO/graphene as better support than carbon black.	[130 - 133]
2.	G-Fe$_3$O$_4$-Au, G-γ-Fe$_2$O$_3$	O-nitroaniline to benzenediamine , triazoles synthesis *via* click chemistry.	Graphene as better support, magnetic separation.	[134, 135]
3.	GO-Ag, G-Ag	Decarboxylative cycloaddition, A3-coupling	High catalyst stability, efficient catalysis	[136, 137]
4.	G-Cu, G-CuO	Arylation	Graphene as a conductive support to stabilize Cu/CuO.	[138, 139]
5.	G-Ru	Arene hydrogenation	High catalyst stability.	[140]
6.	GO	Direct C-C bond formation by Friedel Crafts reaction.	High yield and excellent regioselectivity.	[141 - 143]

(Table 6) cont.....

Serial No.	Composites	Detection Application	Advantage	Refs.
7.	$Fe_3O_4@SiO_2@FeCl_3$	Aza Michael reaction	Easy preparation method, simple recovery, and high efficiency.	[144]
8.	Graphene oxide, Graphene	Ring-opening polymerization	Easy scale- up, fewer side products	[145]

Detection using SERS (Surface-enhanced Raman Scattering)

SERS is a unique and very sensitive method to detect Raman-active molecules. In this technique, the Raman scattering signals obtained from a very small amount of sample can be enhanced to 10^{10} to 10^{11} times; as a result, the detection becomes easier even at a very low concentration level. In this method, Raman active molecules are adsorbed on a rough metal surface or a nanostructure. Nanoparticle-embedded graphene or graphene oxide sheets can also work very well as the base of SERS. Nanoparticles on GO or rGO bases can create electromagnetic hotspots; as graphene has π-electron clouds, it can attract Raman-active molecules by π-π stacking, H-bonding, and by electrostatic attraction. After adsorption, the electromagnetic signal will be enhanced by innumerable folds proving SERS to be a very sensitive technique for the detection of any analyte. The detection limit can reach up to nM or pM level. In 2010, Ling *et al.* for the first time observed that graphene can be used as a surface in SERS. Later several groups have reported the successful roles of graphene, GO, and rGO as active surfaces and summarized in Table **7**.

Table 7. **Few representative examples of the detection of analytes by SERS signalling on graphene-nanoparticles**

Serial No.	Composites	Detection Application	Advantage	Refs.
1.	Graphene/SiO_2/Si	Protoporphyrin IX	Encourages the chemical- enhanced mechanism	[146]
2.	Graphene/AuNR	Rhodamine	Can detect 10^{-11} M	[147]
3.	Graphene/gold nanorods vertical array	Rhodamine	Can detect 10^{-13} M	[148]
4.	Ag grating/graphene/Au	Rhodamine	0.1 nM	[149]
	NPs			
5.	AuNP/G/AuNP or polyethylene	Tetramethylthiur am disulphide (thirum) on orange peel	0.24 ppm	[150]
6.	Graphene-AgNP	Food colorant	0.1 μM	[151]

(Table 7) cont.....

Serial No.	Composites	Detection Application	Advantage	Refs.
7.	Gold-graphene	Hg_{+2}	0.1 nM	[152]
8.	graphene oxide/Gold nanoparticle	Malachite green	2.5 to 100 μmol_L-1	[153]
9.	Ag/Au GO	Different dyes	Very low lod	[154]
10.	Ag@AuNPs modified graphene oxide	Glucose	1.97 ± 0.002 m M in blood	[155]

CONCLUSION

The cost-effective production of graphene plays a pivotal role in unlocking the multifaceted applications of nanoparticle-functionalized graphene [156]. This economical synthesis approach facilitates the incorporation of various nanoparticles into graphene matrices, enabling a plethora of innovative applications across diverse domains. From personalized nanomedicine through enhanced drug delivery and imaging, to advanced energy storage with improved battery and supercapacitor performance, and from efficient environmental remediation using graphene-based composites, to fine-tuned electronics, sensors, and catalysis, the synergy between cost-effective graphene and functional nanoparticles reshapes industries and solutions. This amalgamation not only capitalizes on graphene's exceptional properties but also leverages the specific functionalities of nanoparticles, creating a profound impact on sectors ranging from healthcare and energy to the environment and electronics, fostering innovation and sustainability on a global scale.

Graphene was popularized as the lightest, strongest, safest, and greenest material. The applications are versatile; it starts from making rust-proof durable goods, sports gears, wearable technologies (graphene-based smartphone), transistors, semiconductors, batteries, wind and solar power, supercapacitors, environmental remediation, food waste reduction, crop protection *etc.* Several applications, including water purification, photocatalytic reactions, chemical catalysis, and biosensing, have drawn attention to nanoparticle-incorporated graphene and graphene oxide. But some people thought it to be more of a hype than a hope. Now the future seems more interesting in the multidisciplinary applications of nano-graphene. Fe_3O_4-GO-based soft robots [157] have been possible which is capable of rotating, levitating, capturing, and un-clawing. These have been possible by synergistically using magnetic, ultrasonic, humidity, and light energy.

The journey of graphene is in its infancy. There are myriads of applications in the pipeline that will be in our hands soon. If the flag of graphene keeps on flying high, the future will be named 'the age of graphene'.

LIST OF ABBREVIATIONS

CNT	Carbon nanotubes
OLED	Organic light emitting diode
rGO	Reduced graphene oxide
GO	Graphene oxide
AI	Artificial intelligence
FRET	Forster resonance energy transfer
FET	Field-effect transistor
SERS	Surface enhanced Raman scattering
QD	Quantum dot
UCNP	Upconversion nanoparticles
PEMFC	Proton exchange membrane fuel cells
DMFC	Direct methanol fuel cell
ORR	Oxygen reduction reaction
HER	Hydrogen evolution reaction
MOR	Methanol oxidation reaction
OER	Oxygen evolution reaction

ACKNOWLEDGEMENTS

PS sincerely thanks the Science and Engineering Research Board (SRG/2021/000969) for funding.

REFERENCES

[1] Lee, C.; Wei, X.; Kysar, J.W.; Hone, J. Measurement of the elastic properties and intrinsic strength of monolayer graphene. *Science,* **2008**, *321*(5887), 385-388.
[http://dx.doi.org/10.1126/science.1157996] [PMID: 18635798]

[2] Cao, J.; Liu, M.; Liu, Z. Alternating Current Field Effects in Atomically Ferroelectric Ultrathin Films. *Materials,* **2022**, *15*(7), 2506.
[http://dx.doi.org/10.3390/ma15072506]

[3] Pathak, R.; Punetha, V. D.; Bhatt, S.; Punetha, M. Multifunctional role of carbon dot-based polymer nanocomposites in biomedical applications: A review. *J. Mater. Sci.,* **2023**, *58*(15), 6419-6443.
[http://dx.doi.org/10.1007/s10853-023-08408-4]

[4] Bhatt, S.; Punetha, V.D.; Pathak, R.; Punetha, M. Graphene in nanomedicine: A review on nano-bio factors and antibacterial activity. *Colloids Surf. B Biointerfaces,* **2023**, *226*, 113323.
[http://dx.doi.org/10.1016/j.colsurfb.2023.113323] [PMID: 37116377]

[5] Bhatt, S.; Pathak, R.; Punetha, V. D.; Punetha, M. Recent advances and mechanism of antimicrobial efficacy of graphene-based materials: A review. *J. Mater. Sci.,* **2023**, *58*(19), 7839-7867.
[http://dx.doi.org/10.1007/s10853-023-08534-z]

[6] Bhatt, S.; Punetha, V.D.; Pathak, R.; Punetha, M. Two-dimensional carbon nanomaterial-based

biosensors: Micromachines for advancing the medical diagnosis. *Adv. Struct. Mater.,* **2023**, *190*, 181-225.
[http://dx.doi.org/10.1007/978-3-031-28942-2_9]

[7] Pathak, R.; Punetha, V.D.; Bhatt, S.; Punetha, M. Carbon nanotube-based biocompatible polymer nanocomposites as an emerging tool for biomedical applications. *Eur. Polym. J.,* **2023**, *196*, 112257.
[http://dx.doi.org/10.1016/j.eurpolymj.2023.112257]

[8] Singh, S.; Hasan, M.R.; Sharma, P.; Narang, J. Graphene nanomaterials: The wondering material from synthesis to applications. *Sens. Int.,* **2022**, *3*, 100190.
[http://dx.doi.org/10.1016/j.sintl.2022.100190]

[9] Bhatt, S.; Pathak, R.; Punetha, V.D.; Punetha, M. Shape memory hallmarks and antimicrobial efficacy of polyurethane composites. *React. Funct. Polym.,* **2023**, *191*, 105678.
[http://dx.doi.org/10.1016/j.reactfunctpolym.2023.105678]

[10] Lin, J.; Liu, Y.; Sui, H.; Sagoe-Crentsil, K.; Duan, W. Microstructure of graphene oxide–silica-reinforced OPC composites: Image-based characterization and nano-identification through deep learning. *Cement Concr. Res.,* **2022**, *154*, 106737.
[http://dx.doi.org/10.1016/j.cemconres.2022.106737]

[11] Zhang, J.; Mou, L.; Jiang, X. Surface chemistry of gold nanoparticles for health-related applications. *Chem. Sci.,* **2020**, *11*(4), 923-936.
[http://dx.doi.org/10.1039/C9SC06497D] [PMID: 34084347]

[12] Liu, J.; Fu, S.; Yuan, B.; Li, Y.; Deng, Z. Toward a universal "adhesive nanosheet" for the assembly of multiple nanoparticles based on a protein-induced reduction/decoration of graphene oxide. *J. Am. Chem. Soc.,* **2010**, *132*(21), 7279-7281.
[http://dx.doi.org/10.1021/ja100938r] [PMID: 20462190]

[13] Novoselov, K.S.; Geim, A.K.; Morozov, S.V.; Jiang, D.; Zhang, Y.; Dubonos, S.V.; Grigorieva, I.V.; Firsov, A.A. Electric field effect in atomically thin carbon films. *Science,* **2004**, *306*(5696), 666-669.
[http://dx.doi.org/10.1126/science.1102896] [PMID: 15499015]

[14] Yi, M.; Shen, Z. A review on mechanical exfoliation for the scalable production of graphene. *J. Mater. Chem. A Mater. Energy Sustain.,* **2015**, *3*(22), 11700-11715.
[http://dx.doi.org/10.1039/C5TA00252D]

[15] Zeng, Z.; Yi, L.; He, J.; Hu, Q.; Liao, Y.; Wang, Y.; Luo, W.; Pan, M. Hierarchically porous carbon with pentagon defects as highly efficient catalyst for oxygen reduction and oxygen evolution reactions. *J. Mater. Sci.,* **2020**, *55*(11), 4780-4791.
[http://dx.doi.org/10.1007/s10853-019-04327-5]

[16] Khan, A.; Habib, M.R.; Kumar, R.R.; Islam, S.M.; Arivazhagan, V.; Salman, M.; Yang, D.; Yu, X. Wetting behaviors and applications of metal-catalyzed CVD grown graphene. *J. Mater. Chem. A Mater. Energy Sustain.,* **2018**, *6*(45), 22437-22464.
[http://dx.doi.org/10.1039/C8TA08325H]

[17] Hummers, W.S., Jr; Offeman, R.E. Preparation of graphitic oxide. *J. Am. Chem. Soc.,* **1958**, *80*(6), 1339-1339.
[http://dx.doi.org/10.1021/ja01539a017]

[18] Chang, K.; Chen, W.; Ma, L.; Li, H.; Li, H.; Huang, F.; Xu, Z.; Zhang, Q.; Lee, J-Y. Graphene-like MoS2/amorphous carbon composites with high capacity and excellent stability as anode materials for lithium ion batteries. *J. Mater. Chem.,* **2011**, *21*(17), 6251.
[http://dx.doi.org/10.1039/c1jm10174a]

[19] Tien, H.W.; Huang, Y.L.; Yang, S.Y.; Wang, J.Y.; Ma, C.C.M. The production of graphene nanosheets decorated with silver nanoparticles for use in transparent, conductive films. *Carbon,* **2011**, *49*(5), 1550-1560.
[http://dx.doi.org/10.1016/j.carbon.2010.12.022]

[20]   Su, Y.; Lu, X.; Xie, M.; Geng, H.; Wei, H.; Yang, Z.; Zhang, Y. A one-pot synthesis of reduced graphene oxide–Cu2S quantum dot hybrids for optoelectronic devices. *Nanoscale,* **2013**, *5*(19), 8889-8893.
[http://dx.doi.org/10.1039/c3nr02992a] [PMID: 23907643]

[21]   Hussain Shar, A. Facile synthesis of reduced graphene oxide encapsulated selenium nanoparticles prepared by hydrothermal method for acetone gas sensors. *Chem. Phys. Lett.,* **2020**, *755*, 137797.
[http://dx.doi.org/10.1016/j.cplett.2020.137797]

[22]   Moradi, A.G. One-step electrodeposition synthesis of silver-nanoparticle-decorated graphene on indium-tin-oxide for enzymeless hydrogen peroxide detection. *Carbon,* **2013**, *62*, 405-412.
[http://dx.doi.org/10.1016/j.carbon.2013.06.025.]

[23]   Golinelli, D.L.C.; Machado, S.A.S.; Cesarino, I. Synthesis of silver nanoparticle-graphene composites for electroanalysis applications using chemical and electrochemical methods. *Electroanalysis,* **2017**, *29*(4), 1014-1021.
[http://dx.doi.org/10.1002/elan.201600669]

[24]   Darabdhara, G.; Das, M.R.; Singh, S.P.; Rengan, A.K.; Szunerits, S.; Boukherroub, R. Ag and Au nanoparticles/reduced graphene oxide composite materials: Synthesis and application in diagnostics and therapeutics. *Adv. Colloid Interface Sci.,* **2019**, *271*, 101991.
[http://dx.doi.org/10.1016/j.cis.2019.101991] [PMID: 31376639]

[25]   Hu, W.; Peng, C.; Lv, M.; Li, X.; Zhang, Y.; Chen, N.; Fan, C.; Huang, Q. Protein corona-mediated mitigation of cytotoxicity of graphene oxide. *ACS Nano,* **2011**, *5*(5), 3693-3700.
[http://dx.doi.org/10.1021/nn200021j] [PMID: 21500856]

[26]   Chang, Y.; Yang, S.T.; Liu, J.H.; Dong, E.; Wang, Y.; Cao, A.; Liu, Y.; Wang, H. *In vitro* toxicity evaluation of graphene oxide on A549 cells. *Toxicol. Lett.,* **2011**, *200*(3), 201-210.
[http://dx.doi.org/10.1016/j.toxlet.2010.11.016] [PMID: 21130147]

[27]   Aliabadi, M.; Shagholani, H.; Yunessnia lehi, A. Synthesis of a novel biocompatible nanocomposite of graphene oxide and magnetic nanoparticles for drug delivery. *Int. J. Biol. Macromol.,* **2017**, *98*, 287-291.
[http://dx.doi.org/10.1016/j.ijbiomac.2017.02.012] [PMID: 28167110]

[28]   Horváth, L.; Magrez, A.; Burghard, M.; Kern, K.; Forró, L.; Schwaller, B. Evaluation of the toxicity of graphene derivatives on cells of the lung luminal surface. *Carbon,* **2013**, *64*, 45-60.
[http://dx.doi.org/10.1016/j.carbon.2013.07.005]

[29]   Punetha, V.D.; Rana, S.; Yoo, H.J.; Chaurasia, A.; McLeskey, J.T., Jr; Ramasamy, M.S.; Sahoo, N.G.; Cho, J.W. Functionalization of carbon nanomaterials for advanced polymer nanocomposites: A comparison study between CNT and graphene. *Prog. Polym. Sci.,* **2017**, *67*, 1-47.
[http://dx.doi.org/10.1016/j.progpolymsci.2016.12.010]

[30]   Nabais, J.M.V.; Gomes, J.A.; Suhas, P.J.M.; Carrott, P.J.M.; Laginhas, C.; Roman, S. Phenol removal onto novel activated carbons made from lignocellulosic precursors: Influence of surface properties. *J. Hazard. Mater.,* **2009**, *167*(1-3), 904-910.
[http://dx.doi.org/10.1016/j.jhazmat.2009.01.075] [PMID: 19233559]

[31]   Jiao, T. Facile and Scalable Preparation of Graphene Oxide-Based Magnetic Hybrids for Fast and Highly Efficient Removal of Organic Dyes. *Sci Rep,* **2015**, *5*(1), 12451.
[http://dx.doi.org/10.1038/srep12451]

[32]   Jolivet, J-P.; Chanéac, C.; Tronc, E. Iron oxide chemistry. From molecular clusters to extended solid networks. *Chem. Commun.,* **2004**, (5), 481-487.
[http://dx.doi.org/10.1039/c2jm15544c] [PMID: 14973569]

[33]   Wang, F. Effect of oxygen-containing functional groups on the adsorption of cationic dye by magnetic graphene nanosheets. *Chem. Eng. Res. Des.,* **2017**, *128*, 155-161.
[http://dx.doi.org/10.1016/j.cherd.2017.10.007]

[34]   Zhao, D.; Zhu, H.; Wu, C.; Feng, S.; Alsaedi, A.; Hayat, T.; Chen, C. Facile synthesis of magnetic Fe_3O_4/graphene composites for enhanced U(VI) sorption. *Appl. Surf. Sci.,* **2018**, *444*, 691-698.
[http://dx.doi.org/10.1016/j.apsusc.2018.03.121]

[35]   Anastasiou, E.; Lorentz, K.O.; Stein, G.J.; Mitchell, P.D. Prehistoric schistosomiasis parasite found in the Middle East. *Lancet Infect. Dis.,* **2014**, *14*(7), 553-554.
[http://dx.doi.org/10.1016/S1473-3099(14)70794-7] [PMID: 24953264]

[36]   Kumar Gupta, V.; Agarwal, S.; Asif, M.; Fakhri, A.; Sadeghi, N. Application of response surface methodology to optimize the adsorption performance of a magnetic graphene oxide nanocomposite adsorbent for removal of methadone from the environment. *J. Colloid Interface Sci.,* **2017**, *497*, 193-200.
[http://dx.doi.org/10.1016/j.jcis.2017.03.006] [PMID: 28284073]

[37]   Ali, N.; Zada, A.; Zahid, M.; Ismail, A.; Rafiq, M.; Riaz, A.; Khan, A. Enhanced photodegradation of methylene blue with alkaline and transition-metal ferrite nanophotocatalysts under direct sun light irradiation. *J. Chin. Chem. Soc.,* **2019**, *66*(4), 402-408.
[http://dx.doi.org/10.1002/jccs.201800213]

[38]   Chang, Y.P.; Ren, C.L.; Qu, J.C.; Chen, X.G. Preparation and characterization of Fe_3O_4/graphene nanocomposite and investigation of its adsorption performance for aniline and p-chloroaniline. *Appl. Surf. Sci.,* **2012**, *261*, 504-509.
[http://dx.doi.org/10.1016/j.apsusc.2012.08.045]

[39]   Nehru, S.; Sakthinathan, S.; Tamizhdurai, P.; Chiu, T.W.; Shanthi, K. Reduced graphene oxide/multiwalled carbon nanotube composite decorated with Fe_3O_4 magnetic nanoparticles for electrochemical determination of hydrazine in environmental water. *J. Nanosci. Nanotechnol.,* **2020**, *20*(5), 3148-3156.
[http://dx.doi.org/10.1166/jnn.2020.17379] [PMID: 31635659]

[40]   Dai, K.; Liu, G.; Xu, W.; Deng, Z.; Wu, Y.; Zhao, C.; Zhang, Z. Judicious fabrication of bifunctionalized graphene oxide/$MnFe_2O_4$ magnetic nanohybrids for enhanced removal of Pb(II) from water. *J. Colloid Interface Sci.,* **2020**, *579*, 815-822.
[http://dx.doi.org/10.1016/j.jcis.2020.06.085] [PMID: 32673858]

[41]   Geng, Z.; Lin, Y.; Yu, X.; Shen, Q.; Ma, L.; Li, Z.; Pan, N.; Wang, X. Highly efficient dye adsorption and removal: A functional hybrid of reduced graphene oxide-Fe_3O_4 nanoparticles as an easily regenerative adsorbent. *J. Mater. Chem.,* **2012**, *22*(8), 3527.
[http://dx.doi.org/10.1039/c2jm15544c]

[42]   Li, J.; Zhang, S.; Chen, C.; Zhao, G.; Yang, X.; Li, J.; Wang, X. Removal of Cu(II) and fulvic acid by graphene oxide nanosheets decorated with Fe_3O_4nanoparticles. *ACS Appl. Mater. Interfaces,* **2012**, *4*(9), 4991-5000.
[http://dx.doi.org/10.1021/am301358b] [PMID: 22950475]

[43]   Liu, M.; Chen, C.; Hu, J.; Wu, X.; Wang, X. Synthesis of magnetite/graphene oxide composite and application for cobalt(II) removal. *J. Phys. Chem. C,* **2011**, *115*(51), 25234-25240.
[http://dx.doi.org/10.1021/jp208575m]

[44]   He, H.; Li, S.; Shi, X.; Wang, X.; Liu, X.; Wang, Q.; Guo, A.; Ge, B.; Khan, N.U.; Huang, F. Quantitative nanoscopy of small blinking graphene nanocarriers in drug delivery. *Bioconjug. Chem.,* **2018**, *29*(11), 3658-3666.
[http://dx.doi.org/10.1021/acs.bioconjchem.8b00589] [PMID: 30346721]

[45]   Liu, H.; Xiang, H.; Li, Z.; Meng, Q.; Li, P.; Ma, Y.; Zhou, H.; Huang, W. Flexible and degradable multimodal sensor fabricated by transferring laser-induced porous carbon on starch film. *ACS Sustain. Chem. Eng.,* **2020**, *8*(1), 527-533.
[http://dx.doi.org/10.1021/acssuschemeng.9b05968]

[46]   Sreejith, S.; Ma, X.; Zhao, Y. Graphene oxide wrapping on squaraine-loaded mesoporous silica nanoparticles for bioimaging. *J. Am. Chem. Soc.,* **2012**, *134*(42), 17346-17349.

[http://dx.doi.org/10.1021/ja305352d] [PMID: 22799451]

[47]   Qian, R.; Maiti, D.; Zhong, J.; Xiong, S.S.; Zhou, H.; Zhu, R.; Wan, J.; Yang, K. Multifunctional nano-graphene based nanocomposites for multimodal imaging guided combined radioisotope therapy and chemotherapy. *Carbon,* **2019,** *149,* 55-62.
[http://dx.doi.org/10.1016/j.carbon.2019.04.046]

[48]   Wang, G.; Ma, Y.; Wei, Z.; Qi, M. Development of multifunctional cobalt ferrite/graphene oxide nanocomposites for magnetic resonance imaging and controlled drug delivery. *Chem. Eng. J.,* **2016,** *289,* 150-160.
[http://dx.doi.org/10.1016/j.cej.2015.12.072]

[49]   Baktash, M.S.; Zarrabi, A.; Avazverdi, E.; Reis, N.M. Development and optimization of a new hybrid chitosan-grafted graphene oxide/magnetic nanoparticle system for theranostic applications. *J. Mol. Liq.,* **2021,** *322,* 114515.
[http://dx.doi.org/10.1016/j.molliq.2020.114515]

[50]   Wang, Y.; Shi, L.; Wu, H.; Li, Q.; Hu, W.; Zhang, Z.; Huang, L.; Zhang, J.; Chen, D.; Deng, S.; Tan, S.; Jiang, Z. Graphene oxide-IPDI-Ag/ZnO@Hydroxypropyl cellulose nanocomposite films for biological wound-dressing applications. *ACS Omega,* **2019,** *4*(13), 15373-15381.
[http://dx.doi.org/10.1021/acsomega.9b01291] [PMID: 31572836]

[51]   Kim, T.I.; Kwon, B.; Yoon, J.; Park, I.J.; Bang, G.S.; Park, Y.; Seo, Y.S.; Choi, S.Y. Antibacterial activities of graphene oxide-molybdenum disulfide nanocomposite films. *ACS Appl. Mater. Interfaces,* **2017,** *9*(9), 7908-7917.
[http://dx.doi.org/10.1021/acsami.6b12464] [PMID: 28198615]

[52]   Konwar, A.; Kalita, S.; Kotoky, J.; Chowdhury, D. Chitosan-iron oxide coated graphene oxide nanocomposite hydrogel: A robust and soft antimicrobial biofilm. *ACS Appl. Mater. Interfaces,* **2016,** *8*(32), 20625-20634.
[http://dx.doi.org/10.1021/acsami.6b07510] [PMID: 27438339]

[53]   Insomphun, C.; Chuah, J.A.; Kobayashi, S.; Fujiki, T.; Numata, K. Influence of hydroxyl groups on the cell viability of polyhydroxyalkanoate (PHA) scaffolds for tissue engineering. *ACS Biomater. Sci. Eng.,* **2017,** *3*(12), 3064-3075.
[http://dx.doi.org/10.1021/acsbiomaterials.6b00279] [PMID: 33445351]

[54]   Venu, M.; Venkateswarlu, S.; Reddy, Y.V.M.; Seshadri Reddy, A.; Gupta, V.K.; Yoon, M.; Madhavi, G. Highly sensitive electrochemical sensor for anticancer drug by a zirconia nanoparticle-decorated reduced graphene oxide nanocomposite. *ACS Omega,* **2018,** *3*(11), 14597-14605.
[http://dx.doi.org/10.1021/acsomega.8b02129] [PMID: 30555980]

[55]   Vasanthakumar, A.; Rejeeth, C.; Vivek, R.; Ponraj, T.; Jayaraman, K.; Anandasadagopan, S.K.; Vinayaga Moorthi, P. Design of bio-graphene-based multifunctional nanocomposites exhibits intracellular drug delivery in cervical cancer treatment. *ACS Appl. Bio Mater.,* **2022,** *5*(6), 2956-2964.
[http://dx.doi.org/10.1021/acsabm.2c00280] [PMID: 35620928]

[56]   Zheng, F.F.; Zhang, P.H.; Xi, Y.; Chen, J.J.; Li, L.L.; Zhu, J.J. Aptamer/graphene quantum dots nanocomposite capped fluorescent mesoporous silica nanoparticles for intracellular drug delivery and real-time monitoring of drug release. *Anal. Chem.,* **2015,** *87*(23), 11739-11745.
[http://dx.doi.org/10.1021/acs.analchem.5b03131] [PMID: 26524192]

[57]   Kumari, S.; Sharma, P.; Yadav, S.; Kumar, J.; Vij, A.; Rawat, P.; Kumar, S.; Sinha, C.; Bhattacharya, J.; Srivastava, C.M.; Majumder, S. A novel synthesis of the graphene oxide-silver (GO-Ag) Nanocomposite for unique physiochemical applications. *ACS Omega,* **2020,** *5*(10), 5041-5047.
[http://dx.doi.org/10.1021/acsomega.9b03976] [PMID: 32201790]

[58]   Jovito, B.L.; Paterno, L.G.; Sales, M.J.A.; Gross, M.A.; Silva, L.P.; de Souza, P.; Báo, S.N. Graphene oxide/zinc oxide nanocomposite displaying selective toxicity to glioblastoma cell lines. *ACS Appl. Bio Mater.,* **2021,** *4*(1), 829-843.
[http://dx.doi.org/10.1021/acsabm.0c01369]

[59]   Zhao, R.; Kong, W.; Sun, M.; Yang, Y.; Liu, W.; Lv, M.; Song, S.; Wang, L.; Song, H.; Hao, R. Highly stable graphene-based nanocomposite (GO-PEI-Ag) with broad-spectrum, long-term antimicrobial activity and antibiofilm effects. *ACS Appl. Mater. Interfaces,* **2018**, *10*(21), 17617-17629.
[http://dx.doi.org/10.1021/acsami.8b03185] [PMID: 29767946]

[60]   Li, J.; Xie, J.; Gao, L.; Li, C.M. Au nanoparticles-3D graphene hydrogel nanocomposite to boost synergistically *in situ* detection sensitivity toward cell-released nitric oxide. *ACS Appl. Mater. Interfaces,* **2015**, *7*(4), 2726-2734.
[http://dx.doi.org/10.1021/am5077777] [PMID: 25580718]

[61]   Chen, H.; Wang, Z.; Zong, S.; Wu, L.; Chen, P.; Zhu, D.; Wang, C.; Xu, S.; Cui, Y. SERS-fluorescence monitored drug release of a redox-responsive nanocarrier based on graphene oxide in tumor cells. *ACS Appl. Mater. Interfaces,* **2014**, *6*(20), 17526-17533.
[http://dx.doi.org/10.1021/am505160v] [PMID: 25272041]

[62]   Cao, Y.; Cheng, Y.; Zhao, G. Near-infrared light-, magneto-, and pH-responsive GO–Fe$_3$O$_4$/Poly(N-isopropylacrylamide)/alginate nanocomposite hydrogel microcapsules for controlled drug release. *Langmuir,* **2021**, *37*(18), 5522-5530.
[http://dx.doi.org/10.1021/acs.langmuir.1c00207] [PMID: 33929865]

[63]   Zhao, R.; Lv, M.; Li, Y.; Sun, M.; Kong, W.; Wang, L.; Song, S.; Fan, C.; Jia, L.; Qiu, S.; Sun, Y.; Song, H.; Hao, R. Stable nanocomposite based on PEGylated and silver nanoparticles loaded graphene oxide for long-term antibacterial activity. *ACS Appl. Mater. Interfaces,* **2017**, *9*(18), 15328-15341.
[http://dx.doi.org/10.1021/acsami.7b03987] [PMID: 28422486]

[64]   Chen, M.L.; He, Y.J.; Chen, X.W.; Wang, J.H. Quantum-dot-conjugated graphene as a probe for simultaneous cancer-targeted fluorescent imaging, tracking, and monitoring drug delivery. *Bioconjug. Chem.,* **2013**, *24*(3), 387-397.
[http://dx.doi.org/10.1021/bc3004809] [PMID: 23425155]

[65]   Chen, Z.; Liu, Y.; Hao, L.; Zhu, Z.; Li, F.; Liu, S. Reduced graphene oxide-zirconium dioxide–thionine nanocomposite integrating recognition, amplification, and signaling for an electrochemical assay of protein kinase activity and inhibitor screening. *ACS Appl. Bio Mater.,* **2018**, *1*(5), 1557-1565.
[http://dx.doi.org/10.1021/acsabm.8b00451] [PMID: 34996206]

[66]   Li, D.; Deng, M.; Yu, Z.; Liu, W.; Zhou, G.; Li, W.; Wang, X.; Yang, D.P.; Zhang, W. Biocompatible and stable GO-coated Fe$_3$O$_4$ nanocomposite: A robust drug delivery carrier for simultaneous tumor MR imaging and targeted therapy. *ACS Biomater. Sci. Eng.,* **2018**, *4*(6), 2143-2154.
[http://dx.doi.org/10.1021/acsbiomaterials.8b00029] [PMID: 33435038]

[67]   Zheng, H.; Ji, Z.; Roy, K.R.; Gao, M.; Pan, Y.; Cai, X.; Wang, L.; Li, W.; Chang, C.H.; Kaweeterawat, C.; Chen, C.; Xia, T.; Zhao, Y.; Li, R. Engineered graphene oxide nanocomposite capable of preventing the evolution of antimicrobial resistance. *ACS Nano,* **2019**, *13*(10), 11488-11499.
[http://dx.doi.org/10.1021/acsnano.9b04970] [PMID: 31566947]

[68]   Jannesari, M.; Akhavan, O.; Madaah Hosseini, H.R.; Bakhshi, B. Graphene/CuO$_2$ nanoshuttles with controllable release of oxygen nanobubbles promoting interruption of bacterial respiration. *ACS Appl. Mater. Interfaces,* **2020**, *12*(32), 35813-35825.
[http://dx.doi.org/10.1021/acsami.0c05732] [PMID: 32664715]

[69]   Liu, Y.; Wang, X.; Si, B.; Wang, T.; Wu, Y.; Liu, Y.; Zhou, Y.; Tong, H.; Zheng, X.; Xu, A. Zinc oxide/graphene oxide nanocomposites efficiently inhibited cadmium-induced hepatotoxicity *via* releasing Zn ions and up-regulating MRP1 expression. *Environ. Int.,* **2022**, *165*, 107327.
[http://dx.doi.org/10.1016/j.envint.2022.107327] [PMID: 35667343]

[70]   Jannesari, M.; Akhavan, O.; Madaah Hosseini, H.R.; Bakhshi, B. Oxygen-Rich Graphene/ZnO$_2$-Ag nanoframeworks with pH-Switchable Catalase/Peroxidase activity as O$_2$ Nanobubble-Self generator

for bacterial inactivation. *J. Colloid Interface Sci.,* **2023**, *637*, 237-250.
[http://dx.doi.org/10.1016/j.jcis.2023.01.079] [PMID: 36701869]

[71]   Tade, R.S.; Patil, P.O. Theranostic prospects of graphene quantum dots in breast cancer. *ACS Biomater. Sci. Eng.,* **2020**, *6*(11), 5987-6008.
[http://dx.doi.org/10.1021/acsbiomaterials.0c01045] [PMID: 33449670]

[72]   Zhang, Z.; Su, T.; Han, Y.; Yang, Z.; Wei, J.; Jin, L.; Fan, H. A convergent synthetic platform for dual anticancer drugs functionalized by reduced graphene nanocomposite delivery for hepatocellular cancer. *Drug Deliv.,* **2021**, *28*(1), 1982-1994.
[http://dx.doi.org/10.1080/10717544.2021.1974606] [PMID: 34569406]

[73]   Fathy, R.M.; Mahfouz, A.Y. Eco-friendly graphene oxide-based magnesium oxide nanocomposite synthesis using fungal fermented by-products and gamma rays for outstanding antimicrobial, antioxidant, and anticancer activities. *J. Nanostruct. Chem.,* **2021**, *11*(2), 301-321.
[http://dx.doi.org/10.1007/s40097-020-00369-3]

[74]   Thapa, R.K.; Kim, J.H.; Jeong, J.H.; Shin, B.S.; Choi, H.G.; Yong, C.S.; Kim, J.O. Silver nanoparticle-embedded graphene oxide-methotrexate for targeted cancer treatment. *Colloids Surf. B Biointerfaces,* **2017**, *153*, 95-103.
[http://dx.doi.org/10.1016/j.colsurfb.2017.02.012] [PMID: 28231500]

[75]   Jafarizad, A.; Aghanejad, A.; Sevim, M.; Metin, Ö.; Barar, J.; Omidi, Y.; Ekinci, D. Gold nanoparticles and reduced graphene oxide-gold nanoparticle composite materials as covalent drug delivery systems for breast cancer treatment. *ChemistrySelect,* **2017**, *2*(23), 6663-6672.
[http://dx.doi.org/10.1002/slct.201701178]

[76]   Zamani, M.; Rostami, M.; Aghajanzadeh, M.; Kheiri Manjili, H.; Rostamizadeh, K.; Danafar, H. Mesoporous titanium dioxide zinc oxide-graphene oxide nanocarriers for colon-specific drug delivery. *J. Mater. Sci.,* **2018**, *53*(3), 1634-1645.
[http://dx.doi.org/10.1007/s10853-017-1673-6]

[77]   Shahbazi, R.; Payan, A.; Fattahi, M. Preparation, evaluations and operating conditions optimization of nano TiO_2 over graphene based materials as the photocatalyst for degradation of phenol. *J. Photochem. Photobiol. Chem.,* **2018**, *364*, 564-576.
[http://dx.doi.org/10.1016/j.jphotochem.2018.05.032]

[78]   Ramalingam, G.; Perumal, N.; Priya, A.K.; Rajendran, S. A review of graphene-based semiconductors for photocatalytic degradation of pollutants in wastewater. *Chemosphere,* **2022**, *300*, 134391.
[http://dx.doi.org/10.1016/j.chemosphere.2022.134391] [PMID: 35367486]

[79]   Minale, M.; Gu, Z.; Guadie, A.; Kabtamu, D.M.; Li, Y.; Wang, X. Application of graphene-based materials for removal of tetracyclines using adsorption and photocatalytic-degradation: A review. *J. Environ. Manage.,* **2020**, *276*, 111310.
[http://dx.doi.org/10.1016/j.jenvman.2020.111310] [PMID: 32891984]

[80]   Yu, L.; Xu, W.; Liu, H.; Bao, Y. Titanium dioxide-reduced graphene oxide composites for photocatalytic degradation of dyes in water. *Catalysts,* **2022**, *12*(11), 1340.
[http://dx.doi.org/10.3390/catal12111340]

[81]   Rubio-Govea, R.; Orona-Návar, C.; Hernández, A. Visible light driven photocatalytic nanocomposite for the degradation of Rhodamine B in water. *IOP Conf Ser Earth Environ Sci,* **2020**, *471*(1), 012014.
[http://dx.doi.org/10.1088/1755-1315/471/1/012014]

[82]   Alwan, S.H.; Salem, K.H.; Alshamsi, H.A. Visible light-driven photocatalytic degradation of Rhodamine B dye onto TiO_2/rGO nanocomposites. *Mater. Today Commun.,* **2022**, *33*, 104558.
[http://dx.doi.org/10.1016/j.mtcomm.2022.104558]

[83]   Tayebi, M.; Kolaei, M.; Tayyebi, A.; Masoumi, Z.; Belbasi, Z.; Lee, B.K. Reduced graphene oxide (RGO) on TiO_2 for an improved photoelectrochemical (PEC) and photocatalytic activity. *Sol. Energy,* **2019**, *190*, 185-194.
[http://dx.doi.org/10.1016/j.solener.2019.08.020]

[84]   Alwan, S.H.; Salem, K.H.; Alshamsi, H.A. Visible light-driven photocatalytic degradation of Rhodamine B dye onto TiO$_2$/rGO nanocomposites. *Mater. Today Commun.,* **2022**, *33*, 104558.
[http://dx.doi.org/10.1016/j.mtcomm.2022.104558]

[85]   de Almeida, G.C.; Della Santina Mohallem, N.; Viana, M.M. Ag/GO/TiO$_2$ nanocomposites: the role of the interfacial charge transfer for application in photocatalysis. *Nanotechnology,* **2022**, *33*(3), 035710.
[http://dx.doi.org/10.1088/1361-6528/ac2f24] [PMID: 34638115]

[86]   Huang, J.; Chen, W.; Yu, X.; Fu, X.; Zhu, Y.; Zhang, Y. Fabrication of a ternary BiOCl/CQDs/rGO photocatalyst: The roles of CQDs and rGO in adsorption-photocatalytic removal of ciprofloxacin. *Colloids Surf. A Physicochem. Eng. Asp.,* **2020**, *597*, 124758.
[http://dx.doi.org/10.1016/j.colsurfa.2020.124758]

[87]   Afreen, G.; Shoeb, M.; Upadhyayula, S. Effectiveness of reactive oxygen species generated from rGO/CdS QD heterostructure for photodegradation and disinfection of pollutants in waste water. *Mater. Sci. Eng. C,* **2020**, *108*, 110372.
[http://dx.doi.org/10.1016/j.msec.2019.110372] [PMID: 31924024]

[88]   Shad, N.A.; Sajid, M.M.; Afzal, A.M.; Amin, N.; Javed, Y.; Hassan, S.; Imran, Z.; Razaq, A.; Yousaf, M.I.; Munawar, A.; Sharma, S.K. Facile synthesis of Bi2WO6/rGO nanocomposites for photocatalytic and solar cell applications. *Ceram. Int.,* **2021**, *47*(11), 16101-16110.
[http://dx.doi.org/10.1016/j.ceramint.2021.02.185]

[89]   Ma, J.; Zhao, B.; Fan, X.; Wang, W.; Chen, X.; Shao, N.; Jiang, P. Ag/rGO/Bi2WO6 nanocomposite as a highly efficient and stable photocatalyst for Rhodamine B degradation under visible light irradiation. *Diamond Related Mater.,* **2022**, *127*, 109143.
[http://dx.doi.org/10.1016/j.diamond.2022.109143]

[90]   Margaret, S.M.; Paul Winston, A.J.P.; Muthupandi, S.; Shobha, P.; Sagayaraj, P. Enhanced photocatalytic degradation of phenol using urchin-Like ZnO microrod-reduced graphene oxide composite under visible-light irradiation. *J. Nanomater.,* **2021**, *2021*, 1-11.
[http://dx.doi.org/10.1155/2021/5551148]

[91]   Yuan, A.; Lei, H.; Xi, F.; Liu, J.; Qin, L.; Chen, Z.; Dong, X. Graphene quantum dots decorated graphitic carbon nitride nanorods for photocatalytic removal of antibiotics. *J. Colloid Interface Sci.,* **2019**, *548*, 56-65.
[http://dx.doi.org/10.1016/j.jcis.2019.04.027] [PMID: 30981964]

[92]   Men, X.; Wu, Y.; Chen, H.; Fang, X.; Sun, H.; Yin, S.; Qin, W. Facile fabrication of TiO$_2$/Graphene composite foams with enhanced photocatalytic properties. *J. Alloys Compd.,* **2017**, *703*, 251-257.
[http://dx.doi.org/10.1016/j.jallcom.2017.01.353]

[93]   Shabestari, M.E.; Baselga, J.; Martin, O. Photocatalytic behavior of supported copper double salt: The role of graphene oxide. *J. Chem.,* **2022**, *2022*, 1-9.
[http://dx.doi.org/10.1155/2022/7844259]

[94]   Ismail, A.A.; Faisal, M.; Al-Haddad, A. Mesoporous WO$_3$-graphene photocatalyst for photocatalytic degradation of Methylene Blue dye under visible light illumination. *J. Environ. Sci.,* **2018**, *66*, 328-337.
[http://dx.doi.org/10.1016/j.jes.2017.05.001] [PMID: 29628102]

[95]   Ramar, V.; Balasubramanian, K. Reduced graphene Oxide/WO$_3$ nanorod composites for photocatalytic degradation of methylene blue under sunlight irradiation. *ACS Appl. Nano Mater.,* **2021**, *4*(5), 5512-5521.
[http://dx.doi.org/10.1021/acsanm.1c00863]

[96]   Sahoo, D.P.; Patnaik, S.; Parida, K. Construction of a Z-Scheme Dictated WO 3– X /Ag/ZnCr LDH Synergistically Visible Light-Induced Photocatalyst towards Tetracycline Degradation and H 2 Evolution. *ACS Omega,* **2019**, *4*(12), 14721-14741.
[http://dx.doi.org/10.1021/acsomega.9b01146]

[97]   Govindaraj, T.; Mahendran, C.; Manikandan, V.S.; Archana, J.; Shkir, M.; Chandrasekaran, J. Fabrication of WO3 nanorods/RGO hybrid nanostructures for enhanced visible-light-driven photocatalytic degradation of Ciprofloxacin and Rhodamine B in an ecosystem. *J. Alloys Compd.,* **2021**, *868*, 159091.
[http://dx.doi.org/10.1016/j.jallcom.2021.159091]

[98]   Xu, J.; Cui, Y.; Han, Y.; Hao, M.; Zhang, X. ZnO–graphene composites with high photocatalytic activities under visible light. *RSC Advances,* **2016**, *6*(99), 96778-96784.
[http://dx.doi.org/10.1039/C6RA19622E]

[99]   Frindy, S.; Sillanpää, M. Synthesis and application of novel α- Fe$_2$O$_3$/graphene for visible-light enhanced photocatalytic degradation of RhB. *Mater. Des.,* **2020**, *188*, 108461.
[http://dx.doi.org/10.1016/j.matdes.2019.108461]

[100] Hernández-Majalca, B.C.; Meléndez-Zaragoza, M.J.; Salinas-Gutiérrez, J.M.; López-Ortiz, A.; Collins-Martínez, V. Visible-light photo-assisted synthesis of GO-TiO$_2$ composites for the photocatalytic hydrogen production. *Int. J. Hydrogen Energy,* **2019**, *44*(24), 12381-12389.
[http://dx.doi.org/10.1016/j.ijhydene.2018.10.152]

[101] Li, J.; Tang, Y.; Jin, R.; Meng, Q.; Chen, Y.; Long, X.; Wang, L.; Guo, H.; Zhang, S. Ultrasonic-microwave assisted synthesis of GO/g-C3N4 composites for efficient photocatalytic H$_2$ evolution. *Solid State Sci.,* **2019**, *97*, 105990.
[http://dx.doi.org/10.1016/j.solidstatesciences.2019.105990]

[102] Li, W.; Wang, X.; Li, M.; He, S.; Ma, Q.; Wang, X. Construction of Z-scheme and p-n heterostructure: Three-dimensional porous g-C3N4/graphene oxide-Ag/AgBr composite for high-efficient hydrogen evolution. *Appl. Catal. B,* **2020**, *268*, 118384.
[http://dx.doi.org/10.1016/j.apcatb.2019.118384]

[103] Wang, S.; Zhu, Y.; Jiang, M.; Cui, J.; Zhang, Y.; He, W. TiO$_2$ nanotube/g-C3N4/graphene composite as high performance anode material for Na-ion batteries. *Vacuum,* **2021**, *184*, 109926.
[http://dx.doi.org/10.1016/j.vacuum.2020.109926]

[104] Peng, G.Z.; Hardiansyah, A.; Lin, H-T.; Lee, R-Y.; Kuo, C-Y.; Pu, Y-C.; Liu, T-Y. Photocatalytic degradation and reusable SERS detection by Ag nanoparticles immobilized on g-C3N4/graphene oxide nanosheets. *Surf. Coat. Tech.,* **2022**, *435*, 128212.
[http://dx.doi.org/10.1016/j.surfcoat.2022.128212]

[105] Liu, J.; Wei, X.; Sun, W.; Guan, X.; Zheng, X.; Li, J. Fabrication of S-scheme CdS-g-C$_3$N$_4$-graphene aerogel heterojunction for enhanced visible light driven photocatalysis. *Environ. Res.,* **2021**, *197*, 111136.
[http://dx.doi.org/10.1016/j.envres.2021.111136] [PMID: 33839114]

[106] Lei, Y.; Yang, C.; Hou, J.; Wang, F.; Min, S.; Ma, X.; Jin, Z.; Xu, J.; Lu, G.; Huang, K-W. Strongly coupled CdS/graphene quantum dots nanohybrids for highly efficient photocatalytic hydrogen evolution: Unraveling the essential roles of graphene quantum dots. *Appl. Catal. B,* **2017**, *216*, 59-69.
[http://dx.doi.org/10.1016/j.apcatb.2017.05.063]

[107] Lu, N.; Wang, P.; Su, Y.; Yu, H.; Liu, N.; Quan, X. Construction of Z-Scheme g-C$_3$N$_4$/RGO/WO$_3$ with in situ photoreduced graphene oxide as electron mediator for efficient photocatalytic degradation of ciprofloxacin. *Chemosphere,* **2019**, *215*, 444-453.
[http://dx.doi.org/10.1016/j.chemosphere.2018.10.065] [PMID: 30336321]

[108] Liu, Y.; Xu, X.; Li, A.; Si, Z.; Wu, X.; Ran, R.; Weng, D. A strategy to construct (reduced graphene oxide, γ-Fe$_2$O$_3$)/C$_3$N$_4$ step-scheme photocatalyst for visible-light water splitting. *Catal. Commun.,* **2021**, *157*, 106327.
[http://dx.doi.org/10.1016/j.catcom.2021.106327]

[109] Saquib, M.; Kaushik, R.; Halder, A. Photoelectrochemical activity of Ag Coated 2D-TiO $_2$ /RGO heterojunction for hydrogen evolution reaction and environmental remediation. *ChemistrySelect,* **2020**, *5*(21), 6376-6388.

[http://dx.doi.org/10.1002/slct.202000843]

[110] Nethravathi, P.C.; Suresh, D. Silver-doped ZnO embedded reduced graphene oxide hybrid nanostructured composites for superior photocatalytic hydrogen generation, dye degradation, nitrite sensing and antioxidant activities. *Inorg. Chem. Commun.,* **2021**, *134*, 109051.
[http://dx.doi.org/10.1016/j.inoche.2021.109051]

[111] Victoria Tafoya, J.P.; Doszczeczko, S.; Titirici, M.M.; Jorge Sobrido, A.B. Enhancement of the electrocatalytic activity for the oxygen reduction reaction of boron-doped reduced graphene oxide *via* ultrasonic treatment. *Int. J. Hydrogen Energy,* **2022**, *47*(8), 5462-5473.
[http://dx.doi.org/10.1016/j.ijhydene.2021.11.127]

[112] Ji, Z.; Chen, J.; Guo, Z.; Zhao, Z.; Cai, R.; Rigby, M.T.P.; Haigh, S.J.; Perez-Page, M.; Shen, Y.; Holmes, S.M. Graphene/carbon structured catalyst layer to enhance the performance and durability of the high-temperature proton exchange membrane fuel cells. *J. Energy Chem.,* **2022**, *75*, 399-407.
[http://dx.doi.org/10.1016/j.jechem.2022.08.004]

[113] Velayutham, R.; Palanisamy, K.; Manikandan, R.; Velumani, T.; Kumar AP, S.; Puigdollers, J.; Chul Kim, B. Synergetic effect induced/tuned bimetallic nanoparticles (Pt-Ni) anchored graphene as a catalyst for oxygen reduction reaction and scalable SS-314L serpentine flow field proton exchange membrane fuel cells (PEMFCs). *Mater. Sci. Eng. B,* **2022**, *282*, 115780.
[http://dx.doi.org/10.1016/j.mseb.2022.115780]

[114] Peng, K.J.; Lai, J.Y.; Liu, Y.L. Nanohybrids of graphene oxide chemically-bonded with Nafion: Preparation and application for proton exchange membrane fuel cells. *J. Membr. Sci.,* **2016**, *514*, 86-94.
[http://dx.doi.org/10.1016/j.memsci.2016.04.062]

[115] Chen, J.; Bailey, J.J.; Britnell, L.; Perez-Page, M.; Sahoo, M.; Zhang, Z.; Strudwick, A.; Hack, J.; Guo, Z.; Ji, Z.; Martin, P.; Brett, D.J.L.; Shearing, P.R.; Holmes, S.M. The performance and durability of high-temperature proton exchange membrane fuel cells enhanced by single-layer graphene. *Nano Energy,* **2022**, *93*, 106829.
[http://dx.doi.org/10.1016/j.nanoen.2021.106829]

[116] Devrim, Y.; Bulanık Durmuş, G.N. Composite membrane by incorporating sulfonated graphene oxide in polybenzimidazole for high temperature proton exchange membrane fuel cells. *Int. J. Hydrogen Energy,* **2022**, *47*(14), 9004-9017.
[http://dx.doi.org/10.1016/j.ijhydene.2021.12.257]

[117] Gizem Güneştekin, B.; Medetalibeyoglu, H.; Atar, N.; Lütfi Yola, M. Efficient direct-methanol fuel cell based on graphene quantum dots/multi-walled carbon nanotubes composite. *Electroanalysis,* **2020**, *32*(9), 1977-1982.
[http://dx.doi.org/10.1002/elan.202060074]

[118] Hanifah, M.F.R.; Jaafar, J.; Othman, M.H.D.; Ismail, A.F.; Rahman, M.A.; Yusof, N.; Aziz, F.; Rahman, N.A.A. One-pot synthesis of efficient reduced graphene oxide supported binary Pt-Pd alloy nanoparticles as superior electro-catalyst and its electro-catalytic performance toward methanol electro-oxidation reaction in direct methanol fuel cell. *J. Alloys Compd.,* **2019**, *793*, 232-246.
[http://dx.doi.org/10.1016/j.jallcom.2019.04.114]

[119] Li, Y.; Zhou, Y.; Zhu, C.; Hu, Y.H.; Gao, S.; Liu, Q.; Cheng, X.; Zhang, L.; Yang, J.; Lin, Y. Porous graphene doped with Fe/N/S and incorporating Fe_3O_4 nanoparticles for efficient oxygen reduction. *Catal. Sci. Technol.,* **2018**, *8*(20), 5325-5333.
[http://dx.doi.org/10.1039/C8CY01328D]

[120] Farzaneh, A.; Goharshadi, E.K.; Gharibi, H.; Saghatoleslami, N.; Ahmadzadeh, H. Insights on the superior performance of nanostructured nitrogen-doped reduced graphene oxide in comparison with commercial Pt/C as cathode electrocatalyst layer of passive direct methanol fuel cell. *Electrochim. Acta,* **2019**, *306*, 220-228.
[http://dx.doi.org/10.1016/j.electacta.2019.03.120]

[121]   Forootan Fard, H.; Khodaverdi, M.; Pourfayaz, F.; Ahmadi, M.H. Application of N-doped carbon nanotube-supported Pt-Ru as electrocatalyst layer in passive direct methanol fuel cell. *Int. J. Hydrogen Energy,* **2020**, *45*(46), 25307-25316.
[http://dx.doi.org/10.1016/j.ijhydene.2020.06.254]

[122]   Baronia, R.; Goel, J.; Gautam, G.; Singh, D.; Singhal, S.K. Synthesis and characterization of nitrogen doped reduced graphene oxide (N-rGO) supported PtCu anode catalysts for direct methanol fuel cell. *J. Nanosci. Nanotechnol.,* **2019**, *19*(7), 3832-3843.
[http://dx.doi.org/10.1166/jnn.2019.16301] [PMID: 30764941]

[123]   Jiang, Z.J.; Jiang, Z.; Tian, X.; Luo, L.; Liu, M. Sulfonated Holey Graphene Oxide (SHGO) filled sulfonated poly(ether ether ketone) membrane: The role of holes in the SHGO in improving its performance as proton exchange membrane for direct methanol fuel cells. *ACS Appl. Mater. Interfaces,* **2017**, *9*(23), 20046-20056.
[http://dx.doi.org/10.1021/acsami.7b00198] [PMID: 28535030]

[124]   Holmes, S.M.; Balakrishnan, P.; Kalangi, V.S.; Zhang, X.; Lozada-Hidalgo, M.; Ajayan, P.M.; Nair, R.R. 2D crystals significantly enhance the performance of a working fuel cell. *Adv. Energy Mater.,* **2017**, *7*(5), 1601216.
[http://dx.doi.org/10.1002/aenm.201601216]

[125]   Su, H.; Hu, Y.H. Recent advances in graphene-based materials for fuel cell applications. *Energy Sci. Eng.,* **2021**, *9*(7), 958-983.
[http://dx.doi.org/10.1002/ese3.833]

[126]   Liu, T.; Li, C.; Yuan, Q. Facile synthesis of PtCu alloy/graphene oxide hybrids as improved electrocatalysts for alkaline fuel cells. *ACS Omega,* **2018**, *3*(8), 8724-8732.
[http://dx.doi.org/10.1021/acsomega.8b01347] [PMID: 31459004]

[127]   Raja, A.; Son, N.; Swaminathan, M.; Kang, M. Electrochemical behavior of heteroatom doped on reduced graphene oxide with RuO_2 for HER, OER, and supercapacitor applications. *J. Taiwan Inst. Chem. Eng.,* **2022**, *138*, 104471.
[http://dx.doi.org/10.1016/j.jtice.2022.104471]

[128]   Kumar, R.; Sahoo, S.; Joanni, E.; Singh, R.K. A review on the current research on microwave processing techniques applied to graphene-based supercapacitor electrodes: An emerging approach beyond conventional heating. *J. Energy Chem.,* **2022**, *74*, 252-282.
[http://dx.doi.org/10.1016/j.jechem.2022.06.051]

[129]   Navalon, S.; Dhakshinamoorthy, A.; Alvaro, M.; Garcia, H. Carbocatalysis by graphene-based materials. *Chem. Rev.,* **2014**, *114*(12), 6179-6212.
[http://dx.doi.org/10.1021/cr4007347] [PMID: 24867457]

[130]   Machado, B.F.; Serp, P. Graphene-based materials for catalysis. *Catal. Sci. Technol.,* **2012**, *2*(1), 54-75.
[http://dx.doi.org/10.1039/C1CY00361E]

[131]   Scheuermann, G.M.; Rumi, L.; Steurer, P.; Bannwarth, W.; Mülhaupt, R. Palladium nanoparticles on graphite oxide and its functionalized graphene derivatives as highly active catalysts for the Suzuki-Miyaura coupling reaction. *J. Am. Chem. Soc.,* **2009**, *131*(23), 8262-8270.
[http://dx.doi.org/10.1021/ja901105a] [PMID: 19469566]

[132]   Tang, Z.; Shen, S.; Zhuang, J.; Wang, X. Noble-metalpromoted three- dimensional macroassembly of single-layered graphene oxide. *Angew. Chem.,* **2010**, *122*(27), 4707-4711.
[http://dx.doi.org/10.1002/ange.201000270]

[133]   Li, Y.; Fan, X.; Qi, J.; Ji, J.; Wang, S.; Zhang, G.; Zhang, F. Palladium nanoparticle-graphene hybrids as active catalysts for the Suzuki reaction. *Nano Res.,* **2010**, *3*(6), 429-437.
[http://dx.doi.org/10.1007/s12274-010-0002-z]

[134]   Zeng, T.; Zhang, X.; Ma, Y.; Niu, H.; Cai, Y. A novel Fe_3O_4-graphene-Au multifunctional

nanocomposite: Green synthesis and catalytic application. *J. Mater. Chem.,* **2012**, *22*(35), 18658-18663.
[http://dx.doi.org/10.1039/c2jm34198k]

[135]   Salam, N.; Sinha, A.; Mondal, P.; Roy, A.S.; Jana, N.R.; Islam, S.M. Efficient and reusable graphene-γ-Fe$_2$O$_3$ magnetic nano-composite for selective oxidation and one-pot synthesis of 1,2,3-triazole using a green solvent. *RSC Advances,* **2013**, *3*(39), 18087-18098.
[http://dx.doi.org/10.1039/c3ra43184c]

[136]   Kim, J.D.; Palani, T.; Kumar, M.R.; Lee, S.; Choi, H.C. Preparation of reusable Ag-decorated graphene oxide catalysts for decarboxylative cycloaddition. *J. Mater. Chem.,* **2012**, *22*(38), 20665-20670.
[http://dx.doi.org/10.1039/c2jm35512d]

[137]   Salam, N.; Sinha, A.; Roy, A.S.; Mondal, P.; Jana, N.R.; Islam, S.M. Synthesis of silver-graphene nanocomposite and its catalytic application for the one-pot three-component coupling reaction and one-pot synthesis of 1,4-disubstituted 1,2,3-triazoles in water. *RSC Advances,* **2014**, *4*(20), 10001-10012.
[http://dx.doi.org/10.1039/c3ra47466f]

[138]   Mondal, P.; Sinha, A.; Salam, N.; Roy, A.S.; Jana, N.R.; Islam, S.M. Enhanced catalytic performance by copper nanoparticle-graphene based composite. *RSC Advances,* **2013**, *3*(16), 5615-5623.
[http://dx.doi.org/10.1039/c3ra23280h]

[139]   Honraedt, A.; Le Callonnec, F.; Le Grognec, E.; Fernandez, V.; Felpin, F.X. C-H arylation of benzoquinone in water through aniline activation: synergistic effect of graphite-supported copper oxide nanoparticles. *J. Org. Chem.,* **2013**, *78*(9), 4604-4609.
[http://dx.doi.org/10.1021/jo4004426] [PMID: 23551327]

[140]   Yao, K.X.; Liu, X.; Li, Z.; Li, C.C.; Zeng, H.C.; Han, Y. Preparation of a Ru-nanoparticles/defectiv-graphene composite as a highly efficient arene- hydrogenation catalyst. *ChemCatChem,* **2012**, *4*(12), 1938-1942.
[http://dx.doi.org/10.1002/cctc.201200354]

[141]   Hu, F.; Patel, M.; Luo, F.; Flach, C.; Mendelsohn, R.; Garfunkel, E.; He, H.; Szostak, M. Graphene-catalyzed direct Friedel–Crafts alkylation reactions: Mechanism, selectivity, and synthetic utility. *J. Am. Chem. Soc.,* **2015**, *137*(45), 14473-14480.
[http://dx.doi.org/10.1021/jacs.5b09636] [PMID: 26496423]

[142]   Allahresani, A.; Nasseri, M.A.; Akbari, A.; Nasab, B.Z. Graphene oxide based solid acid as an efficient and reusable nano-catalyst for the green synthesis of diindolyl-oxindole derivatives in aqueous media. *React. Kinet. Mech. Catal.,* **2015**, *116*(1), 249-259.
[http://dx.doi.org/10.1007/s11144-015-0883-7]

[143]   Peng, X.; Zen, Y.; Liu, Q.; Liu, L.; Wang, H. Graphene oxide as a green carbon material for cross-coupling of indoles with ethers *via* oxidation and the Friedel- Crafts reaction. *Org. Chem. Front.,* **2019**, *6*(21), 3615-3619.
[http://dx.doi.org/10.1039/C9QO00926D]

[144]   Parsa, H.; Taghizadeh, M. J.; Mossavi, M.; Hosseini, J. Synthesis, characterization and using Fe$_3$O$_4$@ SiO$_2$@ FeCl3 as a new nanocatalyst for aza- Michael reaction between amines and ethyl crotonate. *Iranian J. Cataly.,* **2020**, *10*(3), 189-194.

[145]   Garg, B.; Ling, Y. C. Versatilities of graphene-based catalysts in organic transformations. *Green Mater. 1,* **2013**, *1*(1), 47-61.
[http://dx.doi.org/10.1680/gmat.12.00008]

[146]   Ling, X.; Zhang, J. First-layer effect in graphene-enhanced Raman scattering. *Small,* **2010**, *6*(18), 2020-2025.
[http://dx.doi.org/10.1002/smll.201000918] [PMID: 20730826]

[147]   Dong, J.; Zhao, X.; Gao, W.; Han, Q.; Qi, J.; Wang, Y.; Guo, S.; Sun, M. Nanoscale vertical arrays of

gold nanorods by self-assembly: Physical mechanism and application. *Nanoscale Res. Lett.,* **2019,** *14*(1), 118.
[http://dx.doi.org/10.1186/s11671-019-2946-6] [PMID: 30941536]

[148] Zhao, X.; Dong, J.; Cao, E.; Han, Q.; Gao, W.; Wang, Y.; Qi, J.; Sun, M. Plasmon-exciton coupling by hybrids between graphene and gold nanorods vertical array for sensor. *Appl. Mater. Today,* **2019,** *14,* 166-174.
[http://dx.doi.org/10.1016/j.apmt.2018.12.013]

[149] Jun, D. Plasmon-exciton coupling for nanophotonic sensing on chip. *Opt. Express,* **2020,** *28*(14), 20817-20829.
[http://dx.doi.org/10.1364/OE.387867]

[150] Dong, J. Flexible and transparent AuNP/G/AuNP 'Sandwich' substrate for surface-enhanced raman scattering. *Mater. Today Nano,* **2019,** (Apr), 100067.
[http://dx.doi.org/10.1016/j.mtnano.2019.100067]

[151] Xie, Y.; Li, Y.; Niu, L.; Wang, H.; Qian, H.; Yao, W. A novel surface-enhanced Raman scattering sensor to detect prohibited colorants in food by graphene/silver nanocomposite. *Talanta,* **2012,** *100,* 32-37.
[http://dx.doi.org/10.1016/j.talanta.2012.07.080] [PMID: 23141308]

[152] Ding, X.; Kong, L.; Wang, J.; Fang, F.; Li, D.; Liu, J. Highly sensitive SERS detection of Hg2+ ions in aqueous media using gold nanoparticles/graphene heterojunctions. *ACS Appl. Mater. Interfaces,* **2013,** *5*(15), 7072-7078.
[http://dx.doi.org/10.1021/am401373e] [PMID: 23855919]

[153] Fu, W.L.; Zhen, S.J.; Huang, C.Z. One-pot green synthesis of graphene oxide/gold nanocomposites as SERS substrates for malachite green detection. *Analyst,* **2013,** *138*(10), 3075-3081.
[http://dx.doi.org/10.1039/c3an00018d] [PMID: 23586069]

[154] Links, D. A. Nanoscale Highly reproducible and sensitive surface-enhanced Raman scattering from colloidal plasmonic nanoparticle *via* stabilization of hot spots in graphene oxide liquid crystal. *Nanoscale,* **2012,** *4,* 6649-6657.
[http://dx.doi.org/10.1039/c2nr31035j]

[155] Gupta, V. K.; Atar, N.; Torul, H. A novel glucose biosensor platform based on Ag@AuNPs modified graphene oxide nanocomposite and SERS application. *J. Coll. Interf. Sci.,* **2013,** *406,* 231-237.
[http://dx.doi.org/10.1016/j.jcis.2013.06.007]

[156] Sahoo, N.G.; Sandeep, M. A process of manufacturing. Indian Patent 352780, 2016.

[157] Han, B.; Gao, Y-Y.; Zhang, Y-L.; Liu, Y-Q.; Ma, Z-C.; Guo, Q.; Zhu, L.; Chen, Q-D.; Sun, H-B. Multi-field-coupling energy conversion for flexible manipulation of graphene-based soft robots. *Nano Energy,* **2020,** *71,* 104578.
[http://dx.doi.org/10.1016/j.nanoen.2020.104578]

CHAPTER 3

Toxicity of Graphene Family and Remediation Approaches

Shalini Bhatt[*, 1], **Rakshit Pathak**[1] and **Neha Faridi**[2]

[1] *Centre of Excellence for Research, P.P. Savani University, Surat-394125, Gujrat, India*

[2] *Defence Institute of Bio-Energy Research, DRDO, Haldwani, Uttarakhand- 263139, India*

Abstract: Graphene family nanomaterials (GFNs) appeared to be extensively exploited in numerous diverse fields predominantly in the biomedical sector, owing to distinctive physical, chemical as well as biological/biocompatible characteristics. With the expanding uses, individuals are now exposed to GFNs more often and through a variety of different routes. Upon exposure, these materials exhibit varying amounts of toxicity in biological systems used for toxicological examinations. Administration by various routes leads to penetration by breaching physical barriers and eventually gets disseminated in various tissues or may accumulate in the cells, and subsequently may get eliminated from the body. The present chapter provides information about the toxic effect of the GFNs in several organs encompassing studies in various animals and cell lines. Different factors including lateral size, functionality, concentration as well as protein corona formation, *etc.* influencing the toxicity status of the GFNs have been elaborated. Furthermore, some representative toxicity mechanisms include mitochondrial as well as DNA impairment, and oxidative damage to name a few. At last, we have provided toxicity remediation approaches for GFNs.

Keywords: Bioimaging, Graphene, Graphene Family Nanomaterials, Mechanism, Remediation, Toxicity.

INTRODUCTION

Graphene (Gr) has turned up as a phenomenal nanocarbon with exceptional characteristics. Its carbon (C) atoms get organized in a two-dimensional (2D) hexagonal and planar pattern. Each of the C-atom is hybridized in sp^2 fashion and contains four bonds, including a single pie bond that is out of the plane and a single sigma bond between each of its three neighbors [1]. Graphene-family nanomaterials (GFN) include Gr and its derivatives like monolayer Gr, few-layer Gr (FLG), Gr nanosheets, Gr nanoplatelets (GNPs), Gr nanoribbons, graphene

---

[*] **Corresponding author Shalini Bhatt:** Centre of Excellence for Research, P.P. Savani University, Surat-394125, Gujrat, India; E-mail: shalini.bhatt@ppsu.ac.in

Vinay Deep Punetha (Ed.)

oxide (GO), and reduced GO (rGO), *etc.* as illustrated in Fig. (**1**) [2, 3]. Like Gr, they are 2D C-based materials having honeycomb patterns that may be further modified by reduction, oxidation, as well as functionalization [4]. GFNs have demonstrated potential uses in diverse multiple fields, including electronics, biomedicine, antioxidant, and antimicrobial activities, sensing, bio-imaging, bioenergy, biodiesel catalysts, and environmental remediation, on account of remarkable mechanical, electrical as well as thermal properties displayed by them [2, 5 - 9].

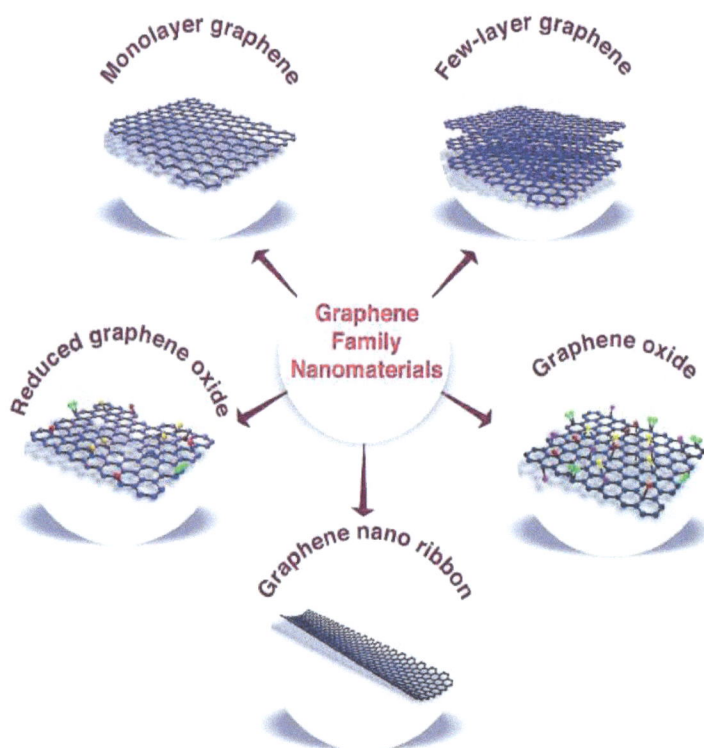

Fig. (1). Schematic illustration of graphene family nanomaterials (GFNs).

Functionalized Gr materials display improved interfacial contact and interactions with bacterial and mammalian cells as well as proteins, lipids, *etc.*, making them a promising nano-system for use in wide applications most importantly biomedicine and bioengineering. Owing to its widespread use in diverse fields, the production of Gr and its derivatives has received huge interest. Extensive production has aroused biocompatibility concerns amongst the environment as well as the population. To assess the potential health and ecological effects of graphene family nanomaterials (GFNs), the Graphene Flagship Project (GFP) has invested significant effort. The GFP in conjunction with the Human Brain Project is the first Future Technology- Flagship Project of the European Commission, whose goal is to direct significant scientific and technical incitement through extensive, interdisciplinary activities of research and development [10]. It started in 2013 as a 10-year plan involving more than 150 educational and research organizations from more than 20 countries, making up the alliance [11]. The growing use of GFN has increased their exposure across the population. Industries as well as production plants employing these materials in their processing must identify and manage potential hazards that could be posed upon its exposure. Therefore, safety must be thoroughly examined for their market integration to be successful. However, risk assessment using control banding, previously used for chemicals, has recently contributed to the field of nanomaterial risk assessment and management [12, 13]. This method is used to ensure the safety of employees exposed to novel products about which not much information is available. In summery, the "hazard bands" of the novel product are assigned following the degree of hazard associated with the corresponding known items. Each band represents a risk management method. In particular, the hazard band of a GFN can be identified depending on the known toxicity of a comparable GBM from the same subfamily, as established *in vitro* or *in vivo* using animal models. Furthermore, a band of exposure will be established depending on the efficiency as well as the amount used. Crossing these bands necessitates some sort of risk management at the workstation, including precautions or controls [14]. Along with the toxicological hazards associated with extensive production and use, the potential of unintended occupational as well as environmental exposure is increasing. Occupational exposure to these materials poses potent toxicity to researchers and workers. They gain access into the body by oral, subcutaneous/intravenous/intraperitoneal injections and intra-tracheal instillations. By breaching various barriers in the body namely, the blood-brain, blood-testis, blood-placenta, and blood-air, they get gathered in several internal sensitive organs such as the liver, spleen, and lungs causing chronic as well as acute tissue damage [15].

GFNs toxicity databases are so lacking that it might be difficult to assign a hazard band to nanomaterials. Additionally, the extent to which nanomaterials are being used for industrial applications is growing quickly [16, 17], and the safety evaluation of these materials becomes a prerequisite, which is intrinsically linked with the development of new technology. Individual assessment of toxicity evaluation became a time-consuming economically expensive task. As a result, additional supplementary alternatives including grouping and quantitative structure–activity connection were proposed for toxicity evaluation [18]. This paved the way for the scientific community to achieve significant advancements in the understanding of specific mechanisms of action in the field of nano-toxicology. In the meantime, a safe-by-design concept related to a control banding method is increasingly seen as incorporating the predicted safety implications of materials into the design and manufacturing stages [19 - 21]. As a result, this method suggests that safety should be a top priority when developing nanomaterials, and the adverse effects of the proposed GFNs should be known or researched from the start of the development. Creating a safer-by-design nanomaterial requires a thorough investigation of its negative effects and the knowledge of physico-chemical property will affect its toxicity.

Reports suggest that the Gr sheet market size will grow from $254.0 million in 2022 to $ 1719.0 million by 2030 [22] due to the growing demand for GFNs with extraordinary properties such as high mechanical strength, tensile strength, elasticity, electrical conductivity, *etc.* With the increasing use of GFN, serious concerns regarding its compatibility stressed the researchers, so that potential toxic effects could be minimized. The present chapter describes the toxicity of GFNs with an emphasis on the toxicity to the population as well as the environment. We discuss the principal exposure routes as well as the major organs affected such as the immune system of the body, lungs, cardiovascular, central nervous system (CNS) and reproductive systems.

TOXICITY OF GRAPHENE FAMILY NANOMATERIALS

The last decade has witnessed extensive research on the toxicological implications of GFNs. This can be witnessed by the availability of an enormous number of research papers as well as review articles and chapters studying the toxicity of GFNs. According to reports, GFNs have been demonstrated to exhibit noticeable cytotoxicity in experiments conducted on several bacteria and animal models both *in vitro* and *in vivo* [23]. Owing to their superior solubility, and stability in water and under physiological conditions than to other GFNs, GO, and rGO have been the subject of the majority of investigations. Nanomaterials can be exposed to the body by inhalation, skin adsorption, and ingestion in environmental or occupational contexts [24]. Entry *via* dermal injections often occurs during tattoo making where some of the tattoo pigments are nanomaterial [25, 26]. Intravenous, subcutaneous, and intraperitoneal entry of the nanomaterials often occurs during medication when nanoparticles are created for particular medicinal use. Nanomaterials may later move through the body and arrive at locations outside of their original entrance site. Eventually, the substances show their biological or toxicological impacts on a particular target organ or multiple organs [27]. Fig. (**2**) depicts the toxicity of GFNs to various organs and their effects on the respective organs.

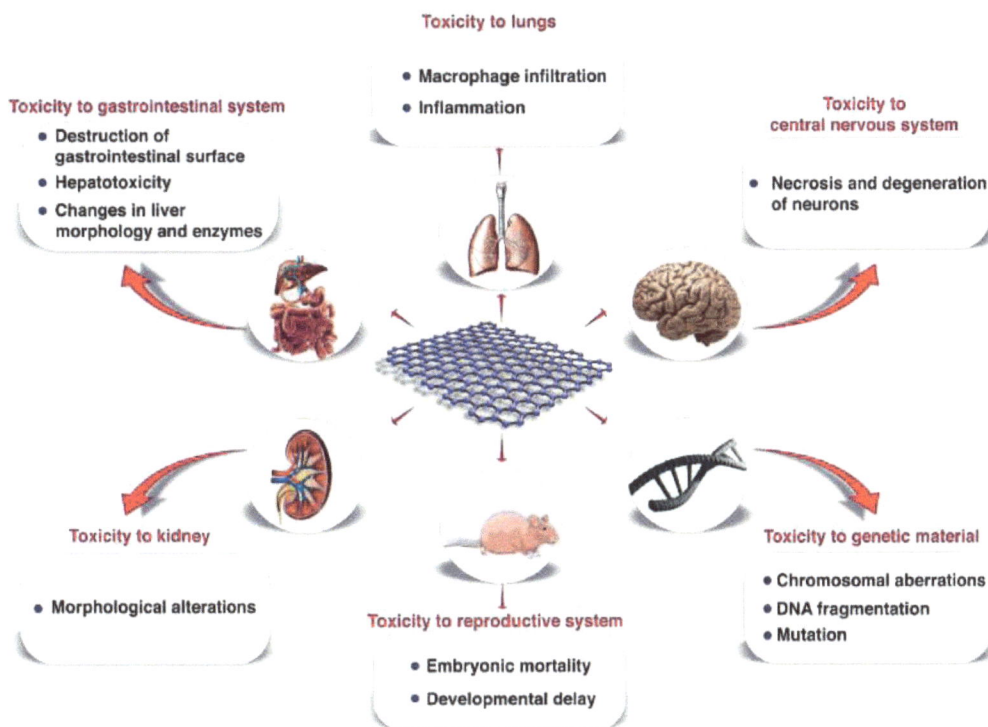

Fig. (2). Toxicity of graphene family nanomaterials to different systems in the body.

Further various *in vivo* as well as *in vitro* studies have been performed by the scientific communities for the assessment of the toxicity potential of GFN. Table **1** and Table **2** outline some of the notable *in vivo* and *in vitro* studies so that a quick overview could be conveyed to the readers. Moreover, these studies have been discussed in the upcoming sections involving the comprehensive toxicological effect of GFNs in the immune system, cardiovascular, gastrointestinal tract, and reproductive system.

Table 1. *In vivo* **studies involving toxicity evaluation.**

Graphene Family Nanomaterial	Dose	Cell Line/Organism	Toxic Effect	Refs.
GO, FLG, sFLG	5 µg/mL	NHBE cells.	Cytotoxic effects such as necrosis and apoptosis, and cell death.	[28]
P-GO, bP-GO	5 and 25 µg/mL	Human dendritic cells.	Suppression of cell differentiation and maturation.	[29]

(Table 1) cont.....

Graphene Family Nanomaterial	Dose	Cell Line/Organism	Toxic Effect	Refs.
G	300-500 µg/mL	Human embryonic kidney cells, human lung cancer cells.	Reproductive toxicity, reduction in brood size	[30]
GO	0, 10, 20, 40, 60, 80, and 100 µg/mL	TM3 (Leydig) and TM4 (Sertoli) Cell lines.	GO exhibits size and dose-dependent cyto- as well as genotoxicity, TM3 cells were more sensitive.	[31]
GO, rGO	10, 50, 100, 200 µg/mL	RPE cells.	GO and rGO induce ROS-dependent genotoxicity *via* DNA damage.	[32]
GO, rGO	GO: 0-250 µg/mL	L5178YTk ± cells, and the Caco-2 cells.	GO was not mutagenic in the MLA; rGO was genotoxic at 250 and 125 µg/mL concentration, respectively after 4 and 24h exposure.	[33]
rGO, G-OH, G-COOH, G-NH$_2$	0.1–10 mg/L	Human neuroblastoma cells.	Studies pertaining to GFNs exhibit neurotoxicity *via* a membrane and oxidative damage which depends upon the functional group attached to the material.	[34]
rGO, c-rGO, pro-rGO, pro- c-rGO	-	Mammary epithelial cells.	c-rGO and pro-c-rGO displayed higher toxicity than rGO and pro-rGO.	[35]
GO, V-rGO	20–100 µg/mL	Human acute monocytic leukemia cell line THP-1 cells.	Treated THP-1 cells exert dose-dependent effects on cell viability, proliferation *via* LDH leakage, ROS stress, and lipid peroxidation leading to DNA damage and elevated pro-apoptotic gene expression.	[31]

Table 2. *In vivo* studies involving toxicity evaluation.

Graphene Family Nanomaterial	Dose	Cell Line/Organism	Toxic Effect	Refs.
GO, rGO	GO: 5 mg/L rGO: 10 mg/L	*C.* elegans	Reduction in reproductive health, impaired fertilization, and egg hatching.	[36]
G	300-500 µg/mL	*C. elegans*, human embryonic kidney cells, human lung cancer cells.	Reproductive toxicity, reduction in brood size.	[30]
GONs	500 mg/kg	Wistar rats	Granulomatous reaction in accumulation of GON in the capsular region of visceral organs, neuronal degeneration and necrosis.	[37]
GO	10.0mg/kg	Winstar rat	GO has toxicity to the mammalian liver.	[38]

(Table 2) cont.....

Graphene Family Nanomaterial	Dose	Cell Line/Organism	Toxic Effect	Refs.
GO, rGO	0, 18, 54 or 162 µg/mouse	Mice	GO causes changes in hepatic lipid homeostasis and induces a more inflammatory response as compared to rGO.	[39]
GO	2 mg/kg and 5 mg/kg	Male mice	GO caused liver toxicity mediated by oxidative stress.	[40]
GO	5mg/L	*C. elegans*	GO causes neurotoxicity.	[41]
pG, GO	pG: 0, 5, 10, 15, 20, and 25 µg/L GO: 0, 0.1, 0.2, 0.3, and 0.4 mg/mL	Zebrafish	pG and GO accumulation leads to multi-organ toxicity.	[42]
GO	0.9 mg/mL	Zebrafish	Small-sized GO exerts a toxic effect on glutamate-based neuronal transmission.	[43]
GO	1, 10 µg/L	*C. elegans*	GO produced multi-generational toxicity which disappeared at parents.	[44]
GO-SH	100µg/L	*C. elegans*	Continued contact with GO-SH causes toxicity in the nematode.	[45]
GO	1, 10, 25, 50, 100, and 200 µg/mL	Mice	The interactions of GOs with various renal compartments, renal excretion mechanisms, and possible kidney damage were all impacted by their lateral size.	[46]
GO	10 mg /L	*C. elegans*	GO have adverse effects on nematodes' growth, survival, and reproduction.	[47]
Micrometric and nanometric-sized GO	-	Mice	GO-induced size and dose-dependent DNA damage.	[48]

Abbreviations: G: Graphene; pG: Pristine graphene; GO: Graphene oxide; rGO: reduced GO; c-rGO: concentrated GO, pro-rGO: Protein corona coated-rGO; pro- c-rGO: Protein corona coated-c- rGO; P-GO: Linear Poly ethyleneglycol GO; bP-GO: Branched Poly ethyleneglycol GO; LDH: Lactate dehydrogenase; ROS: Reactive oxygen species; MLA: Mouse lymphoma assay; MMP: Mitochondrial membrane potential; DNA: Deoxy ribose nucleic acid; FLG: Few-layer graphene; sFLG: small FLG; NHBE: Non-transformed human bronchial epithelial cell. Abbreviations: G: Graphene; pG: Pristine graphene; GO: Graphene oxide; rGO: reduced GO; GONs: GO nanoplatelets; DNA: Deoxyribose nucleic acid; C. elegans: *Caenorhabditis elegans*; GO-SH: Thiolated GO.

Toxicity to the Immune System

The immune system clearly distinguishes between self and non-self-antigens and thus protects the body from foreign invasion. Assessment of the impact of GFNs on the immune system is necessary to determine whether the nanomaterial's presence inside the body could be tackled so that it is eventually eliminated from the body by the immune system or its persistence leads to chronic illness. Dendritic cells of the immune system are significant professional antigen-presenting cells (APCs) with the special ability to trigger the primary immune response and transmit information to the cells of the adaptive/acquired immune response [49]. T-cells display an indispensable effect in the adaptive immune response. Macrophages, another cell of the immune system, have a role in both innate as well as adaptive immune response by phagocytosis. It is a process by which macrophages engulf foreign material, remove dead cells from the body as well as boost immune response. GFNs have been shown to cause dysfunction in these immune cells, leading to the appearance of the toxicological effect of these nanoparticles.

Bio-persistence of GFNs inside the human body exerts an adverse effect on host health by disrupting their physiological and immunological balance [50, 51]. Park *et al.* studied GNPs for their *in vivo* as well as *in vitro*-induced toxicity [52]. They showed that GNPs caused a sub-chronic inflammatory reaction in mice and *in vitro*, it caused mitochondrial damage, which later induced autophagy and death. In a different investigation, scientists aimed to assess the long-term pulmonary persistence of GNPs on both local and systemic health. They conclude that the prolonged presence of GNP in the lung may have harmful consequences on health by interfering with the body's immunological and physiological equilibrium [51]. Surface functionalized GNP administered intratracheally caused a considerable amount of acute neutrophilic inflammation, which eventually went away. However, positively charged GNPs caused somewhat more inflammation than negatively charged ones. Similar, extra-pulmonary translocation patterns into the mediastinal lymph nodes were seen for all kinds of GNPs, and there was clear evidence of extended retention. Consecutive analysis of GO-nanoplatelets and rGO-nanoplatelets reveals that different oxidation degrees lead to toxicity to monocyte cells of the immune system leading to damage in endocytosis as well as phagocytosis [53]. Uzhviyuk *et al.* evaluated the effect of GO on human dendritic cells [29]. They showed that GO-PEG (polyethylene glycol) suppresses the maturation and differentiation of dendritic cells. Gr and its derivatives were examined for their immunotoxicity by Shaohai Xu and co-workers who concentrated on B lymphocytes and the differentiated Ig-secreting plasma cell [54]. They demonstrated that Gr and its derivatives might stimulate B cell activation and affect the expression of B cell surface markers without causing cell death. Additionally, they might interfere with the plasma cell's Ig secretion pathway, indicating a novel possible mechanism employed by GFNs for immunotoxicity. Surface functionalized GNP administered intratracheally caused a considerable amount of acute neutrophilic inflammation, which eventually went away [55]. However, positively charged GNPs caused somewhat more inflammation than negatively charged ones. Similar extra-pulmonary translocation patterns into the mediastinal lymph nodes were seen for all kinds of GNPs.

Toxicity to Internal Organs

Nanomaterials possess the potential to engender damage to various internal organs including the liver, lungs, kidney, intestinal organs, *etc.*, as they have been found to get into the systemic circulation. Studies using theoretical models and animals have been used to monitor and evaluate the toxicity and biocompatibility of GFNs. They may cause immediate inflammation and persistent harm by interfering with the biologically normal functioning of the vital organs. Amrollahi-Sharifabadi and co-workers analyzed the toxicity of GO nanoplatelets (GONs) *in vivo* in male Wistar rats. Histopathological findings indicate that they were toxic to the brain, liver, lungs, spleen, kidney, and intestine [37]. They bring about inflammation and granulomatous reactions in the capsular regions of the visceral organs as well as neuronal degeneration and necrosis in the brain, including the cerebellum. GFNs given orally to mice produced negligible absorption and organ distribution *via* the digestive system, but when administered intravenously or after intra-tracheal instillation, nanomaterials partitioned to the liver and spleen [56, 57]. The long-term liver toxicity of FLG produced by utilizing an arc discharge approach has been documented by Sasidharan and colleagues. They showed that the Kupffer cells of the liver of FLG-treated rats engulfed Gr and reported inflammatory modifications including portal congestion, sporadic sinusoidal congestion, and severe hepatocyte degradation [58]. A similar study to examine the hepato-toxicity of the GO in rats indicated that it exhibits dose-dependent toxicity [38]. The presence of high serum levels of liver function enzymes suggests liver inflammation and damage. Further presence of histopathological lesions supports alteration in the biochemical composition. GO has also been shown to exhibit *in vivo* oxidative stress in the kidneys of male rats [59]. It causes *in vivo* loss of activity associated with antioxidant enzymes, following an upsurge in lipid degradation *via* peroxidation mechanism, and a visible morphological modification. In another study, a histopathological assessment of mice intravenously injected with riboflavin functionalized Gr showed that the nanomaterial was engulfed by macrophages in the liver causing their biodegradation [57]. However, transient deposition of these materials in the kidney leads to necrosis of cells and dilation of tubules. Studies have shown that GO exposure causes acute phase response in the lungs and affects the lipid balance in the liver cells making one more susceptible to the progression of atherosclerosis [39]. Rhazouani *et al.* in an *in vivo* sub-acute toxicity evaluation of GO in mice showed that it causes liver inflammation and induces oxidative stress-induced toxic effects [40].

A toxicology study on orally administered GFN usually in conjunction with anticancer drugs is required for their possible therapeutic uses. GO and GNP aggregates are likely to pass through the stomach with little or no change in their physicochemical characteristics since they are generally inert to physiologically relevant acidic environments. Therefore, it is not necessary to pre-treat pristine GO with acid to examine *in vitro* the possible acute toxicity toward digestive tract cells. However, GO functionalized with drugs or other biomolecules must be carefully evaluated for their toxicity under an acidic environment. Once such *in vitro* studies on the cell lines derived from colon carcinoma showed that high concentrations of GFNs exhibited low acute toxicity [60]. The toxic effect on the lungs has also been studied. Yamany *et al.* studied the effect of intraperitoneal administration of GO in mice [61]. They showed that GO put forth a negative effect on the oxidative state of the lungs leading to oxidative stress in a dose- and time-dependent way. In a different study by Tang and co-workers, it was demonstrated that the surface groups on the Gr affect the activity of digestive enzymes in the gastrointestinal tract [62]. They evaluated the effect of the GO and G modified with hydroxyl (G-OH), carboxy (G-COOH) as well as amino (G-NH$_2$) groups on the enzyme activity and concluded that the activity of α-pancreatic amylase and trypsin was altered while no effect was seen in pepsin activity.

Toxicity in the CNS

In recent years, the treatment of brain malignancies, intracranial and spinal biocompatible devices, and bio-sensing and bio-imaging methods have all made extensive use of GFNs. Steady accumulation and sustained persistence of these materials in the CNS raise serious concerns regarding their chronic toxicity along with aiding in the benefits of drug delivery. Different studies have concluded the adverse effect of GFNs on the cells of the CNS. The potential toxicity of the GO nano-platelets was evaluated *in vivo* in rats [37]. The study's findings demonstrated that GO was harmful to nerve tissue. Microscopic examinations demonstrated that certain neuronal cells in the cerebral and cerebellar cortex exhibited necrosis and degeneration. Specifically, the cytoplasm of the Purkinje cells was constricted, their nuclei vanished, and their form was altered. These alterations were visible in animals receiving a higher dosage of GO. Additionally, animals who had consumed GO showed signs of hemorrhage in their brain tissue. Kim *et al.* studied the neurotoxic capability of GFNs in a nematode, *Caenorhabditis elegans* [41]. After exposure to GO, *C. elegans* showed increased ROS and GO deposition in the head area. Exposure to GO causes a significant decline in the levels of all the neurotransmitters including tyrosine, tryptophan, dopamine, tyramine, and GABA. Moreover, the downregulation of the expression of *ttx-1* and *ceh-14* genes were observed, which is essential for the functioning of AFD thermos-sensory neurons (Amphid neurons with finger-like, AfD, ciliated endings). Manjunatha and co-workers discovered that GO may accumulate in brain tissue at various concentrations and significantly alter the heart rate and survival of young zebrafish [42]. Further, it was demonstrated that the small GO nanosheets selectively block glutamatergic transmission in brain explants taken from animals injected with the nanomaterial and in neuronal cells *in vitro* [43]. Small GO microinjection into the embryonic zebrafish spinal cord inhibits the excitatory synaptic transmission of the spinal network in a targeted manner, without compromising spinal cell viability. This is accompanied by the disruption in the locomotor behaviour induced by the spinal cord's neural network. Similar results were observed during the evaluation of multi-generational toxic effects of the GO *in vivo* model using *C. elegans* [44]. A decrease in locomotion behaviours was observed in the filial generations, which followed the degeneration of neurons involved in the movement.

Toxicity in the Reproduction System

Toxicity evaluation for the reproductive system is an important part of toxicological studies since it not only affects the person but also has the potential to be harmful to future generations and have long-lasting hazardous consequences. Developing biomedical uses of GFNs and the potential for tissue accumulation following prolonged contact strongly implies that the possible reproductive and developmental dangers of these materials require serious consideration. However, the research on this aspect is very limited and needs to be cautiously taken up by the scientific community so that further hazards can be minimized. Xu and colleagues studied the dose-dependent effect of rGO exposure on female mice during pre and post-fertilization [63]. It was observed that the administration of low and intermediate doses of 6.25 and 12.5 mg/kg of rGO in the late gestation period led to abortion, respectively. However, when injected before pregnancy or during early gestation periods may still lead to the birth of healthy offspring. Therefore, a greater comprehension of the effects of rGO nanosheet toxicity on female reproductive fitness and the development of the offspring is required. Further, the inherent risks of GFNs were examined by exposing the worm *C. elegans* to GO and rGO [36]. In a different study, size-dependent toxicity of GO on the reproductive cells such as Leydig and Sertoli cells was studied by Gurunathan and co-workers [31]. Cell viability and proliferation assays revealed that GO nanosheets of 20 and 100nm (GO-20 and GO-100 respectively) sizes exhibit size and dose-dependent toxicity to reproductive cells. G-20 caused increased loss of cell viability as well as cell proliferation, elevated leakage of lactate dehydrogenase (LDH), and production of reactive oxygen species (ROS) as compared to GO-100. Ling Jin *et al.* showed reproductive toxicity of GO in *C. elegans* [44]. The reproductive ability of the nematode was impaired as observed by the reduced brood sizes and oocyte counts, increased germline apoptosis, and abnormal expression of reproductive-related genes in the worms under study. GO had a greater effect on the reproductive ability of the worm than the rGO. The observed GO-induced reproductive failure was due to the crosstalk between the wingless-type MMTV integration site family (Wnt) pathway and the mitogen-activated protein kinase (MAPK) pathway. Wnt signaling pathways, which are evolutionarily conserved signal transduction cascades, regulate a wide array of biological processes during embryonic and larval development, and are also implicated in adult cancer and various disorders related to aging. [64]. Prolonged exposure to thiolated GO (GO-SH) causes adverse effects on the reproductive organs in *C. elegans* [45]. These effects were manifested by the decline in locomotion-based behavior as well as brood size on exposure to GO-SH.

Toxicity to Genetic Material

Genotoxicity is referred to as any changes that alter the genetic constitution of the DNA. GFNs can be a potent genotoxic agent *via* primary or secondary mechanisms [65]. Primary genotoxicity is generated by the direct contact of the GFNs with the genetic material *i.e.* DNA or/and its related proteins, while indirectly by ROS-induced effects on DNA replication or repair machinery. DNA is a polynucleotide molecule consisting of ribose sugar and nitrogenous bases. DNA is negatively charged at physiological pH. Negatively charged oxygen-containing functional groups on these materials such as GO, get attached *via* H-bonding as well as π-π stacking interaction leading to the DNA damage [66, 67]. Molecular dynamics simulation studies have shown that wrinkled Gr surface upon interaction with the DNA leads to structural deformation of DNA while idealized Gr does not affect the DNA structure [68]. DNA binding with the wrinkled Gr leads to zipper-like unfolding of double-stranded DNA, thus exposing additional DNA bases to the region that is wrinkled and speeding up the DNA deformation. Genotoxicity of GO and rGO was studied *in vitro* by Ou and co-workers and demonstrated that these materials induce ROS-dependent DNA damage [32]. Meanwhile, the DNA damage increased as marked by the reduction of saturated C-O bonds present in GO. In a different study, involving exposure of GO and GNPs through the intestine by *in vitro* model stimulating the human intestinal barrier, no oxidative DNA damage was observed [67, 69]. ROS-independent DNA damage was demonstrated by downregulating the genes involved in response to DNA damage and repair mechanisms. Pulmonary exposure to GO and rGO causes variation in the response of the transcriptome in the liver as well as lungs, thus affecting gene expression [39]. However, exposure to GO induces differential expression of the gene at an elevated level when compared with rGO and affects signaling pathways in the liver and lungs.

FACTORS INFLUENCING THE TOXICITY OF GFN

Various characteristics of nanomaterials influence the toxicity potential in biological systems. These factors include surface functionalization, size, concentration, aggregation, and protein corona effect. Maintaining optimum conditions of all these factors may avoid the toxic effects of these materials. But sometimes it may often be seen that these concentrations are incapable of carrying out the desired function for which it is administrated in the biological system. Fig. **(3)** represents various factors affecting the toxicity of GFNs and has been explained thoroughly in the upcoming sections.

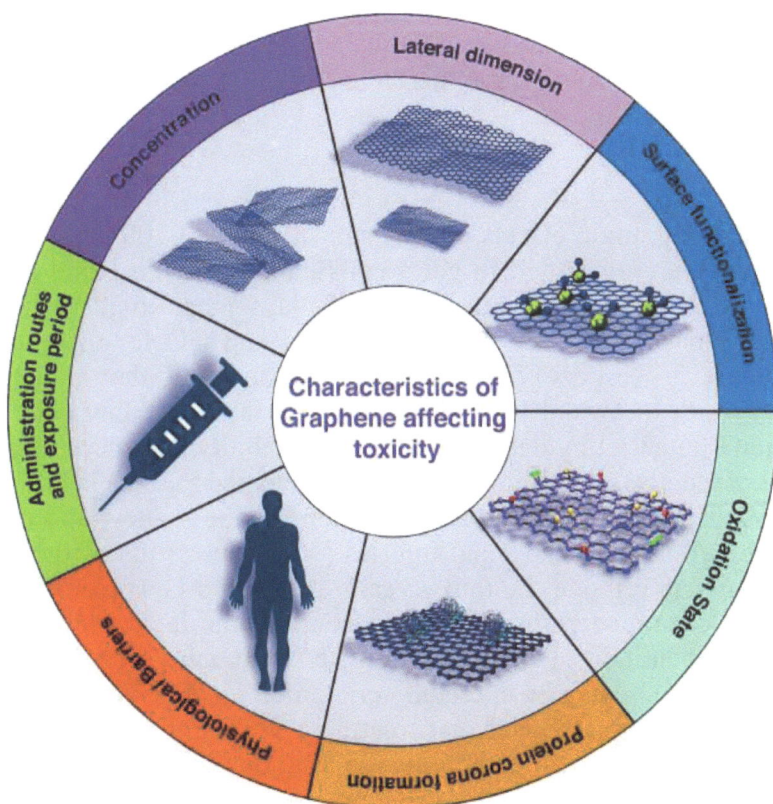

Fig. (3). Different characteristics of graphene family nanomaterials affecting toxicity.

Concentration

The concentration of GFNs also affects the toxicity of the produced material. Various studies are available that support the concentration-dependent effect of toxicity in both cancerous as well as non-cancerous cell lines [31, 40, 70]. In one such study, GO in a concentration range of 0-50µg/mL showed dose-dependent toxic effects. HEK293 cell lines exposed to varying concentrations of GO showed a decline in viability as well as proliferation with increasing concentrations of GO [71]. Similar results were observed during the toxicity assessment of three different-sized GO. GO with low and intermediate doses of 0.1 and 0.25 mg respectively did not show any evident toxicity, while a high of 0.4 mg generated chronic damage [72]. The higher GO concentration gets deposited mostly in the lungs, liver, spleen, and kidneys, making it tough for the kidneys to remove them. Moreover, extended exposure with a decreased concentration of GFNs leads to effective toxicity in the *in vivo* model of *C. elegans* [73]. Long-term exposure affects nematodes' ability to reproduce and move about, shortens their lifespan, and causes oxidative stress. Further genotoxicity of GO and rGO was assessed in a concentration range of 0-250 µg/mL [33]. The results indicated that GO at a concentration of 0-250 µg /mL was not mutagenic, while rGO revealed mutagenic activity at a concentration of 250 g/mL and 125 µg/mL after 4 hours and 24 hours of exposure, respectively.

Lateral Size

In a different study, toxic effects of GO nanosheets, GO-100 and GO-20 of 100 and 20nm size were evaluated in germ cells namely Leyding and Sertoli cells, respectively. Results showed that GO-20 displays more toxicity than GO-100 [74]. Further analysis of three different-sized G and GO for their toxicity was done by Jia and co-workers [75]. This study demonstrated that in comparison to the different sizes of G and GO, the small material at the low conc. exhibited higher toxicity potential by decreasing the cell viability and increasing the DNA impairment. Recently, Chen and co-workers, elucidated that the lateral size of GO influences their interaction with various regions of the kidney, and its excretion pathways as well as potential renal injuries [46]. The results of the study show that the kidney can remove the thin, evenly distributed GOs from the body, but that this renal clearance pathway is influenced by the lateral dimension of the GOs. The large lateral-sized GO gets eliminated by tubular secretion, whereas the tiny lateral-sized GOs may be excreted *via* glomerular filtration. Comprehensive histological evaluations showed that treatment with small and large-sized GO triggers dose-dependent histopathological alterations in the glomeruli and renal tubules, respectively. Additionally, inflammation was observed in the proximal renal tubules. The lateral size of GFNs affects the density of functional groups, and thus further influences their toxicity. Small-sized material has a high density of functional groups on its surface and thus induces enhanced cytotoxicity. An increased number of functional groups on the small-sized material causes ROS-dependent oxidative stress. Further oxidative stress causes mitochondrial dysfunction as well as genomic instability affecting cellular homeostasis.

Surface Functionalization

Surface functionalization is extensively utilized in modifying chemical behavior including attached groups, hydrophobicity, and C-O ratio of the GFNs. Modification of these chemical properties affects their biocompatibility with biological systems. Further toxicity determination of the functionalized nanoparticle becomes essential when it is to be used in biomedicine or any other applications involving human use. Various research works have been conducted to evaluate the resultant toxicity induced or mitigated by the functionalization of GFNs.

Surface functionalization with various polymers has also been shown to decrease the toxicity of GFNs. Functionalization with polyethylene glycol (PEG) [76], polyvinyl alcohol (PVA) [77, 78], PEGylated poly-L-lysine [79], chitosan [80, 81], *etc.* enhances biocompatibility and reduces the toxicity of the nanomaterials. Because of electrostatic charges and non-specific protein binding, the majority of GFNs tend to assemble in physiological solutions. Therefore, functionalized GFNs have increased solubility as well as biocompatibility and reduced cytotoxicity. Various studies have concluded that surface functionalization of pristine Gr and GO is crucial for minimizing the toxic effects. Covalent bonding and non-covalent physisorption processes are the two major methods that are frequently employed for surface modification to create desired functionalized GFNs. Different molecules or polymeric substances that are used for the covalent modification of GFNs include enzymes, various aliphatic and aromatic amines, amino acids, amine-containing biomolecules, and silanes. Hydrophobic interaction and electrostatic forces are examples of noncovalent functionalization techniques that appear to be more adaptable than covalent techniques.

However, GFN toxicity is not always reduced by functionalization. Guo and co-workers evaluated the neurotoxic effects of rGO, carboxylated-G (COOH-G), hydroxylated-G (OH-G), and aminated-G (NH_2-G) on the growth of human neuroblastoma cells (SK-N-SH) [34]. The difference in toxicity was attributed to their surface oxygen content as well as differences in the sizes of these nanoparticles. This shows decreased cytotoxicity. Neurotoxic potential follows the order $G-NH_2 < RGO < G-COOH \approx G-OH$. Lower toxicity of $G-NH_2$ is due to the positively charged amino group present on the Gr that favors neuronal growth and declines the charge transfer to cells and thus exhibits sustained toxicity. Additionally, the neuronal release of lactate dehydrogenase and membrane lipid peroxidation was observed in a similar order. Moreover, it was observed that the longer exposure time, reduces the toxicity of G-COOH, G-OH, and RGO; however, the toxicity of $G-NH_2$ remained constant, indicating that $G-NH_2$ may produce longer-lasting chronic neurotoxicity than the other substances.

Protein Corona Formation

GFNs can be administered intravenously, intramuscularly, or intraperitoneally to reach the bloodstream. In the microenvironment of blood circulation, they get fenced by a multitude of blood components namely blood cells and plasma proteins, which interact as well as inadvertently functionalize the surface of GFNs. This bio-molecular corona formation in the bloodstream was originally described as protein adsorption onto the surface of nanomaterial representing spontaneously random adsorption of biomolecules including proteins, DNA, and metabolites to the surface of GFNs. This bio-molecular corona modifies the particles' surface chemistry and gives the NMs a "biological identity," enabling the body to recognize them as "self." This results in recognition as well as the interaction of the cellular receptors with the corona proteins and small molecules, thus triggering cellular absorption processes and other signaling pathways which might eventually result in harmful effects or toxicity. Studies have been performed on the proteins in the corona of GBMs. The colloidal stability, absorption, and final toxicity of GBMs are affected by fast adsorption to the surface after contact with the biological fluid.

This was experimentally proved by evaluating the toxicity of GO coated with protein bovine serum albumin (BSA) [82]. It was observed that the BSA protein corona of GO lessens the cytotoxicity of GO by reducing the cell membrane penetration. BSA adsorption on the surface causes a decrease in the accessible surface area and negative stearic effect, resulting in a significant reduction in the penetration of protein coronas, thereby a reduction in cell membrane damage and reduced cytotoxicity. Generally, the presence of immunoglobulin (Ig) G and complement proteins in protein corona assist in the recognition of nanoparticles by the immune cells, leading to endocytosis by the reticuloendothelial cells or nonspecific interactions with the receptors on the cell membrane. Yang *et al.* showed that the toxicity of GO was alleviated by protein corona coating on the surface which reduces its endocytosis by the cells [83]. Albumin corona-coated GO has reduced the toxicity in the nematode, *C. elegans* which is manifested by 50 and 100% mitigation of fertility and growth respectively [47]. In a different study, rGO and concentrated rGO (c-rGO), prepared by dispersing in 4 and 2 mg/mL sodium cholate, respectively; as well as protein corona-coated materials namely pro-rGO and pro- c-rGO respectively were evaluated for their toxicity in mammary epithelial cell lines [35]. Pro-r-GO exhibits more severe toxicity as compared to other materials.

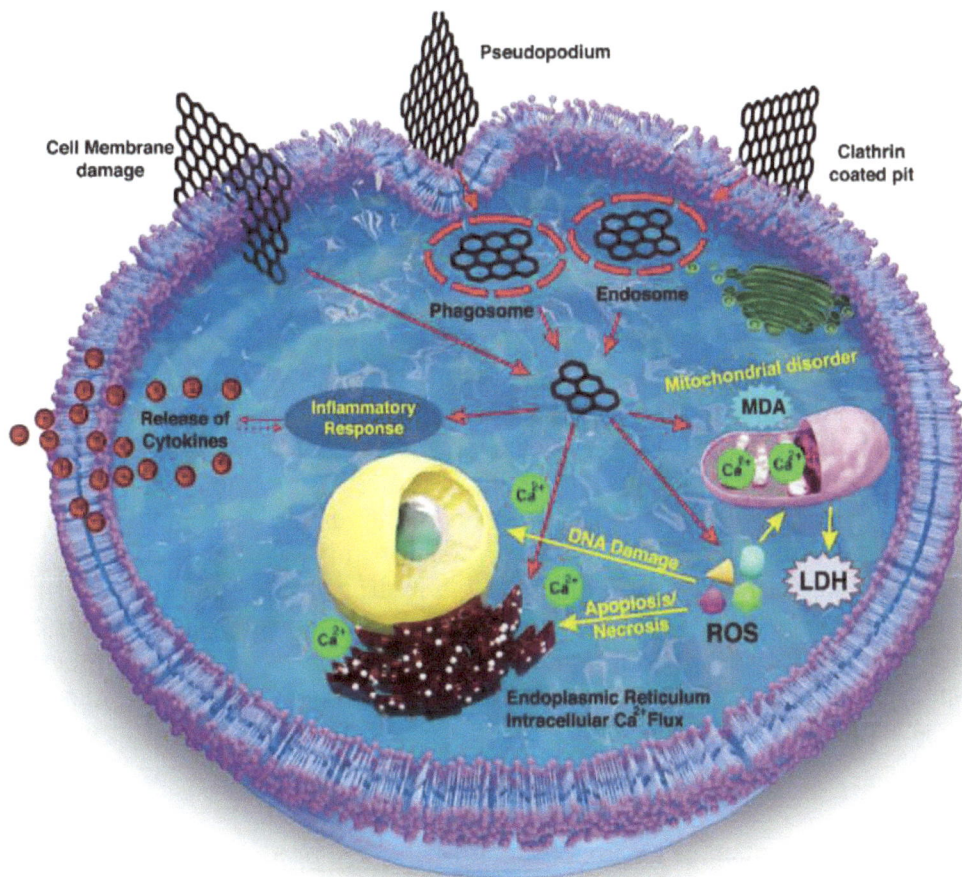

Fig. (4). Toxicity mechanism explored by graphene family nanomaterials.

MECHANISM OF TOXICITY

Various mechanisms by which GFNs exert their toxic effect in the biological system include physical destruction, oxidative stress, damage to cellular organelles like mitochondria and DNA, inflammation, apoptosis, autophagy, necrosis, and epigenetic changes. Specific characteristics such as shape, size, functional group density, and charge transfer capabilities of these materials determine the mechanistic action to induce the cytotoxic effect [84]. The section below describes possible mechanisms undertaken by GFNs to manifest potential toxic action inside the biological systems. GFNs enter the cell by a variety of routes where they cause the release of Ca2+, an increase in LDH and MDA, and the formation of ROS. GFNs then result in many types of cell harm, such as cellular membrane destruction, inflammation, DNA and/or mitochondrial dysfunction/damage, apoptosis, and necrosis. Fig. (4) represents various mechanisms followed by GFNs to induce toxicity and have been explained in the upcoming section.

Physical Damage

GFNs cause physical destruction-mediated toxicity in the biological system. Sharp edges of these materials have the potential to physically damage living organisms by piercing the cell membrane. Membrane damage insinuates the release of a cellular component. LDH and protease are secreted into the medium by the damaged cell membrane and their measurement determines the level of damage [31]. The edges are reactive too, due to the presence of reactive functional groups. The adsorption of GO on RBCs causes hemolysis by causing the cell membrane to rupture [85]. Large-sized sheets have been shown to wrap the cell and thus induce microbial toxicity [86]. Studies using molecular dynamics simulations have revealed that Gr may be found in the hydrophobic core of biological membranes and has a significant affinity for phospholipids [87]. Further, GNP aggregate because of their high hydrophobicity and low solubility, which has been shown to bind to proteins on cell surfaces and compromise membrane integrity [88]. Gr enters and removes a sizable number of phospholipids from cell membranes due to the strong interactions it has with lipids, which causes cytotoxicity.

Oxidative Stress

The primary mechanism of Gr toxicity is linked to the production of intracellular ROS, which adversary affects proteins and DNA. It arises when increased ROS levels overtake the action of enzymes like catalase, superoxide dismutase (SOD), or glutathione peroxidase (GSH) [89, 90]. Gr may enter cells either passively by the process of endocytosis; or actively by actin-dependent macro-pinocytosis or clathrin-mediated energy-dependent endocytosis. Once inside the cell, it hampers the electron transport system, thus inducing the excessive production of ROS including H_2O_2 and hydroxyl radicals. ROS acts a secondary messenger in several intracellular signaling pathways and leads to the destruction of various cellular organelles such as oxidation of membrane lipids, DNA damage, protein denaturation, and mitochondrial dysfunction. Key ROS scavengers inside the cells are unable to restore the GFN-mediated oxidative stress [91, 92]. Excessive ROS production upon the interaction of GO can induce carcinogenesis, ageing, and mutations inside the cell. Moreover, oxidative stress plays a key part in the acute lung damage caused by GO [93, 94]. After being exposed to GO, SOD, and GSH-PX (Plasma glutathione peroxidase) activity declined which is dependent on the exposure time and dosage. Likewise, Human Lung Fibroblasts (HLF) cells exposed to GO showed oxidative stress as the primary factor in apoptosis and DNA damage. In pristine Gr-treated cells, the production of ROS led to the activation of two pro-apoptotic members of the Bcl-2 family of protein namely, Bim and Bax; in addition to the stimulation of the mitogen-activated protein kinase (MAPK) and Transforming growth factor beta (TGF-β) [95, 96].

Mitochondrial Impairment

Mitochondrial activity is essential for maintaining cellular energy balance, regulating calcium ion signaling, managing metabolism, and controlling cell death. In addition, their activity serves as a principal source of ROS pool in the cell. Once inside the cell, GFNs get internalized by the mitochondria present in the cytoplasm, which later affects the structure as well as the functioning of the mitochondria. Further, it is demonstrated to decrease the mitochondrial membrane potential (MMP), which later results in the Ca^{2+} imbalance [97]. The key mechanism for mitochondrial impairment is the lowering of MMP. Eventually, excessive ROS generation marks mitochondrial-induced apoptosis [98]. Mitochondria are the powerhouse of the cell and generate adenosine triphosphate (ATP), therefore cells GFN treated cells *in vitro* show a reduced level of ATP, indicative of mitochondrial toxicity induced by the nanomaterials [71].

DNA Damage

GFNs may have substantial genotoxic potential and result in severe DNA damage because of their tiny size, large surface area, and surface charge. Their effects encompass chromosomal disintegration, DNA breaks, mutations, DNA aberrations, and the formation of oxidative adducts [99, 100]. GFN interacts with DNA *via* π- π stacking interaction as well as H-bonding bringing about DNA deformation and cleavage. Excessive ROS production may further aid in oxidative DNA damage causing double or single-stranded DNA breaks, DNA cross-links, and base modification [32, 39, 101]. Various *in vitro* and *in vivo* studies have proved that GFNs induce DNA damage [102 - 104] by inducing DNA fragmentation, mutation as well as a chromosomal aberration [105, 106]. Chronic as well as long-term exposure can cause persistent DNA damage eventually leading to carcinogenic events [48, 107]. Nanometric (small) sized GO sheets do not cause long-lasting *i.e.* persistent DNA damage independent of the dose and time of exposure. On the other hand, repeated exposure with micrometric (large) sized GO sheets induces long-term double-stranded breaks in the DNA [48]. However, following just one exposure to the equivalent high dosage of these larger GO sheets, fast recovery was seen, demonstrating the crucial role of exposure chronicity in the establishment of long-lasting DNA damage. It was identified that inflammatory reaction and oxidative stress in the exposed cells lead to the recovery or persistence of lung DNA damage.

Inflammatory Response

GFNs can result in a severe inflammatory response that includes infiltration of inflammatory cells, pulmonary edema, and the development of granulomas when administered intravenously or intratracheally at large dosages [65, 103]. Upon exposure to these materials, cytokines and chemokines are released as a part of the immune response of the cell [31]. Less than toxic concentrations of pristine Gr have been shown to trigger *in vivo* secretion of various inflammatory molecules including cytokines and chemokines [98, 108]. Various studies have proved that exposure to GFNs induces lung inflammation [109, 110]. Li and co-workers stated that GFN with higher oxidation displays an increased level of inflammation [92]. Li *et al.* showed that the agglomeration of GO sheets in the lungs causes acute lung injury leading to chronic inflammation and pulmonary fibrosis [111, 112].

TOXICITY REMEDIATION

Gr and its base materials have influenced several disciplines and new uses are being discovered daily. So, as with other developing technologies, it is essential to take their safety into account. The Health and Environment Work Package of the Graphene Flagship program is committed to the development and research of how these novel materials affect our health and the environment. This team looks at whether there are any concerns associated with Gr and multilayer materials to human health and the environment, and if so, how those risks might be reduced. Following are some strategies that are aimed at toxicity mitigation imposed by exposure to GFNs.

Assessment Strategies

GFNs must be thoroughly assessed for the risks and uncertainties involved at various structural levels before being released onto the market. Derived materials and products must have minimal potential hazards to reduce their development costs, regulatory requirements, and environmental impact. It is critical to assess the hazard and safety of GBNs for the consumer, occupational employees, and environment as a health and safety component due to their widespread usage in a wide range of applications. Moreover, GFNs have considerable differences in their physicochemical properties, which vary depending upon the intended user application, who fabricates it according to the requirement. So the toxicity assessment of GFNs for every product might be extremely difficult and must be assessed on a case-by-case basis. The properties of the nanomaterial must be evaluated by the establishment of a safe design strategy to minimize toxicity.

These strategies thus pin down the potential risks and hazards at the early stages of production. A critical component of ensuring user safety, repeatability, and formulation quality, is the standardization of Gr.

Computational Approaches

Experimental procedures involved in assessing the possible toxicity caused by GFNs are time-intensive owing to the intricacy of the physico-chemical characteristics that affect their activity. Experimental toxicity testing is further complicated by the national and international regulatory limits that govern the use of experimental animals in clinical trials [113, 114]. For these reasons, various, non-experimental methods of completing risk assessment studies of NMs generally have been offered. These are computational tools that use data-driven modeling techniques or physics-based modeling methodologies. The development of safety protocols is the major area of concern owing to their potential toxicity. Data-operated computing strategies include quantitative structure-activity relationship (QSAR) also known as QNARs (quantitative nanostructure-activity relationship approaches) or nano-QSAR [115 - 117]. Further read-across approaches utilized statistical and machine learning (ML) algorithms to gain insight into the co-relation between physico-chemical, structural property of the material along with their biological effects. Molecular dynamic (MD) simulations can reveal the interaction of GFNs with the cellular organelles (lipid bilayer membrane, protein, nucleic-acids). Nano-informatics techniques are still in the initial phases of supervisory acceptability since it is difficult to validate such models [118]. Computational approaches provide insights into the toxicity status of materials that have been introduced to the market or those that are potentially being developed. As a result, the perils to human fitness and the environment may be eradicated at an early stage in a logical design phase without engaging experimental testing processes, until the intended attributes of a planned NM are obtained [119, 120].

Determination of Ideal Physicochemical Properties

Physicochemical properties of the GFNs have a significant role in the toxicity level. These properties such as lateral size, shape, charge, and aggregation/dispersion, as discussed in the previous section, have a dose and time-dependent effect on the potential toxicity manifested by the material. Less oxidized Gr generates a high level of ROS, which further induces toxicity. Small-sized nanomaterials may readily penetrate cell membranes and tissue junctions, causing structural damage to mitochondria or nuclei, where they result in significant DNA damage that causes cell death [121]. Therefore, the optimization of these properties must be done before the development of the engineered material, so that toxicity could be minimized. Recently, a highly stable and concentrated dispersion of Gr in the aqueous media was prepared which showed no toxicity in HeLa and Raw 264.7 cells [57]. Studies conducted *in vivo* revealed no hemato-toxicity compared to controls for biochemical and hematological markers that remained within the normal reference range. Furthermore, immune-phenotyping and activation levels of the cells isolated from the lymph nodes and spleen revealed no observable symptoms of inflammation.

Effective Functionalization

Functionalization affects the observed toxicity in the GFNs. Functionalization of GFNs with organic and inorganic molecules reduces toxicity in biological systems. Effective functionalization governs ROS production as well as the net charge on the material, which further determines the toxicity level. The bulky hydroxyl group on the surface of GO leads to the induction of a large number of differentially expressed genes *in vivo* as compared to rGO [39]. The presence of more hydroxyl groups on the surface of GO may pose more toxicity [39, 112]. In comparison to rGO exposure, GO exposure was often more effective at causing the expression of the altered gene and affecting regulatory functions in these organs. Surface modification with organic/inorganic groups affects the coating of endogenous components onto the nanoparticles and thus hinders translocation by enlarging the hydrodynamic size. Studies have shown that positively charged GFNs are somewhat more inflammatory than negatively charged ones. The surface modification affects the net surface charge as well as the production of ROS, which forms one of the major factors promoting immunogenicity with poorly-soluble nanomaterials. Graphene's photothermal characteristics can facilitate targeted toxicity mitigation through controlled heating [122, 123], while cost-effective synthesis methods enhance its accessibility for sustainable remediation [124]. Integrating these aspects underscores the potential for harnessing graphene's unique attributes to address toxicity challenges and develop environmentally conscious remediation approaches.

CONCLUSION

With the present chapter, we have attempted to provide its readers with in-depth information about the toxicity profiling of GFNs. It covers the effects of toxicity to different organs including the immune system, internal organs, CNS, and reproductive system as well as toxicity to genetic material; factors affecting toxicity as well as underlying mechanisms have also been expounded. We have emphasized the significance of comprehending the structure-activity link that underlies the potential toxicity of these compounds. We summarize by emphasizing the importance of utilizing reliable and established assays for toxicity evaluation to safeguard mankind as well as the environment. Additionally, studies on GFNs should focus on topics important for risk assessment, studies that report the major facets of bio-interactions are also required.

ACKNOWLEDGEMENTS

We would like to acknowledge Ms. Himani Pant, ChemX VV for her valuable editorial support in the preparation of this document.

REFERENCES

[1] Li, Z.; Li, X.; Jian, M. Two-dimensional layered nanomaterial-based electrochemical biosensors for detecting microbial toxins. *Toxins,* **2020**, *12*(1), 20.
[http://dx.doi.org/10.3390/toxins12010020]

[2] Li, J.; Zeng, H.; Zeng, Z.; Zeng, Y.; Xie, T. Promising graphene-based nanomaterials and their biomedical applications and potential risks: A comprehensive review. *ACS Biomater. Sci. Eng.,* **2021**, *7*(12), 5363-5396.
[http://dx.doi.org/10.1021/acsbiomaterials.1c00875] [PMID: 34747591]

[3] Su, S.; Sun, Q.; Gu, X. Two-dimensional nanomaterials for biosensing applications. *TrAC Trends in Analytical Chemistry,* **2019**, *121*, 115668.
[http://dx.doi.org/10.1016/j.trac.2019.115668]

[4] Karaca, B.; Karataş, Y.; Cakar, A.B.; Gülcan, M.; Şen, F. Carbon-based nanostructures and nanomaterials. *Nanoscale Proc.,* **2021**, (Jan), 103-130.
[http://dx.doi.org/10.1016/B978-0-12-820569-3.00004-9]

[5] Yaragalla, S.; Bhavitha, K.B.; Athanassiou, A. A review on graphene-based materials and their antimicrobial properties. *Mater. Sci,* **2021**, *11*(10), 1197.
[http://dx.doi.org/10.3390/coatings11101197]

[6] Pathak, R.; Guleria, K.; Kumari, A.; Mehta, S.P.S. Deacidification of *Camelina sativa L.* seed oil by Physisorption method and characterization of produced biodiesel. *J. Appl. Nat. Sci.,* **2021**, *13*(1), 287-294.
[http://dx.doi.org/10.31018/jans.v13i1.2555]

[7] Zhao, H.; Wang, Y.; Bao, L.; Chen, C. Engineering nano–bio interfaces from nanomaterials to nanomedicines. *Accounts of Materials Research,* **2022**, *3*(3), 263-275.
[http://dx.doi.org/10.1021/accountsmr.2c00072]

[8] Bhatt, S.; Faridi, N.; Raj, S.P.M.; Pathak, D.; Agarwal, A. Cloning, expression and specificity evaluation of type III effector, Rip4, from Ralstonia solanacearum. *Ecol. Environm. Conserv.,* **2021**, *27*, S390-S397.

[9] Punetha, V.D.; Dhali, S.; Rana, A.; Karki, N.; Tiwari, H.; Negi, P.; Basak, S.; Sahoo, N.G. Recent advancements in green synthesis of nanoparticles for improvement of bioactivities: A review. *Curr. Pharm. Biotechnol.,* **2022**, *23*(7), 904-919.
[http://dx.doi.org/10.2174/1389201022666210812115233] [PMID: 34387160]

[10] Aicardi, C.; Reinsborough, M.; Rose, N. The integrated ethics and society programme of the Human Brain Project: Reflecting on an ongoing experience. *J. Responsib. Innov.,* **2018**, *5*(1), 13-37.
[http://dx.doi.org/10.1080/23299460.2017.1331101]

[11] Graphene Flagship|Shaping Europe's digital future. Available from:https://digital-strategy.ec.europa.eu/en/activities/graphene-flagship (Accessed Feb. 27, 2023).

[12]   Adam, G.O.; Sharker, S.M.; Ryu, J.H. Emerging biomedical applications of carbon dot and polymer composite materials. *Appl. Sci,* **2022,** *12*(20), 10565.
[http://dx.doi.org/10.3390/app122010565]

[13]   The State of Queensland. **2023.** Available from:https://www.worksafe.qld.gov.au/safety-an--prevention/hazards/hazardous-exposures/nanotechnology/nanomaterial-control-banding-risk-assessment (Accessed: Feb. 09, 2023).

[14]   Alberto, A.R.; Matos, C.; Carmona-Aparicio, G.; Iten, M. Nanomaterials, a new challenge in the workplace. *Adv. Exp. Med. Biol.,* **2022,** *1357,* 379-402.
[http://dx.doi.org/10.1007/978-3-030-88071-2_15] [PMID: 35583652]

[15]   Koyyada, A.; Orsu, P. Safety and toxicity concerns of graphene and its composites. *Compr. Anal. Chem.,* **2020,** *91,* 327-353.
[http://dx.doi.org/10.1016/bs.coac.2020.08.011]

[16]   Punetha, V.D.; Rana, S.; Yoo, H.J.; Chaurasia, A.; McLeskey, J.T., Jr; Ramasamy, M.S.; Sahoo, N.G.; Cho, J.W. Functionalization of carbon nanomaterials for advanced polymer nanocomposites: A comparison study between CNT and graphene. *Prog. Polym. Sci.,* **2017,** *67,* 1-47.
[http://dx.doi.org/10.1016/j.progpolymsci.2016.12.010]

[17]   Farjadian, F.; Ghasemi, A.; Gohari, O.; Roointan, A.; Karimi, M.; Hamblin, M.R. Nanopharmaceuticals and nanomedicines currently on the market: Challenges and opportunities. *Nanomedicine,* **2019,** *14*(1), 93-126.
[http://dx.doi.org/10.2217/nnm-2018-0120] [PMID: 30451076]

[18]   Achawi, S.; Feneon, B.; Pourchez, J.; Forest, V. Structure-activity relationship of graphene-based materials: Impact of the surface chemistry, surface specific area and lateral size on their *in vitro* toxicity. *Nanomaterials,* **2021,** *11*(11), 2963.
[http://dx.doi.org/10.3390/nano11112963] [PMID: 34835726]

[19]   Kraegeloh, A.; Suarez-Merino, B.; Sluijters, T.; Micheletti, C. Implementation of safe-by-design for nanomaterial development and safe innovation: Why we need a comprehensive approach. *Nanomaterials,* **2018,** *8*(4), 239.
[http://dx.doi.org/10.3390/nano8040239] [PMID: 29661997]

[20]   Schwarz-Plaschg, C.; Kallhoff, A.; Eisenberger, I. Making nanomaterials safer by design? *NanoEthics,* **2017,** *11*(3), 277-281.
[http://dx.doi.org/10.1007/s11569-017-0307-4]

[21]   Córdoba, M.; Zambon, A. How to handle nanomaterials? The re-entry of individuals into the philosophy of chemistry. *Found. Chem.,* **2017,** *19*(3), 185-196.
[http://dx.doi.org/10.1007/s10698-017-9283-6]

[22]   Globe Newswire. Graphene Sheet Market is projected to reach $1719.0 million. **2022.** Available from:https://www.globenewswire.com/news-release/2022/10/13/2533528/0/en/Graphene-She-t-Market-is-projected-to-reach-1719-0-million-by-2030-Globally-at---CAGR-of-27-says-MarketsandMarkets.html (Accessed: Feb. 27, 2023).

[23]   Naikoo, G.A.; Arshad, F.; Almas, M.; Hassan, I.U.; Pedram, M.Z.; Aljabali, A.A.A.; Mishra, V.; Serrano-Aroca, Á.; Birkett, M.; Charbe, N.B.; Goyal, R.; Negi, P.; El-Tanani, M.; Tambuwala, M.M. 2D materials, synthesis, characterization and toxicity: A critical review. *Chem. Biol. Interact.,* **2022,** *365,* 110081.
[http://dx.doi.org/10.1016/j.cbi.2022.110081] [PMID: 35948135]

[24]   Stone, V.; Miller, M.R.; Clift, M.J.D.; Elder, A.; Mills, N.L.; Møller, P.; Schins, R.P.F.; Vogel, U.; Kreyling, W.G.; Alstrup Jensen, K.; Kuhlbusch, T.A.J.; Schwarze, P.E.; Hoet, P.; Pietroiusti, A.; De Vizcaya-Ruiz, A.; Baeza-Squiban, A.; Teixeira, J.P.; Tran, C.L.; Cassee, F.R. Nanomaterials *versus* ambient ultrafine particles: An opportunity to exchange toxicology knowledge. *Environ. Health Perspect.,* **2017,** *125*(10), 106002.
[http://dx.doi.org/10.1289/EHP424] [PMID: 29017987]

[25]   Laux, P.; Tralau, T.; Tentschert, J.; Blume, A.; Dahouk, S.A.; Bäumler, W.; Bernstein, E.; Bocca, B.; Alimonti, A.; Colebrook, H.; de Cuyper, C.; Dähne, L.; Hauri, U.; Howard, P.C.; Janssen, P.; Katz, L.; Klitzman, B.; Kluger, N.; Krutak, L.; Platzek, T.; Scott-Lang, V.; Serup, J.; Teubner, W.; Schreiver, I.; Wilkniß, E.; Luch, A. A medical-toxicological view of tattooing. *Lancet,* **2016**, *387*(10016), 395-402.
[http://dx.doi.org/10.1016/S0140-6736(15)60215-X] [PMID: 26211826]

[26]   Schreiver, I. Synchrotron-based ν-XRF mapping and μ-FTIR microscopy enable to look into the fate and effects of tattoo pigments in human skin. *Sci. Reports,* **2017**, *7*(1), 1-12.
[http://dx.doi.org/10.1038/s41598-017-11721-z]

[27]   Raftis, J.B.; Miller, M.R. Nanoparticle translocation and multi-organ toxicity: A particularly small problem. *Nano Today,* **2019**, *26*, 8-12.
[http://dx.doi.org/10.1016/j.nantod.2019.03.010] [PMID: 31217806]

[28]   Frontiñan-Rubio, J.; González, V. J.; Vázquez, E.; Durán-Prado, M. Rapid and efficient testing of the toxicity of graphene-related materials in primary human lung cells. *Sci. Reports,* **2022**, *12*(1), 1-13.
[http://dx.doi.org/10.1038/s41598-022-11840-2]

[29]   Uzhviyuk, S.V.; Bochkova, M.S.; Timganova, V.P.; Khramtsov, P.V.; Shardina, K.Y.; Kropaneva, M.D.; Nechaev, A.I.; Raev, M.B.; Zamorina, S.A. Interaction of human dendritic cells with graphene oxide nanoparticles *in vitro. Bull. Exp. Biol. Med.,* **2022**, *172*(5), 664-670.
[http://dx.doi.org/10.1007/s10517-022-05451-0] [PMID: 35353288]

[30]   Pattammattel, A.; Pande, P.; Kuttappan, D.; Puglia, M.; Basu, A.K.; Amalaradjou, M.A.; Kumar, C.V. Controlling the graphene-bio interface: Dispersions in animal sera for enhanced stability and reduced toxicity. *Langmuir,* **2017**, *33*(49), 14184-14194.
[http://dx.doi.org/10.1021/acs.langmuir.7b02854] [PMID: 29144756]

[31]   Gurunathan, S.; Kang, M.H.; Jeyaraj, M.; Kim, J.H. Differential immunomodulatory effect of graphene oxide and vanillin-functionalized graphene oxide nanoparticles in human acute monocytic leukemia cell line (THP-1). *Int. J. Mol. Sci.,* **2019**, *20*(2), 247.
[http://dx.doi.org/10.3390/ijms20020247] [PMID: 30634552]

[32]   Ou, L.; Lv, X.; Wu, Z.; Xia, W.; Huang, Y.; Chen, L.; Sun, W.; Qi, Y.; Yang, M.; Qi, L. Oxygen content-related DNA damage of graphene oxide on human retinal pigment epithelium cells. *J. Mater. Sci. Mater. Med.,* **2021**, *32*(2), 20.
[http://dx.doi.org/10.1007/s10856-021-06491-0] [PMID: 33638700]

[33]   Cebadero-Dominguez, Ó.; Medrano-Padial, C.; Puerto, M.; Sánchez-Ballester, S.; Cameán, A.M.; Jos, Á. Genotoxicity evaluation of graphene derivatives by a battery of *in vitro* assays. *Chem. Biol. Interact.,* **2023**, *372*, 110367.
[http://dx.doi.org/10.1016/j.cbi.2023.110367] [PMID: 36706891]

[34]   Guo, Z.; Zhang, P.; Chetwynd, A.J.; Xie, H.Q.; Valsami-Jones, E.; Zhao, B.; Lynch, I. Elucidating the mechanism of the surface functionalization dependent neurotoxicity of graphene family nanomaterials. *Nanoscale,* **2020**, *12*(36), 18600-18605.
[http://dx.doi.org/10.1039/D0NR04179C] [PMID: 32914812]

[35]   Coreas, R.; Castillo, C.; Li, Z.; Yan, D.; Gao, Z.; Chen, J.; Bitounis, D.; Parviz, D.; Strano, M.S.; Demokritou, P.; Zhong, W. Biological impacts of reduced graphene oxide affected by protein corona formation. *Chem. Res. Toxicol.,* **2022**, *35*(7), 1244-1256.
[http://dx.doi.org/10.1021/acs.chemrestox.2c00042] [PMID: 35706338]

[36]   Chatterjee, N.; Kim, Y.; Yang, J.; Roca, C.P.; Joo, S.W.; Choi, J. A systems toxicology approach reveals the Wnt-MAPK crosstalk pathway mediated reproductive failure in *Caenorhabditis elegans* exposed to graphene oxide (GO) but not to reduced graphene oxide (rGO). *Nanotoxicology,* **2017**, *11*(1), 76-86.
[http://dx.doi.org/10.1080/17435390.2016.1267273] [PMID: 27901397]

[37]   Amrollahi-Sharifabadi, M.; Koohi, M.K.; Zayerzadeh, E.; Hablolvarid, M.H.; Hassan, J.; Seifalian, A.M. *In vivo* toxicological evaluation of graphene oxide nanoplatelets for clinical application. *Int. J.*

Nanomedicine, **2018**, *13*, 4757-4769.
[http://dx.doi.org/10.2147/IJN.S168731] [PMID: 30174424]

[38]   Nirmal, N.K.; Awasthi, K.K.; John, P.J. Hepatotoxicity of graphene oxide in Wistar rats. *Environ. Sci. Pollut. Res. Int.,* **2021**, *28*(34), 46367-46376.
[http://dx.doi.org/10.1007/s11356-020-09953-0] [PMID: 32632678]

[39]   Poulsen, S.S.; Bengtson, S.; Williams, A.; Jacobsen, N.R.; Troelsen, J.T.; Halappanavar, S.; Vogel, U. A transcriptomic overview of lung and liver changes one day after pulmonary exposure to graphene and graphene oxide. *Toxicol. Appl. Pharmacol.,* **2021**, *410*, 115343.
[http://dx.doi.org/10.1016/j.taap.2020.115343] [PMID: 33227293]

[40]   Rhazouani, A.; Gamrani, H.; Ed-Day, S.; Lafhal, K.; Boulbaroud, S.; Gebrati, L.; Fdil, N.; Aziz, F. Sub-acute toxicity of graphene oxide (GO) nanoparticles in male mice after intraperitoneal injection: Behavioral study and histopathological evaluation. *Food Chem. Toxicol.,* **2023**, *171*, 113553.
[http://dx.doi.org/10.1016/j.fct.2022.113553] [PMID: 36521574]

[41]   Kim, M.; Eom, H.J.; Choi, I.; Hong, J.; Choi, J. Graphene oxide-induced neurotoxicity on neurotransmitters, AFD neurons and locomotive behavior in Caenorhabditis elegans. *Neurotoxicology,* **2020**, *77*, 30-39.
[http://dx.doi.org/10.1016/j.neuro.2019.12.011] [PMID: 31862286]

[42]   Manjunatha, B.; Seo, E.; Park, S.H.; Kundapur, R.R.; Lee, S.J. Pristine graphene and graphene oxide induce multi-organ defects in zebrafish (Danio rerio) larvae/juvenile: an *in vivo* study. *Environ. Sci. Pollut. Res. Int.,* **2021**, *28*(26), 34664-34675.
[http://dx.doi.org/10.1007/s11356-021-13058-7] [PMID: 33656705]

[43]   Cellot, G.; Vranic, S.; Shin, Y.; Worsley, R.; Rodrigues, A.F.; Bussy, C.; Casiraghi, C.; Kostarelos, K.; McDearmid, J.R. Graphene oxide nanosheets modulate spinal glutamatergic transmission and modify locomotor behaviour in an *in vivo* zebrafish model. *Nanoscale Horiz.,* **2020**, *5*(8), 1250-1263.
[http://dx.doi.org/10.1039/C9NH00777F] [PMID: 32558850]

[44]   Jin, L.; Dou, T.T.; Chen, J.Y.; Duan, M.X.; Zhen, Q.; Wu, H.Z.; Zhao, Y.L. Sublethal toxicity of graphene oxide in Caenorhabditis elegans under multi-generational exposure. *Ecotoxicol. Environ. Saf.,* **2022**, *229*, 113064.
[http://dx.doi.org/10.1016/j.ecoenv.2021.113064] [PMID: 34890989]

[45]   Ding, X.; Wang, J.; Rui, Q.; Wang, D. Long-term exposure to thiolated graphene oxide in the range of μg/L induces toxicity in nematode Caenorhabditis elegans. *Sci. Total Environ.,* **2018**, *616-617*, 29-37.
[http://dx.doi.org/10.1016/j.scitotenv.2017.10.307] [PMID: 29107776]

[46]   Chen, W. Renal clearance of graphene oxide: glomerular filtration or tubular secretion and selective kidney injury association with its lateral dimension. *J. Nanobiotechnol.,* **2023**, *21*(1), 51.
[http://dx.doi.org/10.1186/s12951-023-01781-x]

[47]   Côa, F.; Delite, F.S.; Strauss, M.; Martinez, D.S.T. Toxicity mitigation and biodistribution of albumin corona coated graphene oxide and carbon nanotubes in Caenorhabditis elegans. *NanoImpact,* **2022**, *27*, 100413.
[http://dx.doi.org/10.1016/j.impact.2022.100413] [PMID: 35940564]

[48]   de Luna, L. A. V. Lung recovery from DNA damage induced by graphene oxide is dependent on size, dose and inflammation profile. *Part Fibre Toxicol.,* **2022**, *19*(1), 1-21.
[http://dx.doi.org/10.1186/s12989-022-00502-w]

[49]   Svadlakova, T.; Holmannova, D.; Kolackova, M.; Malkova, A.; Krejsek, J.; Fiala, Z. Immunotoxicity of carbon-based nanomaterials, starring phagocytes. *Int. J. Mol. Sci.,* **2022**, *23*(16), 8889.
[http://dx.doi.org/10.3390/ijms23168889] [PMID: 36012161]

[50]   Kiew, S.F.; Kiew, L.V.; Lee, H.B.; Imae, T.; Chung, L.Y. Assessing biocompatibility of graphene oxide-based nanocarriers: A review. *J. Control. Release,* **2016**, *226*, 217-228.
[http://dx.doi.org/10.1016/j.jconrel.2016.02.015] [PMID: 26873333]

[51]   Park, E.J.; Lee, S.J.; Lee, K.; Choi, Y.C.; Lee, B.S.; Lee, G.H.; Kim, D.W. Pulmonary persistence of graphene nanoplatelets may disturb physiological and immunological homeostasis. *J. Appl. Toxicol.,* **2017**, *37*(3), 296-309.
[http://dx.doi.org/10.1002/jat.3361] [PMID: 27440207]

[52]   Park, E.J.; Lee, G.H.; Han, B.S.; Lee, B.S.; Lee, S.; Cho, M.H.; Kim, J.H.; Kim, D.W. Toxic response of graphene nanoplatelets *in vivo* and *in vitro. Arch. Toxicol.,* **2015**, *89*(9), 1557-1568.
[http://dx.doi.org/10.1007/s00204-014-1303-x] [PMID: 24980260]

[53]   Yan, J.; Chen, L.; Huang, C.C.; Lung, S.C.C.; Yang, L.; Wang, W.C.; Lin, P.H.; Suo, G.; Lin, C.H. Consecutive evaluation of graphene oxide and reduced graphene oxide nanoplatelets immunotoxicity on monocytes. *Colloids Surf. B Biointerfaces,* **2017**, *153*, 300-309.
[http://dx.doi.org/10.1016/j.colsurfb.2017.02.036] [PMID: 28285061]

[54]   Xu, S.; Xu, S.; Chen, S.; Fan, H.; Luo, X.; Yang, X.; Wang, J.; Yuan, H.; Xu, A.; Wu, L. Graphene oxide modulates B cell surface phenotype and impairs immunoglobulin secretion in plasma cell. *J. Nanosci. Nanotechnol.,* **2016**, *16*(4), 4205-4215.
[http://dx.doi.org/10.1166/jnn.2016.11712] [PMID: 27451788]

[55]   Lee, J.K.; Jeong, A.Y.; Bae, J.; Seok, J.H.; Yang, J.Y.; Roh, H.S.; Jeong, J.; Han, Y.; Jeong, J.; Cho, W.S. The role of surface functionalization on the pulmonary inflammogenicity and translocation into mediastinal lymph nodes of graphene nanoplatelets in rats. *Arch. Toxicol.,* **2017**, *91*(2), 667-676.
[http://dx.doi.org/10.1007/s00204-016-1706-y] [PMID: 27129695]

[56]   Mao, L.; Hu, M.; Pan, B.; Xie, Y.; Petersen, E.J. Biodistribution and toxicity of radio-labeled few layer graphene in mice after intratracheal instillation. *Part. Fibre Toxicol.,* **2015**, *13*(1), 7.
[http://dx.doi.org/10.1186/s12989-016-0120-1] [PMID: 26864058]

[57]   Ruiz, A.; Lucherelli, M.A.; Murera, D.; Lamon, D.; Ménard-Moyon, C.; Bianco, A. Toxicological evaluation of highly water dispersible few-layer graphene *in vivo. Carbon,* **2020**, *170*, 347-360.
[http://dx.doi.org/10.1016/j.carbon.2020.08.023]

[58]   Sasidharan, A.; Swaroop, S.; Koduri, C.K.; Girish, C.M.; Chandran, P.; Panchakarla, L.S.; Somasundaram, V.H.; Gowd, G.S.; Nair, S.; Koyakutty, M. Comparative *in vivo* toxicity, organ biodistribution and immune response of pristine, carboxylated and PEGylated few-layer graphene sheets in Swiss albino mice: A three month study. *Carbon,* **2015**, *95*, 511-524.
[http://dx.doi.org/10.1016/j.carbon.2015.08.074]

[59]   Patlolla, A. K.; Randolph, J.; Kumari, S. A.; Tchounwou, P. B. Toxicity evaluation of graphene oxide in kidneys of sprague-dawley rats. *IJERPH,* **2016**, *13*(4), 380.
[http://dx.doi.org/10.3390/ijerph13040380]

[60]   Kucki, M.; Rupper, P.; Sarrieu, C.; Melucci, M.; Treossi, E.; Schwarz, A.; León, V.; Kraegeloh, A.; Flahaut, E.; Vázquez, E.; Palermo, V.; Wick, P. Interaction of graphene-related materials with human intestinal cells: An *in vitro* approach. *Nanoscale,* **2016**, *8*(16), 8749-8760.
[http://dx.doi.org/10.1039/C6NR00319B] [PMID: 27064646]

[61]   El-Yamany, N.A.; Mohamed, F.F.; Salaheldin, T.A.; Tohamy, A.A.; Abd El-Mohsen, W.N.; Amin, A.S. Graphene oxide nanosheets induced genotoxicity and pulmonary injury in mice. *Exp. Toxicol. Pathol.,* **2017**, *69*(6), 383-392.
[http://dx.doi.org/10.1016/j.etp.2017.03.002] [PMID: 28359838]

[62]   Tang, H.; Yang, T.; Chen, L.; Zhang, Y.; Zhu, Y.; Wang, C.; Liu, D.; Guo, Q.; Cheng, G.; Xia, F.; Zhong, T.; Wang, J. Surface chemistry of graphene tailoring the activity of digestive enzymes by modulating interfacial molecular interactions. *J. Colloid Interface Sci.,* **2023**, *630*(Pt B), 179-192.
[http://dx.doi.org/10.1016/j.jcis.2022.10.030] [PMID: 36327721]

[63]   Xu, S.; Zhang, Z.; Chu, M. Long-term toxicity of reduced graphene oxide nanosheets: Effects on female mouse reproductive ability and offspring development. *Biomaterials,* **2015**, *54*, 188-200.
[http://dx.doi.org/10.1016/j.biomaterials.2015.03.015] [PMID: 25907052]

[64]   Mehta, S.; Hingole, S.; Chaudhary, V. The emerging mechanisms of Wnt secretion and signaling in development. *Front. Cell Dev. Biol.,* **2021**, *9,* 714746.
[http://dx.doi.org/10.3389/fcell.2021.714746] [PMID: 34485301]

[65]   Wu, K.; Zhou, Q.; Ouyang, S. Direct and indirect genotoxicity of graphene family nanomaterials on DNA-A review. *Nanomaterials,* **2021**, *11*(11), 2889.
[http://dx.doi.org/10.3390/nano11112889]

[66]   Liu, B.; Zhao, Y.; Jia, Y.; Liu, J. Heating drives DNA to hydrophobic regions while freezing drives DNA to hydrophilic regions of graphene oxide for highly robust biosensors. *J. Am. Chem. Soc.,* **2020**, *142*(34), 14702-14709.
[http://dx.doi.org/10.1021/jacs.0c07028] [PMID: 32786801]

[67]   Domenech, J.; Hernández, A.; Demir, E.; Marcos, R.; Cortés, C. Interactions of graphene oxide and graphene nanoplatelets with the *in vitro* Caco-2/HT29 model of intestinal barrier. *Sci. Reports,* **2020**, *10*(1), 1-15.
[http://dx.doi.org/10.1038/s41598-020-59755-0]

[68]   Li, B.; Zhang, Y.; Meng, X.Y.; Zhou, R. Zipper-like unfolding of dsDNA caused by graphene wrinkles. *J. Phys. Chem. C,* **2020**, *124*(5), 3332-3340.
[http://dx.doi.org/10.1021/acs.jpcc.9b08778]

[69]   Domenech, J.; Rodríguez-Garraus, A.; de Cerain, A. L.; Azqueta, A.; Catalán, J. Genotoxicity of graphene-based materials. *Nanomaterials,* **1795**, *12*(11), 1795.
[http://dx.doi.org/10.3390/nano12111795]

[70]   Cho, E. S.; Ruminski, A. M.; Aloni, S.; Liu, Y. S.; Guo, J.; Urban, J. J. Graphene oxide/metal nanocrystal multilaminates as the atomic limit for safe and selective hydrogen storage. *Nat. Commun.,* **2016**, *7*(1), 1-8.
[http://dx.doi.org/10.1038/ncomms10804]

[71]   Gurunathan, S.; Kang, M.H.; Jeyaraj, M.; Kim, J.H. Differential cytotoxicity of different sizes of graphene oxide nanoparticles in leydig (TM3) and Sertoli (TM4) Cells. *Nanomaterials,* **2019**, *9*(2), 139.
[http://dx.doi.org/10.3390/nano9020139] [PMID: 30678270]

[72]   Wang, K.; Ruan, J.; Song, H.; Zhang, J.; Wo, Y.; Guo, S.; Cui, D. Biocompatibility of graphene oxide. *Nanoscale Res. Lett.,* **2010**, *6*(1), 8.
[http://dx.doi.org/10.1007/s11671-010-9751-6] [PMID: 27502632]

[73]   Tsai, M.H.; Chao, H-R.; Jiang, J-J.; Su, Y-H.; Cortez, M.P.; Tayo, L.L.; Lu, I-C.; Hsieh, H.; Lin, C-C.; Lin, S-L.; Wan Mansor, W.N.; Su, C-K.; Huang, S-T.; Hsu, W-L. Toxicity of low-dose graphene oxide nanoparticles in an *in vivo* wild type of caenorhabditis elegans model. *Aerosol Air Qual. Res.,* **2021**, *21*(5), 200559.
[http://dx.doi.org/10.4209/aaqr.200559]

[74]   Gurunathan, S.; Arsalan Iqbal, M.; Qasim, M.; Park, C.H.; Yoo, H.; Hwang, J.H.; Uhm, S.J.; Song, H.; Park, C.; Do, J.T.; Choi, Y.; Kim, J.H.; Hong, K. Evaluation of graphene oxide induced cellular toxicity and transcriptome analysis in human embryonic kidney cells. *Nanomaterials,* **2019**, *9*(7), 969.
[http://dx.doi.org/10.3390/nano9070969] [PMID: 31269699]

[75]   Jia, P.P.; Sun, T.; Junaid, M.; Yang, L.; Ma, Y.B.; Cui, Z.S.; Wei, D.P.; Shi, H.F.; Pei, D.S. Nanotoxicity of different sizes of graphene (G) and graphene oxide (GO) *in vitro* and *in vivo. Environ. Pollut.,* **2019**, *247,* 595-606.
[http://dx.doi.org/10.1016/j.envpol.2019.01.072] [PMID: 30708322]

[76]   Kazempour, M.; Namazi, H.; Akbarzadeh, A.; Kabiri, R. Synthesis and characterization of PEG-functionalized graphene oxide as an effective pH-sensitive drug carrier. *Artif. Cells Nanomed. Biotechnol.,* **2019**, *47*(1), 90-94.
[http://dx.doi.org/10.1080/21691401.2018.1543196] [PMID: 30663418]

[77] Rahman, M.; Ahmad, M.; Ahmad, J.; Firdous, J.; Ahmad, F.; Mushtaq, G.; Kamal, M.; Akhter, S. Role of graphene nano-composites in cancer therapy: Theranostic applications, metabolic fate and toxicity issues. *Curr. Drug Metab.,* **2015**, *16*(5), 397-409.
[http://dx.doi.org/10.2174/1389200215666141125120633] [PMID: 25429670]

[78] Li, T.; Liu, X.; Li, L.; Wang, Y.; Ma, P.; Chen, M.; Dong, W. Polydopamine-functionalized graphene oxide compounded with polyvinyl alcohol/chitosan hydrogels on the recyclable adsorption of cu(II), Pb(II) and cd(II) from aqueous solution. *J. Polym. Res.,* **2019**, *26*(12), 281.
[http://dx.doi.org/10.1007/s10965-019-1971-6]

[79] Zheng, M.; Pan, M.; Zhang, W.; Lin, H.; Wu, S.; Lu, C.; Tang, S.; Liu, D.; Cai, J. Poly(α-l-lysin-)-based nanomaterials for versatile biomedical applications: Current advances and perspectives. *Bioact. Mater.,* **2021**, *6*(7), 1878-1909.
[http://dx.doi.org/10.1016/j.bioactmat.2020.12.001] [PMID: 33364529]

[80] Kazempour, M.; Namazi, H.; Akbarzadeh, A.; Kabiri, R. Synthesis and characterization of PEG-functionalized graphene oxide as an effective pH-sensitive drug carrier. *Artif. Cells Nanomed. Biotechnol.,* **2019**, *47*(1), 90-94.
[http://dx.doi.org/10.1080/21691401.2018.1543196]

[81] Gholami, A.; Emadi, F.; Amini, A.; Shokripour, M.; Chashmpoosh, M.; Omidifar, N. Functionalization of graphene oxide nanosheets can reduce their cytotoxicity to dental pulp stem cells. *J. Nanomater.,* **2020**, *2020*, 1-14.
[http://dx.doi.org/10.1155/2020/6942707]

[82] Duan, G.; Kang, S.; Tian, X.; Garate, J.A.; Zhao, L.; Ge, C.; Zhou, R. Protein corona mitigates the cytotoxicity of graphene oxide by reducing its physical interaction with cell membrane. *Nanoscale,* **2015**, *7*(37), 15214-15224.
[http://dx.doi.org/10.1039/C5NR01839K] [PMID: 26315610]

[83] Yang, Y.; Han, P.; Xie, X.; Yin, X.; Duan, G.; Wen, L. Protein corona reduced graphene oxide cytotoxicity by inhibiting endocytosis. *Colloid Interface Sci. Commun.,* **2021**, *45*, 100514.
[http://dx.doi.org/10.1016/j.colcom.2021.100514]

[84] Rhazouani, A.; Gamrani, H.; El Achaby, M.; Aziz, K.; Gebrati, L.; Uddin, M.S.; Aziz, F. Synthesis and toxicity of graphene oxide nanoparticles: A literature review of *in vitro* and *in vivo* studies. *BioMed Res. Int.,* **2021**, *2021*, 1-19.
[http://dx.doi.org/10.1155/2021/5518999] [PMID: 34222470]

[85] Wang, T.; Zhu, S.; Jiang, X. Toxicity mechanism of graphene oxide and nitrogen-doped graphene quantum dots in RBCs revealed by surface-enhanced infrared absorption spectroscopy. *Toxicol. Res.,* **2015**, *4*(4), 885-894.
[http://dx.doi.org/10.1039/C4TX00138A]

[86] Xin, Q.; Shah, H.; Nawaz, A.; Xie, W.; Akram, M.Z.; Batool, A.; Tian, L.; Jan, S.U.; Boddula, R.; Guo, B.; Liu, Q.; Gong, J.R. Antibacterial carbon-based nanomaterials. *Adv. Mater.,* **2019**, *31*(45), 1804838.
[http://dx.doi.org/10.1002/adma.201804838] [PMID: 30379355]

[87] Chen, Y.; Pandit, S.; Rahimi, S.; Mijakovic, I. Interactions between graphene-based materials and biological surfaces: A review of underlying molecular mechanisms. *Adv. Mater. Interfaces,* **2021**, *8*(24), 2101132.
[http://dx.doi.org/10.1002/admi.202101132]

[88] Gold, K.; Slay, B.; Knackstedt, M.; Gaharwar, A.K. Antimicrobial activity of metal and metal-oxide based nanoparticles. *Adv. Ther.,* **2018**, *1*(3), 1700033.
[http://dx.doi.org/10.1002/adtp.201700033]

[89] Wen, T.; Liu, J.; He, W.; Yang, A. Nanomaterials and Reactive Oxygen Species (ROS). In: *Nanotechnology in Regenerative Medicine and Drug Delivery Therapy*; Springer, **2020**; pp. 361-387.
[http://dx.doi.org/10.1007/978-981-15-5386-8_8]

[90]   Jaworski, S. Degradation of mitochondria and oxidative stress as the main mechanism of toxicity of pristine graphene on U87 glioblastoma cells and tumors and HS-5 cells. *IJMS,* **2019**, *20*(3), 650.
[http://dx.doi.org/10.3390/ijms20030650]

[91]   Zare, P.; Aleemardani, M.; Seifalian, A.; Bagher, Z.; Seifalian, A. M. Graphene oxide: Opportunities and challenges in biomedicine. *Nanomaterials,* **2021**, *11*(5), 1083.
[http://dx.doi.org/10.3390/nano11051083]

[92]   Li, R.; Guiney, L.M.; Chang, C.H.; Mansukhani, N.D.; Ji, Z.; Wang, X.; Liao, Y.P.; Jiang, W.; Sun, B.; Hersam, M.C.; Nel, A.E.; Xia, T. Surface oxidation of graphene oxide determines membrane damage, lipid peroxidation, and cytotoxicity in macrophages in a pulmonary toxicity model. *ACS Nano,* **2018**, *12*(2), 1390-1402.
[http://dx.doi.org/10.1021/acsnano.7b07737] [PMID: 29328670]

[93]   Zhang, L.; Ouyang, S.; Zhang, H.; Qiu, M.; Dai, Y.; Wang, S.; Wang, Y.; Ou, J. Graphene oxide induces dose-dependent lung injury in rats by regulating autophagy. *Exp. Ther. Med.,* **2021**, *21*(5), 462.
[http://dx.doi.org/10.3892/etm.2021.9893] [PMID: 33747194]

[94]   Tabish, T.A.; Pranjol, M.Z.I.; Jabeen, F.; Abdullah, T.; Latif, A.; Khalid, A.; Ali, M.; Hayat, H.; Winyard, P.G.; Whatmore, J.L.; Zhang, S. Investigation into the toxic effects of graphene nanopores on lung cancer cells and biological tissues. *Appl. Mater. Today,* **2018**, *12*, 389-401.
[http://dx.doi.org/10.1016/j.apmt.2018.07.005]

[95]   Ou, L.; Lin, S.; Song, B.; Liu, J.; Lai, R.; Shao, L. The mechanisms of graphene-based materials-induced programmed cell death: A review of apoptosis, autophagy, and programmed necrosis. *Int. J. Nanomedicine,* **2017**, *12*, 6633-6646.
[http://dx.doi.org/10.2147/IJN.S140526] [PMID: 28924347]

[96]   Li, Y.; Liu, Y.; Fu, Y.; Wei, T.; Le Guyader, L.; Gao, G.; Liu, R.S.; Chang, Y.Z.; Chen, C. The triggering of apoptosis in macrophages by pristine graphene through the MAPK and TGF-beta signaling pathways. *Biomaterials,* **2012**, *33*(2), 402-411.
[http://dx.doi.org/10.1016/j.biomaterials.2011.09.091] [PMID: 22019121]

[97]   Zhang, J.; Cao, H.Y.; Wang, J.Q.; Wu, G.D.; Wang, L. Graphene oxide and reduced graphene oxide exhibit cardiotoxicity through the regulation of lipid peroxidation, oxidative stress, and mitochondrial dysfunction. *Front. Cell Dev. Biol.,* **2021**, *9*, 616888.
[http://dx.doi.org/10.3389/fcell.2021.616888] [PMID: 33816465]

[98]   Liu, L.; Zhang, M.; Zhang, Q.; Jiang, W. Graphene nanosheets damage the lysosomal and mitochondrial membranes and induce the apoptosis of RBL-2H3 cells. *Sci. Total Environ.,* **2020**, *734*, 139229.
[http://dx.doi.org/10.1016/j.scitotenv.2020.139229] [PMID: 32450398]

[99]   Xu, L.; Zhao, J.; Wang, Z. Genotoxic response and damage recovery of macrophages to graphene quantum dots. *Sci. Total Environ.,* **2019**, *664*, 536-545.
[http://dx.doi.org/10.1016/j.scitotenv.2019.01.356] [PMID: 30759415]

[100] Fasbender, S. The low toxicity of graphene quantum dots is reflected by marginal gene expression changes of primary human hematopoietic stem cells. *Sci. Reports,* **2019**, *9*(1), 1-13.
[http://dx.doi.org/10.1038/s41598-019-48567-6]

[101] Flores-López, L.Z.; Espinoza-Gómez, H.; Somanathan, R. Silver nanoparticles: Electron transfer, reactive oxygen species, oxidative stress, beneficial and toxicological effects. Mini review. *J. Appl. Toxicol.,* **2019**, *39*(1), 16-26.
[http://dx.doi.org/10.1002/jat.3654] [PMID: 29943411]

[102] Zhang, C.; Wei, K.; Zhang, W.; Bai, Y.; Sun, Y.; Gu, J. Graphene oxide quantum dots incorporated into a thin film nanocomposite membrane with high flux and antifouling properties for low-pressure nanofiltration. *ACS Appl. Mater. Interfaces,* **2017**, *9*(12), 11082-11094.
[http://dx.doi.org/10.1021/acsami.6b12826] [PMID: 28244726]

[103] Rafieian-Kopaei, M.; Ghazimoradi, M.M.; Azad, F.V.; Jalali, F. The neurotoxic mechanisms of graphene family nanomaterials at the cellular level: A solution-based approach review. *Curr. Pharm. Des.,* **2022**, *28*(44), 3572-3581.
[http://dx.doi.org/10.2174/1381612829666221202093813] [PMID: 36464882]

[104] Wang, X.; Zeng, Z.; Yang, T.; Zhang, P.; Feng, B.; Qing, T. DNA damage caused by light-driven graphene oxide: A new mechanism. *Environ. Sci. Nano,* **2023**, *10*(2), 519-527.
[http://dx.doi.org/10.1039/D2EN00948J]

[105] Akhavan, O.; Ghaderi, E.; Akhavan, A. Size-dependent genotoxicity of graphene nanoplatelets in human stem cells. *Biomaterials,* **2012**, *33*(32), 8017-8025.
[http://dx.doi.org/10.1016/j.biomaterials.2012.07.040] [PMID: 22863381]

[106] Mohamed, H.R.H.; Welson, M.; Yaseen, A.E.; EL-Ghor, A.A. Estimation of genomic instability and mutation induction by graphene oxide nanoparticles in mice liver and brain tissues. *Environ. Sci. Pollut. Res. Int.,* **2020**, *27*(1), 264-278.
[http://dx.doi.org/10.1007/s11356-019-06930-0] [PMID: 31786761]

[107] Hojo, M.; Maeno, A.; Sakamoto, Y.; Ohnuki, A.; Tada, Y.; Yamamoto, Y.; Ikushima, K.; Inaba, R.; Suzuki, J.; Taquahashi, Y.; Yokota, S.; Kobayashi, N.; Ohnishi, M.; Goto, Y.; Numano, T.; Tsuda, H.; Alexander, D.B.; Kanno, J.; Hirose, A.; Inomata, A.; Nakae, D. Two-year intermittent exposure of a multiwalled carbon nanotube by intratracheal instillation induces lung tumors and pleural mesotheliomas in F344 rats. *Part. Fibre Toxicol.,* **2022**, *19*(1), 38.
[http://dx.doi.org/10.1186/s12989-022-00478-7] [PMID: 35590372]

[108] Zhou, H.; Zhao, K.; Li, W.; Yang, N.; Liu, Y.; Chen, C.; Wei, T. The interactions between pristine graphene and macrophages and the production of cytokines/chemokines *via* TLR- and NF-κB-related signaling pathways. *Biomaterials,* **2012**, *33*(29), 6933-6942.
[http://dx.doi.org/10.1016/j.biomaterials.2012.06.064] [PMID: 22796167]

[109] Duch, M.C.; Budinger, G.R.S.; Liang, Y.T.; Soberanes, S.; Urich, D.; Chiarella, S.E.; Campochiaro, L.A.; Gonzalez, A.; Chandel, N.S.; Hersam, M.C.; Mutlu, G.M. Minimizing oxidation and stable nanoscale dispersion improves the biocompatibility of graphene in the lung. *Nano Lett.,* **2011**, *11*(12), 5201-5207.
[http://dx.doi.org/10.1021/nl202515a] [PMID: 22023654]

[110] Loret, T.; de Luna, L.A.V.; Fordham, A.; Arshad, A.; Barr, K.; Lozano, N.; Kostarelos, K.; Bussy, C. Innate but not adaptive immunity regulates lung recovery from chronic exposure to graphene oxide nanosheets. *Adv. Sci.,* **2022**, *9*(11), 2104559.
[http://dx.doi.org/10.1002/advs.202104559] [PMID: 35166457]

[111] Li, B.; Yang, J.; Huang, Q.; Zhang, Y.; Peng, C.; Zhang, Y.; He, Y.; Shi, J.; Li, W.; Hu, J.; Fan, C. Biodistribution and pulmonary toxicity of intratracheally instilled graphene oxide in mice. *NPG Asia Mater.,* **2013**, *5*(4), e44.
[http://dx.doi.org/10.1038/am.2013.7]

[112] Bengtson, S.; Knudsen, K.B.; Kyjovska, Z.O.; Berthing, T.; Skaug, V.; Levin, M.; Koponen, I.K.; Shivayogimath, A.; Booth, T.J.; Alonso, B.; Pesquera, A.; Zurutuza, A.; Thomsen, B.L.; Troelsen, J.T.; Jacobsen, N.R.; Vogel, U. Differences in inflammation and acute phase response but similar genotoxicity in mice following pulmonary exposure to graphene oxide and reduced graphene oxide. *PLoS One,* **2017**, *12*(6), e0178355.
[http://dx.doi.org/10.1371/journal.pone.0178355] [PMID: 28570647]

[113] Trinh, T.X.; Ha, M.K.; Choi, J.S.; Byun, H.G.; Yoon, T.H. Curation of datasets, assessment of their quality and completeness, and nanoSAR classification model development for metallic nanoparticles. *Environ. Sci. Nano,* **2018**, *5*(8), 1902-1910.
[http://dx.doi.org/10.1039/C8EN00061A]

[114] Fontana, F.; Figueiredo, P.; Martins, J.P.; Santos, H.A. Requirements for animal experiments: Problems and challenges. *Small,* **2021**, *17*(15), 2004182.

[http://dx.doi.org/10.1002/smll.202004182] [PMID: 33025748]

[115] Varsou, D.D.; Afantitis, A.; Melagraki, G.; Sarimveis, H. Read-across predictions of nanoparticle hazard endpoints: A mathematical optimization approach. *Nanoscale Adv.,* **2019**, *1*(9), 3485-3498.
[http://dx.doi.org/10.1039/C9NA00242A] [PMID: 36133569]

[116] Papadiamantis, A.G.; Afantitis, A.; Tsoumanis, A.; Valsami-Jones, E.; Lynch, I.; Melagraki, G. Computational enrichment of physicochemical data for the development of a ζ-potential read-across predictive model with Isalos Analytics Platform. *NanoImpact,* **2021**, *22*, 100308.
[http://dx.doi.org/10.1016/j.impact.2021.100308] [PMID: 35559965]

[117] Afantitis, A.; Tsoumanis, A.; Melagraki, G. Enalos suite of tools: Enhancing cheminformatics and nanoinfor - matics through KNIME. *Curr. Med. Chem.,* **2020**, *27*(38), 6523-6535.
[http://dx.doi.org/10.2174/0929867327666200727114410] [PMID: 32718281]

[118] Winkler, D.A. Role of artificial intelligence and machine learning in nanosafety. *Small,* **2020**, *16*(36), 2001883.
[http://dx.doi.org/10.1002/smll.202001883]

[119] Qiu, T.A.; Clement, P.L.; Haynes, C.L. Linking nanomaterial properties to biological outcomes: Analytical chemistry challenges in nanotoxicology for the next decade. *Chem. Commun.,* **2018**, *54*(91), 12787-12803.
[http://dx.doi.org/10.1039/C8CC06473C] [PMID: 30357136]

[120] Casalini, T. Not only *in silico* drug discovery: Molecular modeling towards *in silico* drug delivery formulations. *J. Control. Release,* **2021**, *332*, 390-417.
[http://dx.doi.org/10.1016/j.jconrel.2021.03.005] [PMID: 33675875]

[121] Kang, Y.; Liu, J.; Yin, S.; Jiang, Y.; Feng, X.; Wu, J.; Zhang, Y.; Chen, A.; Zhang, Y.; Shao, L. Oxidation of reduced graphene oxide *via* cellular redox signaling modulates actin-mediated neurotransmission. *ACS Nano,* **2020**, *14*(3), 3059-3074.
[http://dx.doi.org/10.1021/acsnano.9b08078] [PMID: 32057235]

[122] Punetha, V.D.; Ha, Y.M.; Kim, Y.O.; Jung, Y.C.; Cho, J.W. Interaction of photothermal graphene networks with polymer chains and laser-driven photo-actuation behavior of shape memory polyurethane/epoxy/epoxy-functionalized graphene oxide nanocomposites. *Polymer,* **2019**, *181*, 121791.
[http://dx.doi.org/10.1016/j.polymer.2019.121791]

[123] Punetha, V.D.; Ha, Y.M.; Kim, Y.O.; Jung, Y.C.; Cho, J.W. Rapid remote actuation in shape memory hyperbranched polyurethane composites using cross-linked photothermal reduced graphene oxide networks. *Sens. Actuators B Chem.,* **2020**, *321*, 128468.
[http://dx.doi.org/10.1016/j.snb.2020.128468]

[124] Sahoo, N.G.; Sandeep, M. A process of manufacturing. Indian Patent 352780, 2016.

CHAPTER 4

Graphene and Waste Management A Roadmap for Cost-Effective Graphene Production

Mamoona Hayat[1,2]**, Junaid Ali**[*, 1]**, Saira Arif**[*, 2]**, Khaled Pervez**[3]**, Ghayas Uddin Siddiqui**[4] **and Cinzia Casiraghi**[3]

[1] *OptoElectronics Research Laboratory(OERL), Department of Physics, COMSATS University Islamabad, Pakistan*

[2] *Department of Chemistry, COMSATS University Islamabad, Pakistan*

[3] *School of Chemistry, Manchester University, Manchester M13 9PL, United Kingdom*

[4] *Department of Chemical Engineering, Jeju National University, 63243, Jeju, South Korea*

Abstract: The sustainable development goals have provided a boost and economic appeal to recycling and reusing waste. Waste materials like plastic, industrial, and biomass can be exploited as a foundation to produce valuable products, including wonder materials like graphene. It is utilized in almost every field of life, from environmental sustainability to smart clothing. Waste material contains a variety of organic polymers which can be converted into graphene and its derivatives. It uses various methods like metal catalysis, laser ablation techniques, flash Joule heating, and pyrolysis. These methods may produce 3D, 2D, 1D, and 0D graphene. The obtained products have exclusive properties like thermal, optoelectronic, and electrical properties. The potential for removing and converting waste into the revolutionary material of the century opens possibilities for a sustainable and progressive yet less hazardous world for our future generations. Some approaches promise the fabrication of graphene and its spin-offs from biowaste like sugarcane bagasse, dog feces, and grass. Similarly, liquid phase exfoliation of graphene provides less hazardous and sustainable graphene production from materials without using toxic materials or burdening the earth with waste products. The carbon-negative approach proves an environmentally friendly alternative to prevalent waste-burning practices to dispose of such waste. The obtained graphene and related products have distinctive properties and tremendous applications at a fraction of the cost. The potential for removing and converting waste into the revolutionary material of the century opens possibilities for a sustainable and progressive yet less hazardous world for our future generations. This chapter reviews the efficient methods for synthesizing graphene from waste products and its various applications.

---

[*] **Corresponding authors Saira Arif and Junaid Ali:** Department of Chemistry, COMSATS University Islamabad, Pakistan; Department of Chemistry, COMSATS University Islamabad, Pakistan; E-mails: saira.arif@comsasts.edu.pk and junaid_ali@comsats.edu.pk

Vinay Deep Punetha (Ed.)

Keywords: Cost, Environmental, Flash graphene, Graphene quantum dots, Laser-Induced graphene, LIG, Laser ablation, Production, Pyrolysis, Supercapacitor sustainability, Toxicity, Waste.

INTRODUCTION

The world's population has become highly concerned about the environment due to the rising consumption of reliable and non-reliable consumer products and the waste materials from these activities [1]. The concerns are thoughtful because the accumulation of waste threatens the environment and public safety. Recent assessments indicate that landfills were used to discard more than 80% of municipal solid waste. Instead of being disposed of away, this trash can be reprocessed and used in factories to lessen the effects of this issue [2]. The production of graphene from waste materials presents an exciting opportunity to not only tackle the waste problem but also to obtain a valuable and versatile material for numerous industrial applications. By utilizing waste materials as a precursor for graphene production, the aim is to establish a cost-effective and sustainable approach to graphene synthesis, thereby reducing the environmental impact and enhancing resource efficiency. The roadmap for cost-effective graphene production in waste management encompasses a comprehensive exploration of waste sources that can serve as potential graphene precursors. It includes strategies for waste collection, segregation, and processing, followed by innovative methods for the extraction and synthesis of graphene. Additionally, the roadmap outlines the characterization techniques and quality control measures required to ensure the viability and effectiveness of the produced graphene material [3]. The chapter delves into the discussion of various precursors and methods used for graphene production in the context of waste management. It explores different types of waste materials that can serve as potential precursors for graphene synthesis, including biomass, industrial waste, and carbonaceous materials. Additionally, the chapter explores innovative approaches and techniques for the cost-effective production of graphene from these waste sources. It also emphasizes the importance of optimizing the production process to achieve high-quality graphene while minimizing energy consumption and environmental impact. The chapter serves as a comprehensive guide that explores the potential of graphene in waste management, while also outlining the various precursors and methods available for cost-effective graphene production.

TYPES OF WASTE

Plastic waste

The large-scale fabrication and consumption of plastic artifacts for the betterment of human life have posed a severe threat to the environment, as it is nonbiodegradable. The scientific community faces severe challenges regarding solid plastic waste disposal and its conversion to valuable products to mitigate its environmental effects. To overcome the problem, the scientific community is framing strategies for solid plastic waste management that can potentially convert the trash into treasure. Technical backwardness in reprocessing technology plays a significant role in overcoming the solid litter problem, where the fascinating utilization of Nano chemistry provides a benefit over conventional recycling methodologies. The proliferation of carbon nanomaterials underpins solid waste management technology to achieve successful value-added recycling. The critical technology has been mainly driven by the selective transformation of toxic solid waste into graphene nanosheets [4].

Rubber Waste

Used synthetic and natural rubber is discarded in vast quantities each year. Rubber is difficult to recycle and degrade naturally due to its intricate three-dimensional structure. Even though managing rubber waste has become a global issue, there is always a room for improvement. Rubbers with broken bonds or polymer backbones are created *via* vulcanization or degradation. Rubber that has been devulcanized or decomposed can be recycled to create new products or used in various applications for energy recovery [5]. Luoung *et al.* recently designed a method to fabricate turbostratic flash graphene (tFG) using the Flash Joule Heating procedures, where electricity is used to synthesize high-quality tFG [6]. The procedure is less expensive than traditional synthetic methods and consistent with raw waste rubber resources [7].

Electronic Waste

Around the globe, the electric and electronic market is expanding speedily, producing larger quantities of electronic waste (e-waste). E-waste is being dumped openly without considering environmental safety, as e-waste eliminates lethal and noxious environmental effluents [8]. Proper e-waste management is necessary to keep the environment clean. Among the electronic waste management options, incineration and landfilling are the least favored. Instead, the primary stress is upon nature-friendly waste recycling to reduce waste production and its conversion into products with significant value and optimize the usage of resources [9].

The production of high-quality graphene and other carbon nanostructures now uses solid polymeric wastes like polyethylene, polypropylene, and bio-organic waste in exciting and creative new ways [10]. Given the effectiveness of these ground-breaking investigations, the parts of e-waste containing polymers may be used as a substitute originator to synthesize graphene nanosheets and other materials with pore sizes in the nanometer range. Bajpai *et al.* reported that transforming the polymers in e-waste to graphene using Pulsed Laser Ablation (PLA) in a solvent media is a practical, accessible, and computationally efficient process. In this process, chemicals and high vacuum apparatus are not required. Therefore, the process is scalable, repeatable, and ecologically friendly. Moreover, the e-waste, *i.e.*, utilized as a precursor, may be purchased in large amounts. When conducting, ink is made to print a circuit, and the produced graphene nanosheets also exhibit good conductivity. The procedure is helpful from an ecological and economic standpoint because it results in manufacturing a valued product with high scientific value, such as high-quality graphene [11].

Biomass Waste

Due to the exponential increase in energy demand that our society has experienced in the past few years, advanced energy storage technologies must be developed. These technologies must be able to store as much energy as possible while also being made of components that are reasonably priced, safe, and environmentally friendly [12]. Carbon-containing materials have been extensively used for energy storage because of their affordable price, abundance, range of polymorphs and transitions, exceptional physical/chemical stability, and low cost [13]. Usually, synthesis procedures are expensive, time-consuming, and challenging to scale up. Carbons from biomass are more environmentally friendly and are frequently produced by simple thermal treatment of agricultural and forestry waste at elevated temperatures. The resultant products, which exhibit significant electrochemical activity, are exciting substances for application in dual Carbon Lithium-ion Capacitors (LICs) as either positive or negative electrodes [14]. The conductance and connection of the different carbon particles are crucial for Li-insertion anodes. In this aspect, graphene can provide the electrode with mechanical stability, high-speed connections with the current collector, and a 3D conducting network along the electrode to speed up electron transport and transmission [15]. Gómez-Urbano *et al.* reported the fabrication of customized graphene-carbon composites for use in both the electrodes of a LIC by pyrolyzing graphene oxide (GO) and coffee waste powder. It investigated how the physicochemical properties of the materials influenced the electrochemical behavior of both electrodes [16].

METHODS USED FOR THE SYNTHESIS OF GRAPHENE AND ITS DERIVATIVES

Pyrolysis

Pyrolysis is an old technique that can be used to control plastic waste. This process injects waste materials into the reactor at a higher temperature while keeping the environment inert (between 400 and 600 C). The products from the reactor are transported *via* an inert gas, such as nitrogen or argon, which may also maintain the vacuum. As a result, the polymer molecules in the waste materials are broken down into smaller molecules that may be further processed and used. However, the hazardous hydrocarbons generated by pyrolysis are used to create graphene that has a significant value. There are two major types of pyrolysis:

Catalytic Pyrolysis

Catalytic pyrolysis is an efficient method of recycling plastic waste in which lower energy is required along with a catalyst to form valuable products with improved quality. It also reduces the reaction time as compared to thermal pyrolysis. The output and process of catalytic pyrolysis can be improved by utilizing a variety of catalysts. Fig. (**1**) shows the process of catalytic pyrolysis.

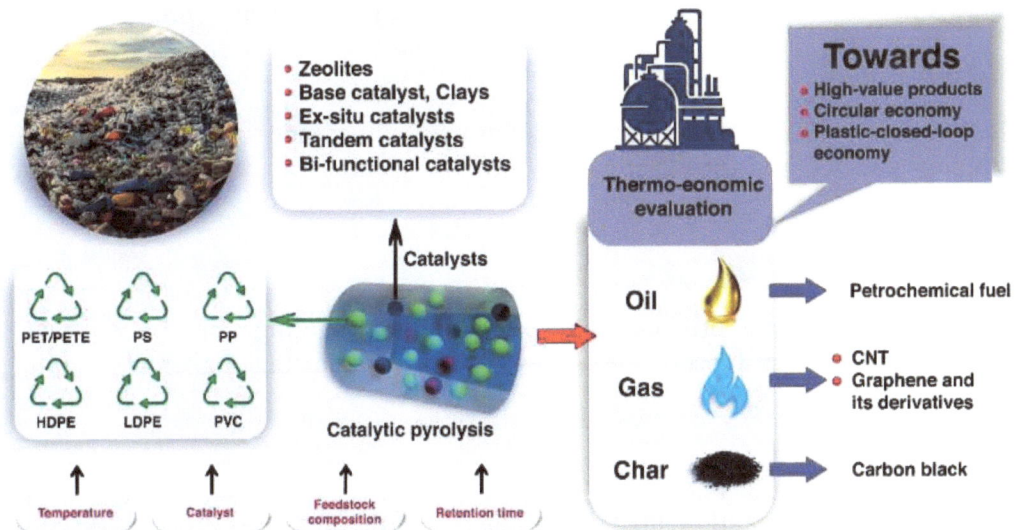

Fig. (1). Catalytic pyrolysis process for e-waste to GS conversion.

3.1.1.1 Factors Affecting Catalysts Activity:

a. The surface area of the catalyst
b. Size of pores
c. Volume of pores
d. Acidity

Thermal Pyrolysis

1. It is an endothermic process requiring high energy to decompose polymers in plastic waste. It converts plastic waste to a significant fraction of products that are low quality and contain impurities.

Factors Affecting Pyrolysis

Temperature

The most crucial variable that impacts yield amount and quality is temperature. Because it influences the cracking processes, which could alter the result, it also impacts the products formed. Long-chain hydrocarbons are created at low temperatures, whereas short-chain hydrocarbons are created at high temperatures due to the C-C bond breaking. The secondary process reaction, which results in the synthesis of aromatic chemicals, is started at a high temperature.

Retention time

It shows a negligible effect on the product while the temperature is kept constant. Studies reported that the exact composition of products was obtained at different retention times at the same temperature.

Feedstock composition

The feedstock being utilized requires different operating conditions, such as polyethylene and polypropylene types of plastic requiring higher temperatures than polystyrene.

Catalyst

The catalyst significantly enhances the quality of the pyrolysis product. It reduces the energy demand and temperature. It accelerates the cracking reactions, synthesizing a more significant proportion of gases than the yield of liquid oil. However, some heavier hydrocarbon chains in the liquid are either accumulated on the catalyst or split into shorter chains, enhancing the purity of the liquid oil produced [17 - 23]. A summary of various critical factors affecting the pyrolysis is illustrated in Fig. (2).

Pyrolysis of Plastic waste

Trashed plastics were collected from a thrift store to begin making graphene nanosheets. The collected plastics were crushed into tiny, flake-like particles with dimensions of 11.21 mm (about 0.44 in) in length and 5.75 mm (about 0.23 in) in width using the Crusher. After being cleaned in a washer with a detergent solution to get rid of the greasy and oily pollutants, the shredded plastics were dried. Following these basic washing and drying procedures, the plastic waste and disk-shaped bentonite nano clay were mixed adequately in a predetermined ratio with a sample mixture unit. Bentonite and the material we must degrade must be thoroughly blended in the mixer.

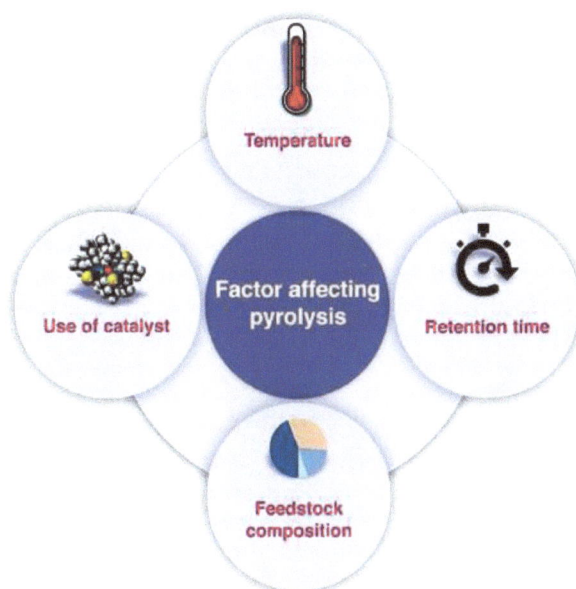

Fig. (2). Illustration of the factors affecting pyrolysis.

Consequently, a horizontal hollow cylindrical feeder unit with a capacity of 0.41 m^3 was used to prepare the sample combination, which was then gradually pyrolyzed at 400 °C in an inert environment of N_2 gas. The framework for forming graphene nanosheets is created by separating all fuels, waxes, and gases with added value. This procedure calls for slow pyrolysis at a temperature of up to 400 °C and a heating rate of 5 °C/min to remove oil-containing hydrocarbons from the polymers. This process employs a slow pyrolysis rate to verify that all oil-containing hydrocarbons and gases in the C1-C4 range have been removed from the sample mixture. The slow rate of pyrolysis also lowers the likelihood of developing waxy components in the sample combination, even if the pyrolysis process produces waxy compounds. The last step also raises the graphene nanosheets' percentage purity level. During slow pyrolysis, a dark-colored charred residue of the amorphous carbon type was created. A basic ignition test was also performed on a few arbitrarily selected specimens from the bundles of the black residue produced following the slow pyrolysis to establish that all the oily hydrocarbons and waxy chemicals had been eliminated during the process. An amorphous, porous, shiny black, charred residue formed after the pyrolysis unit was cooled to ambient temperature and allowed to pyrolyze there slowly. The amorphous black charred residue obtained from the first step is added and processed into an ultra-fine powder in the ball mill unit.

The graphene nanosheets in the system will be more productive as a result. Afterward, this secondary heating process is completed in an inert environment. A secondary reactor with a vertical cylindrical feeder unit and a 0.06 m^3 capacity was utilized for fast pyrolysis. It works at 750 °C with a heating rate of 10 °C/min in an inert atmosphere of N_2 (20 ml/min). Grinding amorphous carbon residue into an ultra-fine powder creates black graphene nanosheets. To maintain the product's purity, nano clay was removed. This was done by washing with distilled water and using 5% HCl to make a moderately acidic treatment. We refined these two temperatures through multiple trials to create a few graphene layers [4]. Fig. (3) explains the process of pyrolysis of plastic waste.

Fig. (3). Pyrolysis of plastic waste.

Pyrolysis of Biowaste

Graphene Quantum Dots from Spent Tea

Short-range ordered Carbon from wasted tea pyrolysis was chopped thermochemically with the help of a microwave to create GQDs. Wasted tea was cleaned, dried in an oven for 12 hours at 80 °C, and ground into a fine powder (90 m). The powder was then heated to 500 °C in an inert environment at 10 °C per minute in a VCTF4 furnace and three hours of pyrolysis time resulted in the production of carbon-rich biochar. After being cleaned with DI water, the biochar precursor was cooked in 0.1 M HCl to remove contamination and lower the ash level. To achieve the purified product, the specimens were cleaned once more with DI water before being dried in an oven at 60 °C. To determine the purity of the precursor, an XPS examination was carried out. According to the findings, the raw precursor was primarily composed of C, O, and N, with residues of Ca. The carbon-rich sample was then injected as 20 mg into a reactor holding 10 ml of DI water. To create an acidic environment for the oxidative cutting of carbon domains, 2 ml of HNO_3 was added to the reactor. The reactor was heated in the microwave for 15–180 minutes at 100–900 W while in reflux. Using 100 ml of DI water, the resultant brown dispersion was diluted. 0.1 um polyvinylidene difluoride (PVDF) filtration membranes were utilized to extract the larger unreacted particles. The produced pale-yellow filtrate, which displayed brilliant luminescence, included GQDs. The solution's acidity made it necessary to purify the produced GQDs.

The GQDs dispersion was dialyzed for one day after being neutralized with NaOH solution till pH 7. It will improve the GQDs solution's purity. By freeze-drying, dry GQDs were produced with a yield of up to 84.5 weight percent. Around 50 mL of the GQDs solution was put into a Teflon-lined autoclave for hydrothermal treatment prior to purification. The dispersion underwent one day of dialysis purification after 8 hours of hydrothermal treatment at 200 °C. Then, it was filtered using a 0.1 m PVDF filtration membrane. The extra water was eliminated using a rotary evaporator. The concentrated solution produced lights brilliantly [14]. In Fig. (**4**) pyrolysis of spent tea is explained. For the synthesis of a few layers' graphenes from KPW, exfoliation is preceded by the pre-carbonization processes. KPW was bought from a fruit market close by. Double distilled water was used to purify the collected KPW, removing impurities like dust. The KPW was cleaned, sun-dried for two days, and then oven-dried for four hours at 90 °C. The KPW was cleaned, dried, and processed *via* a typical blender into a fine powder. Using a Tube furnace apparatus, the produced sample combination was primarily pyrolyzed for 1 hour at 440 °C in an inert environment of N2 gas. At this stage, the sample is subjected to advanced pyrolysis at a heating rate of 5.0 °C/min to eliminate all liquid and gaseous components.

Because of the sample's pre-carbonization, the main chain for the nucleation of GNs was created in this phase. The black-colored amorphous solid was formed after an hour of pyrolysis. Before the second heating operation, this black amorphous solid is ground into a fine black powder and branded as KPWC (pre-carbonized Kinnow Peel Waste Carbon). The powdered KPWC and KCl were mixed entirely in a blender in a 1:2 ratio, with KCl serving as an activator. The reaction mixture was

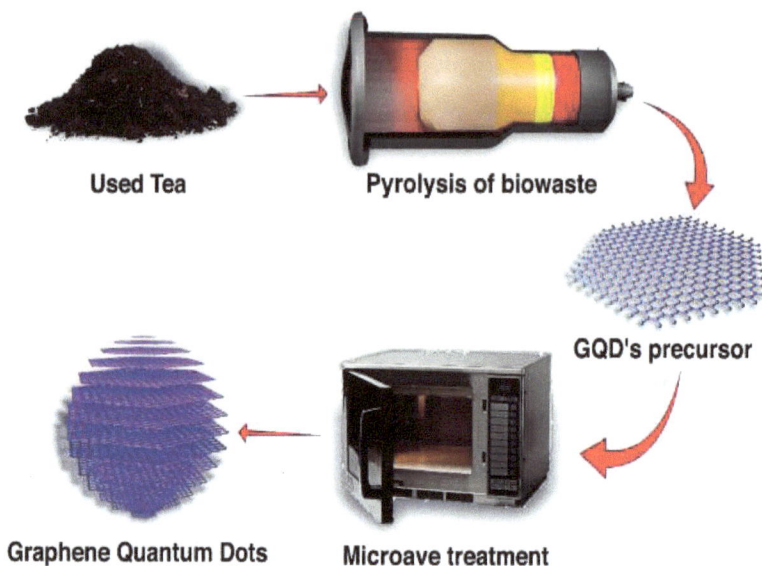

Fig. (4). Pyrolysis of spent tea to synthesize GQDs.

Few Layer Graphene Nanosheets from Kinnow Peel Waste (KPW)

heated a second time at 940 °C with a fast-heating rate of 10 °C/min in an inert environment using the same apparatus. The black charred residue is removed after the instrument has cooled, and any undesired organic contaminants are cleaned off with isopropyl alcohol and double-distilled water. The cleansed black material is now subjected to an hour-long ultrasonic treatment with 30% power and 10 pulse rates. Hence, to obtain the few-layer graphene nanosheet, the ultrasonicated sample is dried and filtered [24]. The mechanism of KPW pyrolysis is shown in Fig. (**5**).

Fig. (5). Pyrolysis of KPW employed to get Graphene sheets from sustainable resources.

Flash Joule Heating

Gram-Scale Bottom-Up Flash Graphene Synthesis

A quartz tube with two loosely fitting electrodes is part of the flash joule heating setup. These electrodes force carbonaceous materials into touch with the carbon sources using two copper wool switches or graphite fillers, allowing volatile compounds to escape. Controlling the compressing force with a modified small clamp to lower sample resistance to 1-1,000 is essential for obtaining a decent flash reaction (0.004-4 S cm^{-1}). A mechanical relay with programmable millisecond-level delay time controls the discharge timing. The entire sample reaction chamber is enclosed inside a low-pressure container (plastic vacuum desiccator) for safety and to facilitate degassing. Yet, the FJH method works just as well at 1 atm. A total of 20 capacitors with a combined capacitance of 0.22 F make up the capacitor bank. A DC The capacitor bank is charged by an output supply with a maximum of 400 V. The prototype system is comfortably transported using a single plastic mobile cart. With the FJH method, we can synthesize 1 g of FG in each batch using a substantial quartz tube with a diameter of 15 mm. The highest voltage and estimated maximum current that each capacitor will supply to the FJH discharge based on a 50-200 ms discharge time are suitable for the voltage and current ratings of the circuit breakers. With maximum currents of 0.7 A and 0.1 A, respectively, the maximum charging and bleeding voltages of 400 V were used. During 100 milliseconds, a pulse can discharge 400 V and up to 1000 A of current to the sample. The mechanical relay was equipped with a 24-mH inductor to guard against surges in current. The mechanical relay may be vulnerable to high-current arcing if the circuit is occasionally switched off without the inductor. To shield the inductor from the spike voltage that happens when the current is cut off, a diode and a low-resistance resistor with the appropriate ratings are connected to the inductor.

A suitable diode is positioned parallel to the capacitor bank to protect the capacitor against reverse polarity in the event of oscillatory decay, which might happen during a fast discharge. In a water-Pluronic (F-127) solution (1%), FG was dissolved at 1 to 10 gL^{-1}. The mixture was sonicated in an ultrasonicator for 40 minutes to form a dark suspension. The dispersion was centrifuged at 1,500 rpm for 30 minutes to separate the larger particles. UV-VIS spectroscopy was used to analyze the supernatant. The dispersions were 500 times diluted before the absorbance at 660 nm was measured [6].

Synthesis of Holey and Wrinkled Flash Graphene

PW (0.65 g), Ca (OAc)$_2$ (1.95 g), and amorphous carbon black BP-2000 (0.165 g) or metallurgical coke (0.125 g) were mixed with finely crushed PW. The resulting mixture of grey powder (2.77 g) was then poured into a quartz tube with an internal diameter of 12 mm. To maintain compression, a snug-fitting spring was then fitted around the tube. At the ends of the tube, graphite electrodes hold the powder, which is subsequently connected to the FJH system by being placed between metal screw electrodes. Never was the sample in contact with the metal. Four tiny (1 mm^2) holes were drilled into one of the graphite electrodes to allow volatiles to escape during the FJH technique and keep the quartz tube from breaking. A resistance of 1.5 k was measured across the powder sample after it had been compressed between the graphite electrodes. After that, during 50 s, a constant 208 V LC discharge was supplied with currents varying from 0.1 A to 25 A. Because dissolving salt and combustible plastic outgas significantly during carbonization, care should be taken to ventilate the reaction chamber quickly. As gases change during the FJH process as a result of polymers volatilizing and salt degrading, the electrodes must have frits to balance the pressure. It stops internal tube expansion from interrupting the circuit and stopping the current discharge.

The electrodes are pressed inward by springs during the FJH discharge to keep in touch with the sample, power source, and electrodes. Instead of using a vacuum desiccator like in traditional FJH processes, an air purge was used to eliminate the volatiles outgassing through the fume hood. The sample was then subjected to an HC flash, which used 140 F capacitors that had been charged to 120 V and discharged for 250 ms to create a short-duration current discharge. The CaO residue can be eliminated by immersing the quartz tube in 100 mL of diluted acid once the HWFG product (around 0.4 g) has been created (1 M acetic or HCl). The HWFG was then filtered using a PVDF filter membrane with a 0.45 m pore size in a vacuum filtering flask. After that, the HWFG powder was dried overnight in a 150 °C oven. Different amounts of HWFG were produced depending on the plastic feedstock, ranging from 0.325 g for virgin HDPE to as little as 0.251 g for waste PVC [25].

Flash Graphene from Rubber Waste

The source of rubber was located between two Carbon electrodes, which were then in contact with copper electrodes. The material is heated by an electrical pulse, which causes it to reach 3000 K in under 0.5 seconds. High-quality graphene is produced due to the dissolution and rearrangement of carbon-carbon bonds. Instead of graphene with an AB-stacked shape, the system forms the kinetically stable turbostratic Flash Graphene (tFG) upon rapid cooling. When carbon black is made from tires (TCB) or carbon black made from shredded rubber tires (CB: SRT) as the carbon source, respectively, about 70% or 47% of the product is collected as tFG. During this process, non-carbon elements dissolve [7].

Laser Ablation Technique

Fabrication of Graphene from E-Waste Using Pulsed Laser Ablation (PLA)

Bajpai *et al.* has shown a sustainable, reproducible, and efficient technique for producing graphene from one of the main components of e-waste, namely the polymeric component, *via* PLA [11]. The PLA method uses technology for laser-assisted solid metal degradation and can limit chemical waste generation, making it environmentally sustainable and able to minimize any potential safety concerns [26]. The PCBs utilized in this research were salvaged from several computer mice (model-HP X500). Manual removal from the PCBs was required for the boards, connectors, wires, and circuitry. Following that, they were manually cut into 4-6 mm-sized fragments.

Recovery of the Polymeric Parts from PCBs

Because mechanical separation involves no chemical change during processing, it has emerged as the most environmentally friendly preliminary processing method for recycling a range of PCB components. The components are separated mechanically based on their shape, size, and density. The disassembled PCBs were ground for 7 hours in an Argon environment using a specially designed cryomill that was heated to 160 ± 10 °C and chilled with liquid nitrogen [27]. The metallic and ceramic parts of PCBs, heavier than polymers and settled at the bottom of the flask, were separated from the polymeric portion of PCBs using continuous filtration and magnetic stirring in distilled water. This procedure was performed several times to produce significant amounts of powdered polymeric e-waste.

Synthesis of Graphene Through PLA

Graphene was synthesized using nanosecond PLA in deionized water at room temperature (28 °C) and a Q-switched Nd-YAG laser at a wavelength of 1064 nm. The laser pulse was repeatedly irradiated at a rate of 10 Hz, with a pulse duration of 10 ns, and at various intensities ranging from 80 to 200 mJ. The e-waste pellet was exposed to the laser pulse for 10 minutes, equal to 6000 shots. The polymeric e-waste powder was cold compressed into pellets before laser ablation using a cylindrical die with a load of 30 kN. The laser ablation media used in this experiment was deionized water, and the laser ablation target was an e-waste pellet with a 4 mm thickness and 15 mm diameter. According to the theoretical density of the polymer powder, each pellet weighed about 1.5 g. Using a reflective mirror, the horizontal laser beam was rotated at an angle of 90 degrees. Using a convex lens with a focal length of 70 mm, the vertical laser ray was concentrated to a spot size of over 1 mm^2 on the target's surface. After laser exposure, the generated colloidal solution was centrifuged at 1000 rpm to separate the graphene sheets [11]. Fig. (6) describes the mechanism of PLA.

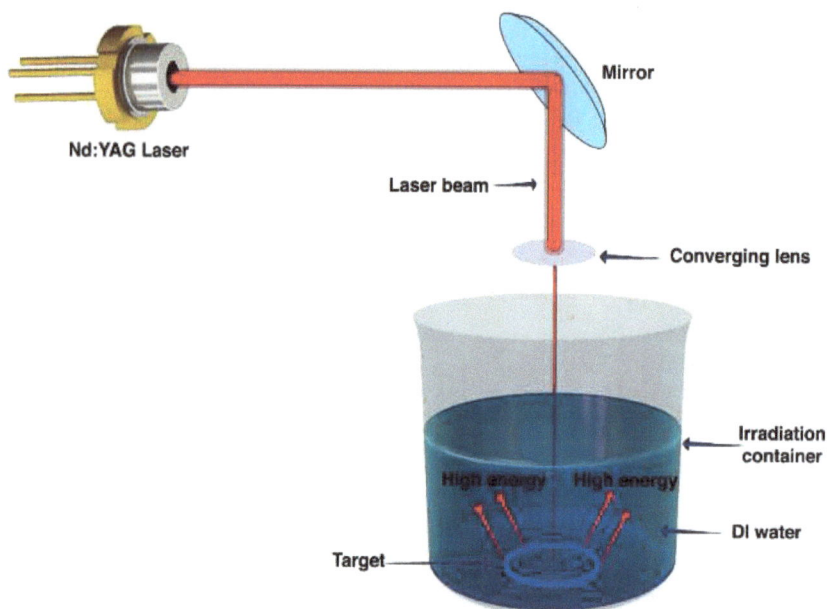

Fig. (6). Pulsed laser ablation similar to the one shown is valid for the synthesis of graphene sheets.

Chemical Vapor Deposition

Synthesis of Graphene Crystals from Solid Plastic Waste

A horizontal AP-CVD system with two split furnaces and a quartz tube of 90 cm in length and 50 mm in diameter was chosen as the reactor. Cu foil with a 20 um thickness and a purity of 99.99% was utilized as the substrate to produce graphene. Plastic waste from product wrapping is employed as a source of Carbon in these investigations [28]. The copper foil was cleaned with acetone using an ultrasonication bath for 30 minutes before being directly inserted into the growth furnace. Before adding the feedstock, the Copper foil was treated thermally in a tube furnace for 30 minutes at 1020 °C in 100 standard cubic centimeters per minute (sccm) of H_2. Annealing is a crucial step to increase the size of the Cu particle, decrease lattice misfit, and eliminate defects, all of which contribute to lowering the number of nucleation sites. The lower-temperature furnace had a ceramic boat containing 3 milligrams of waste plastic as a carbon source. Ar and H_2 were used in a gas combination with flow rates of 98 and 2.5 sccm, respectively, to produce graphene. Throughout the development process, the precursor supply was gradually increased to provide an adequate amount of carbon atoms for the graphene crystal to continue growing. The rate of evaporation changed the injection rate of polymeric components identified in the waste plastic. The SiO_2/Si substrate received individual graphene crystals by applying a poly (methyl methacrylate) (PMMA) supporting layer. A concentrated $Fe (NO_3)_3$ solution with a 50 mg/ml concentration was used to engrave the Cu foil beneath. The PMMA/graphene stack layer was then moved to a SiO_2/Si substrate before being removed using acetone. Graphene was then dried in the air after being processed with a dilute nitric acid solution to eliminate any remaining $Fe (NO_3)_3$.

PROPERTIES OF GRAPHENE AND ITS DERIVATIVES OBTAINED FROM WASTE MATERIALS

Optoelectronic Properties

Minimum Quantum Conductivity and High Mobility

The electron energy and wave vector of graphene are different from the parabolic dispersion of conventional semiconductors due to the material's unique energy band structure. They have found that mass-less Dirac fermions with Fermi velocities up to 1.0×10^6 m/s exhibit a linear dispersion relationship with the wave vector for low-energy electrons close to the energy valley of graphene [29]. Since acoustic phonons are difficult to scatter with carriers close to the Dirac point, graphene has exceptionally elevated carrier mobility. At ambient temperature, the carrier mobility of graphene transistor architectures can approach 1.5×10^4 cm^2/V•s [30].

According to the research, carrier mobility is not temperature dependent. Graphene can both have high carrier mobility and high carrier concentrations (>1012 cm^2). As a result, the resistivity of graphene can be significantly less than other conventional semiconductors. The resistivity of ideal graphene can reach 10^{-6} •cm without flaws or doping [31].

Ballistic Transport

When the carrier density was less than 5×10^9 cm^{-2}, researchers examined suspended graphene's carrier mobility as a temperature function. The carrier mobility reaches 2×10^5 cm^{-2} V^{-1} s^1 as the temperature steadily decreases, demonstrating that the conductivity of suspended graphene near the Dirac point is closely related to temperature and exhibits graphene's ballistic transport features [32].

Ambipolar Electric Field Effects and Adjustable Fermi Energy Levels

Graphene has symmetric valence and conduction bands concerning the Dirac point. Electrons and holes are two distinct types of carriers in the system. Berger *et al.* confirmed that holes and electrons in graphene had the same mobility [33]. In intrinsic single-layer graphene, where the Fermi energy level is located at the Dirac point, the bipolar electric field effect is created by electrons passing through interband transitions from the valence band to the conduction band. The concentrations of electrons and holes can be adjusted to alter the carrier mobility [29].

Absorption Properties of Graphene

The fine structure constant determines how opaque suspended graphene is $\alpha = e^2/\hbar c \approx 1/137$. Graphene is only 0.335 nm thick. However, it absorbs white incoming light at a rate of $\pi\alpha = 2.3\%$ [34]. This is caused by a single sheet of graphene, which has a highly low-energy electron structure that results in a cone of holes and electrons colliding at the Dirac point (K) [35].

Saturable Absorption

Graphene can absorb a wide range of light, from ultraviolet to terahertz. Saturable absorption, a nonlinear optical property of graphene, happens when the intensity of the input light increases over a particular critical value. Graphene has a lower saturation absorption intensity than typical semiconductor materials when used as a saturable absorber of laser mode-locking to create an ultrafast laser [36].

The inter-band and intra-band carrier absorption of graphene is primarily responsible for its optical absorption. The Drude model can fit the absorption characteristics of graphene in the far-infrared band, where intra-band transitions are the primary driver of light absorption. In this band, the subwavelength structure of graphene produces surface plasmon resonance. Graphene's inter-band transitions are primarily responsible for its light absorption in the visible and near-infrared spectrum. The wavelength of the incident light has little bearing on the absorptive capacity band of graphene. 2.3% of light is absorbed by single-layer graphene. The light absorption intensity and the number of graphene layers are positively associated with fewer layers [37].

Inter-Band Transition and Ultrafast Relaxation of Carriers

The Pauli mismatch of carriers saturates graphene with bright light, which researchers used to investigate the relationship between the optical transmission and the pumping power of long-lived graphene. The findings demonstrated that the light transmittance is nearly constant when the pump's soft power is low [38].

The outcomes demonstrated thus that even when the pump soft power is low, the optical penetration remains constant. The optical transmittance improves along with the pump light output. The saturation absorption of graphene can be explained by the complex interactions of the material's photonic carriers (interband, in-band transgression and phonon relaxing of pages). The transition of graphene's carriers between bands during rapid ultrafast light pulse pumping results in nonequilibrium carrier distributions in the conduction and forbidden bands. The nonequilibrium carriers arrive at a thermal equilibrium state through scattering collisions (in-band collisions between runners and phonon radiation), which happen very quickly (†1, 100 fs). The thermal distribution of the carriers is further decreased by the clash of the carriers with lattice vibrations and the relatively slow recombination of the electrons and holes. The interband crossover and relaxation of the carriers in graphene have been investigated using optical pump detection time resolution with ultra-base laser pulses. Using differential transmittance tests, researchers established typical busy time values for carriers of numerous graphene materials at varied pump light intensities [39].

APPLICATIONS OF GRAPHENE

Electrochemical Sensors

The vast surface area and strong conductivity of graphene allow for the detection of trace objects. The large pores boost the identification of binding targets by increasing analyte movement. Functional biological molecules can be found at lower concentrations when graphene is incorporated into direct sensing devices.

Electrochemical Examination

Since electrochemical sensors can sense materials without harming neighboring systems, they have many advantages over biosensor transducers. The redox centers and electrode surface should be as close together as possible for effective detection. Typically, nanoparticles, carbon nanotubes, or nanowires/fibers are added to this process. Graphene has vigorous catalytic activity for detecting various molecular analytes and outstanding electron transport properties [40]. High electrical conductivity and a vast surface area of graphene promote their use for very sensitive enzyme-based electrochemical sensing. Additionally, it is nontoxic and biocompatible, making it suitable for biomedical uses [41]. It has always been difficult for electrochemical electrodes to work on a two-dimensional electrode surface, like those found on metal and glassy carbon electrodes. These difficulties can be reduced by using graphene. It should be mentioned that reduced graphene oxide always has imperfections and functional groups that limit conductivity and functionality [42].

Mediated Detection

This sort of detection relies on the analyte changing the redox properties of a substance in solution to increase the electrode's electrochemical detection. A study employed $K_3[Fe(CN)_6]$ solution, as an illustration to contrast the traits of glassy Carbon with chemical vapor deposition. According to the results of the experiment, graphene had meager charge transfer resistance and series resistance. Increased surface area improved mediated CVD-G performance [43].

Direct Electron Transfer

Graphene research has also been conducted to make graphene perfect as sensors through electrochemical media. An enzyme makes up the sensing region for direct electron transfers. The locus and number of possible active sites for the enzymes affect efficiency. When graphene was applied, it also showed up in the redox behavior of biological molecules, which improved detection. Equipped with surface-bound enzymes for H_2O_2 detection, horseradish per-oxidized was utilized to assess the surface efficiency [43]. Nafion was used to adhere the peroxidized horseradish to the surface. The 2D graphene horseradish peroxidase (HRP) sensors displayed the best efficiency. The graphene's enormous surface area influenced the performance observed throughout the experiment. A dimensional framework for enzyme-less detection has also been evaluated using dopamine detection as a benchmark [44].

Absorbers

High surface area is shown by three-dimensional graphene. They are perfect for absorbing activities because of their qualities. Compared to other materials, graphene has good organic liquid absorption and is also favorable regarding reusability. Again, it has been demonstrated that graphene sheets can absorb certain gases even at low concentrations. Even in that business, the functional group of graphene can reduce gas exposure by capturing and converting undesirable gases.

Gas Absorption Detection

Due to its efficiency, graphene oxide can act as a standpoint for reduction processes. Due to these qualities, the gases may be detected physically and resistively [45]. This is quite prevalent, mainly when graphene oxide (GO) is employed to accelerate the oxidation of SO_2 to SO_3 absorbed at ambient temperature. During this process, the brown-colored GO turns dark. When water is exposed, the trapped SO_3 turns back into sulfuric acid. This makes it simple to filter out the decreased graphene oxide contaminants. The technique helped with the visual detection and storage of any undesired industrial gas [46].

Organic Liquids

The majority of spill cleanup is done with natural microporous absorbers. For instance, zeolites, wool, and sawdust have higher water content and lower oil loading. This behavior makes extracting and recycling the oil very challenging [47]. Some hydrophobic microporous polymers can enhance their absorption rate in oils and organic solvents by 5% of their weight. Nearly 80% of the oil absorbed by the polymer can be recovered by squeezing [48]. Due to its superior loading capacity of more than 83 times its initial weight, expanded graphite is the ideal solution for less expensive oil removal. According to research, 70% of the accumulated oil may be eliminated using vacuum filtration. Despite these advantages, the particles do not indicate that the organic solvents are absorbed more significantly. Again, only a limited number of times can these materials be recycled [49]. Effective and reversible removal of organic solvents and oil spill contaminants is crucial for the future of oil spill cleanup. Therefore, graphene will make a substantial contribution to this area. Research has been done on using graphene as an absorber for various organic compounds [50].

Energy Storage/Conversion Devices

Manufacturing energy storage devices is another use for graphene. This is possible, especially if they have encountered metal oxide, which leads to a limited amount of sheet restacking. The highly conductive linked networks of graphene increase their appeal as potential energy storage applications. Graphene exhibits benefits such as a porous structure, electrochemical stability, and solid mechanical stability when used as an energy storage device.

Fuel Cells

Fuel cells are scientific breakthroughs that efficiently and with promising results in a wide range of applications convert the chemical energy of a fuel directly into electricity. In fuel cells, Platinum and its alloys are generally used as electrodes. However, due to its scarcity and high prices, some other nonprecious metal/metal oxides are being investigated along with nitrogen-doped metal catalysts. These catalysts are less stable and active than platinum catalysts, which makes them less efficient. Active carbons can overcome the issue, but these also have some limitations. Active Carbon typically has a large surface area, but the instability of its support creates serious questions unless they are combined with some other substance. Recent scientific advancements have rewritten the narrative, positioning activated carbons as a stronger alternative to platinum. This shift is attributed to their expansive surface area, superior conductivity, and strong adherence to catalyst particles [51]. Because GO has more active groups, it is suitable for nucleation and attaching catalyst nanoparticles to the surface [52]. Graphene is widely employed in fuel cells, primarily as anode catalyst support material, support, and even as a replacement for the cathode catalyst, as well as in composite and independent electrolyte membranes. It is also utilized in bipolar plates.

At Anode

Two significant applications of graphene at the anode are:

Reducing the Pt and Pt Alloys Catalyst's Load: Graphene is best suited for Pt catalysts that reduce the catalyst's loading or even replace the Pt with other Pt catalysts since it has a larger surface area than other carbon materials. The graphene-based Pt catalyst increased activity by increasing electrical conductivity and ensuring that the Pt was equally distributed throughout. Because of the improved activity, the Pt catalyst used in the fuel cell's anode required less loading, which also improved the catalyst's stability [53]. Furthermore, it is asserted that the addition of heterogeneous atoms such as nitrogen, sulfur, or boron to graphene enhanced the Pt's activity towards alcohol oxidation, either due to an increase in conductivity brought on by the nitrogen atoms' lone electron pair or as a result of an enhancement in wettability that made it simpler for fuel molecules to bind and increased the number of active sites [54].

Pt and Pt alloys are substituted with a nonprecious catalyst: Graphene also supports Ni, the best-known nonprecious catalyst for the anode of low-temperature FCs like methanol in an alkaline environment [55]. Ni supported on graphene displayed great activity compared to carbon nanoparticle or nanofiber support, and the efficiency increased with increasing the Ni loading up to 6% wt. The intercalation of Ni nanoparticles and their well-dispersion on the surface of the graphene sheet is what the researchers believe is responsible for the increase in activity [56]. In general, it was discovered that the high efficiency of nonprecious catalysts over graphene was related to the functional groups on the borders and surface of graphene and its better dispersion [57].

At Cathode

As Support for The Pt Catalyst

Much research has been done on graphene as a catalytic support for Pt catalysts in oxygen reduction processes. The stability of the HO*, which is believed to be the rate-limiting step for the oxygen reduction reaction, is demonstrated to decrease due to the graphene flaws, lowering the activation energy required for oxygen molecule dissociation from 0.37 to 0.16 eV. The graphene support is also involved in reducing the size of the Pt nanoparticles, and the Pt nanoparticles must be equally distributed throughout the graphene surface [58].

As the Electrolyte Membrane of the Fuel Cells

Ions are transferred between the cathode and anode *via* the membrane. The electrolyte membrane requires minimal to no electrical conductivity, strong chemical, thermal, and mechanical stability, and high ionic conductivity.

The development of a composite membrane using graphene oxide has shown to be a potential solution to the electrolyte membrane.

As A Standalone Membrane

In many fuel cells, the electrolyte membrane is a nafion membrane. However, it has drawbacks, including fuel crossover, which reduces energy density, cell voltage, and high cost. It also has poor ionic conductivity in dry environments, brought on by operating the cell at a high temperature [59]. Oxygen and carboxylic functional groups on the surface and margins of GO increased their ionic conductivity and water retention capacity [60]. In PEMFC, GO paper has been effectively utilized as an electrolyte membrane, yielding a power output of $13 mWcm^{-2}$, compared to $9 mWcm^{-2}$ when the nafion membrane was used under the same operating circumstances [61].

At Flow Plates

The bipolar plate (BP) makes up 95% of the FCs' stack's overall volume. The mechanical, thermal, and electrical conductivity of bipolar plates is thought to be strengthened and improved by graphene, a potential candidate. Fuel cells are the perfect technology for using many clean energy sources, including solar, geothermal, ocean, and wind [62]. A carbon composite material made of graphite, carbon black, carbon fibers, and graphene was used by Kakatietal *et al.* to create a BP plate. It was discovered that adding 1% graphene led to higher through- and in-plane conductivities [63].

Microbial Fuel Cells (MFCs)

MFCs are a new class of FCs that are utilized for both the simultaneous production of energy and wastewater treatment. The primary issues preventing their use on a commercial scale are the cost-intensive Pt catalyst at the cathode and the slow electron transmission between the anode and the MFCs' microorganisms [64]. Recent investigations have shown that graphene can enhance the transport of electrons between bacteria and anodes. Zhou *et al.* created carbon fiber paper (CFP) coated with reduced graphene oxide (rGO) either electrochemically (E-rGO) or utilizing a natural extract from eucalyptus leaves (EL), *i.e.*, EL-rGO. In both instances, the performance of the graphene-coated fiber is superior to the bare CFP. The increased electron transport between the microorganisms and the anode surface was responsible for better performance [65].

Fabrication of Dye-Sensitized Solar Cells (DSSC)

Outstandingly plentiful and environmentally friendly renewable energy sources, solar energy has shown the great capability to substitute fossil fuels and thereby shield the atmosphere from the adverse effects of their use [66]. Dye-sensitized solar cells have immense power for direct solar energy harvesting. They comprise three components: an iodide electrolyte, a counter electrode, and a dye-adsorbed TiO_2 photoanode. Under the influence of a light beam on a TiO_2 photo anode, the iodide is converted into triiodide. As a result, one electron is released into TiO_2. The triiodide is converted to iodide when electrons migrate from the external circuit to the counter electrode. A platinum catalyst is necessary to hasten the process. With acceptable performance and stability, graphene is believed to replace such pricey platinum catalysts [67]. (Fig. 7) clearly explains the working principle of DSSC. The photoelectrode is the essential component of the DSSC because it is responsible for absorbing light and converting the photons of solar energy into electrical energy. A compelling photo anode is necessary to guarantee the power functioning of the DSSCs in terms of efficacy, fill factor, and IPCE. The photoanode is supposed to deliver good electron injections and collection and good light utilization [68]. The application of graphene at the photoanode, electrolyte, and counter electrode was found to increase the effectiveness and strength of the DSSCs.

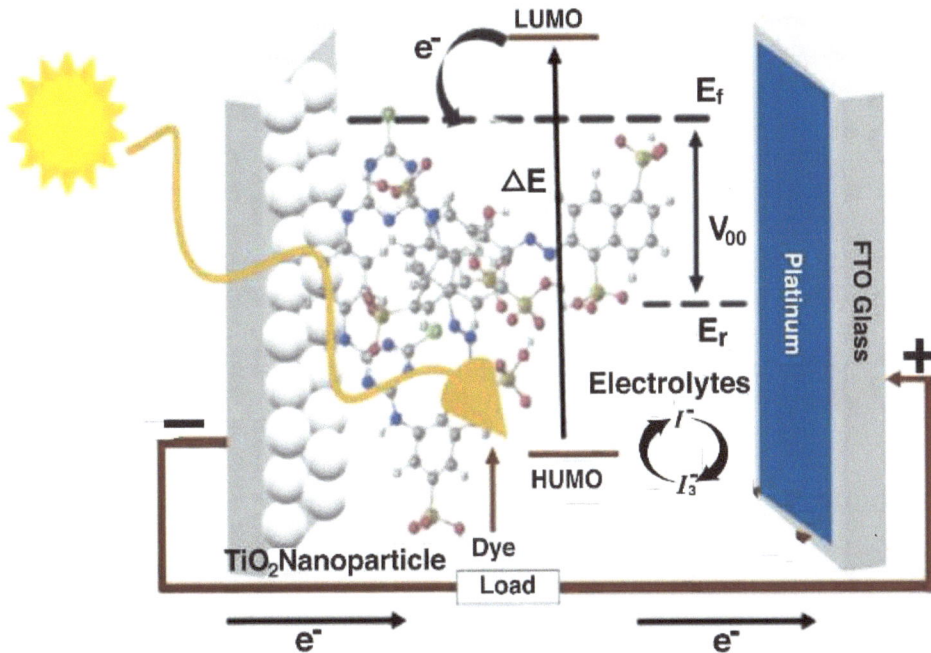

Fig. (7). A DSSC can convert solar energy to electrical energy more efficiently if graphene is added between TiO_2 and ITO electrode.

At Photoanode

The photoanode of DSSC is constructed from a glass-like opaque conduction material coated with TiO_2. Excellent thermal stability, low impedance, high transparency, and high-temperature carrier mobility of graphene are all characteristics of this material. Wang *et al.* created the first graphene electrode for DSSC, but due to flaws in the graphene film created in this paper, the DSSC using graphene performed poorly compared to the gold electrode [69]. As a counter electrode for DSSC, the finest layer of graphene films was produced on a SiO_2 substrate *via* chemical vapor deposition at ambient temperature. The achieved power conversion efficiency was 4.25%, which was not significantly distinct from those of FTO counter electrodes [70]. Since graphene's work function lies between the conduction bands of TiO_2 and ITO, adding it to the semiconductor will boost the charge collection at the DSSC anode [71]. However, as graphene content rises, transmittance falls, so the ideal level is needed. The efficiency of DSSC improved by 39% when a thin layer of graphene was synthesized and employed as a photoanode on top of TiO^{2-} [72]. Using UV-assisted photocatalytic reduction, a layer of graphene/TiO_2 nanoparticles was employed to create an interface between monocrystalline TiO_2 and FTO. The PCE is 5.26%. Good adhesion between FTO and TiO_2 was made possible by the improved PCE related to graphene/TiO_2. Due to graphene's ability to stop charges from recombining, the DSSC's efficiency increased significantly when it was spin-coated on FTO [73]. The performance of DSSC using graphene-TiO_2 was superior to that of DSSC using TiO_2-ZnO because graphene improved the charge transit rate by boosting light absorption and minimizing charge recombination. The produced graphene nanosheets are employed as a photoanode in DSSC. They are coupled with TiO_2 nanoparticles due to graphene's involvement in increasing dye loading and improving electronic transport, adding graphene to TiO_2-enhanced photocurrent. Utilizing Nafion for the initial graphene dispersion, heterogeneous coagulation was used to integrate graphene into TiO_2 particles. Because there were more accessible sites and less charge recombination, introducing graphene to TiO_2 improved the efficiency of DSSC, where the short circuit current density improved by more than 70% [74].

At Electrolyte

Utilizing liquid electrolytes has several benefits, including excellent wettability, simplicity in formation, less viscosity, and high efficiency. However, they could be a better choice due to the limited stability of the liquid electrolyte brought on by outflow, evaporation, and electrode deterioration by the electrolyte [75]. These difficulties can be fixed with polymer electrolytes. Graphene was added to polymer electrolytes, and the findings in terms of stability and effectiveness were encouraging. For example, a GO aerogel containing 3-methoxypropionitrile (MPN) was effectively used as an electrolyte in a DSSC with a performance of 6.70 percent as opposed to 7.18% for liquid electrolytes [76].

At Counter Electrode

The biggest obstacles to Pt's industrial adoption in DSSCs are its high cost and linear resources, despite the metal's excellent activity for I^{3-} reduction and corrosion resistance [77]. A high level of I^{3-} reduction activity was seen in graphene. A counter electrode in DSSC that combines thermal reduction (TR) and chemical reduction (CR) utilizing hydriodic acid (HI) performed exceptionally well (DSSC efficiency 7.8%), coming very near to the performance of Pt catalyst (DSSC efficiency 8.6%). This performance is noticeably better than that obtained by thermally reducing the rGO, whether paired with chemically reducing it with hydrazine hydrate (HH) or when mixed with both (HHI) [78]. The combined TR and CR with HI performed better because of increased charge transfer, supported by the impedance measurement. The elimination of hydroxyl and epoxide groups by HI was a factor in the increased charge transfer [79].

Batteries

Graphitic Carbon is used to make anodic materials perfect for lithium-ion batteries. The maximum intercalation state of the battery is created when LiC_6 is produced from graphite carbon. Between two sheets of storage, where the Li and Carbon are located, there occurs an electron transfer. Consequently, the energy capacity is 372 mA h g^{-1} [80]—long-range crystal organization in graphitic material results in enormous diffusion distances. Several studies have been undertaken over the past few decades to improve the specific energy capacity. According to research, adding a Li_2 covalent site can enable carbons to store more ions than their theoretical maximum. Li is retained on both sides of the six-membered ring, *i.e.*, benzene. The occurrence of edges and vacancies also supports lithium retention [81]. The trade-off between low conductivity and a shorter lifetime is the primary constraint on graphite anode performance. Therefore, reduced graphene oxide is the best option to achieve equilibrium. Enhanced capacities result from the merit of the two-dimensionally formed material [52].

Li Air Battery

Compared to current lithium-ion batteries, lithium-air batteries can store more energy. They are available in prototypes with a lithium anode, a membrane, and a porous electrode for the cathodic area. The cathodic region of a lithium-air battery is where the main problem is visible. Lithium reduces oxygen during operation to produce lithium oxide. This shortens life expectancy by blocking the flow of oxygen and electrolytes, respectively. Low O_2 diffusion rates caused by carbon accumulation and tightly bound binder particles impact rate performance. This suggests that a highly effective carbon cathode with a porous structure is required to boost performance. Graphene's enormous surface area and open pore structure make it a feasible solution to these problems. Their powerful conduction can guarantee the proper use of the porous electrode [82].

Fabrication of Supercapacitors

The development of graphene ensures the future of double-layered electrochemical devices. Graphene has been used in several research projects, with capacitance readings between 100 and 200 F/g [83]. According to research, assembling graphene into a supercapacitor can increase the electrode surface area of the material and the network conductivity. These phenomena produce large macropore channels, facilitating ionic transport through the electrolyte. This demonstrates that the freestanding graphene network can reduce the mass of inactive electrodes made of polymer binders, additives, and current collectors [84].

The 2D graphene sheet is lightweight, has a large surface area, is highly electrically conductive, and has an adjustable thickness and good mechanical properties. For thin, flexible supercapacitors, 2D graphene is thought to be the best material [85].

3D graphene structures, including graphene scaffolds and gels, were also explored for the fabrication of supercapacitors because their distinctive formation arises due to the presence of multiple types of interconnected pores, large surface areas, and quick carrier transit lines. The fusion electrode based on MnO^{2-}graphene foam, built on 3D graphene structures, has a specific capacitance of 389 F/g, a power density of 25 KW/kg, and an energy density of 44 Wh/kg [86].

CONCLUSION

After the complete utilization of a product, people are used to dumping the waste materials openly in the environment posing high alerts to the environment. Waste is a complex mixture of polymers, organic compounds, precious metals, and heavy metals. Landfilling and incineration of waste produce toxic gases that are lethal for the environment and human beings. Another problem associated with this is the need for primary resources. The solution to this problem is the environmentally sustainable recycling of waste materials. Recycling may be a beneficial option to protect the environment and primary sources. People have started working on it and are producing valuable products such as graphene and its derivatives from waste materials. Now researchers are trying to make graphene and related products from the carbonaceous materials present in the waste using several techniques. Graphene and its derivatives have remarkable properties and high-end applications.

REFERENCES

[1]   Hoornweg, D.; Bhada-Tata, P.; Kennedy, C. Environment: Waste production must peak this century. *Nature,* **2013**, *502*(7473), 615-617.
        [http://dx.doi.org/10.1038/502615a] [PMID: 24180015]

[2]   Sahoo, N.G.; Esteves, R.J.; Punetha, V.D.; Pestov, D.; Arachchige, I.U.; McLeskey, J.T., Jr Schottky diodes from 2D germanane. *Appl. Phys. Lett.,* **2016**, *109*(2), 023507.
        [http://dx.doi.org/10.1063/1.4955463]

[3]   Ikram, R.; Mohamed Jan, B.; Nagy, P.B.; Szabo, T. Recycling waste sources into nanocomposites of graphene materials: Overview from an energy-focused perspective. *Nanotechnol. Rev.,* **2023**, *12*(1), 20220512.
        [http://dx.doi.org/10.1515/ntrev-2022-0512]

[4]   Sahoo, N.G.; Sandeep, M. A process of manufacturing. Indian Patent 352780, **2016**.

[5]   Chittella, H.; Yoon, L.W.; Ramarad, S.; Lai, Z.W. Rubber waste management: A review on methods, mechanism, and prospects. *Polym. Degrad. Stabil.,* **2021**, *194*, 109761.
        [http://dx.doi.org/10.1016/j.polymdegradstab.2021.109761]

[6]   Luong, D.X.; Bets, K.V.; Algozeeb, W.A.; Stanford, M.G.; Kittrell, C.; Chen, W.; Salvatierra, R.V.;

Ren, M.; McHugh, E.A.; Advincula, P.A.; Wang, Z.; Bhatt, M.; Guo, H.; Mancevski, V.; Shahsavari, R.; Yakobson, B.I.; Tour, J.M. Gram-scale bottom-up flash graphene synthesis. *Nature,* **2020,** *577*(7792), 647-651.
[http://dx.doi.org/10.1038/s41586-020-1938-0] [PMID: 31988511]

[7]   Advincula, P.A.; Luong, D.X.; Chen, W.; Raghuraman, S.; Shahsavari, R.; Tour, J.M. Flash graphene from rubber waste. *Carbon,* **2021,** *178,* 649-656.
[http://dx.doi.org/10.1016/j.carbon.2021.03.020]

[8]   Iqbal, A.; Jan, M.R.; Shah, J.; Rashid, B. Dispersive solid phase extraction of precious metal ions from electronic wastes using magnetic multiwalled carbon nanotubes composite. *Miner. Eng.,* **2020,** *154,* 106414.
[http://dx.doi.org/10.1016/j.mineng.2020.106414]

[9]   Kaya, M. Recovery of metals and nonmetals from electronic waste by physical and chemical recycling processes. *Waste Manag.,* **2016,** *57,* 64-90.
[http://dx.doi.org/10.1016/j.wasman.2016.08.004] [PMID: 27543174]

[10] Liu, X.; Ma, C.; Wen, Y.; Chen, X.; Zhao, X.; Tang, T.; Holze, R.; Mijowska, E. Highly efficient conversion of waste plastic into thin carbon nanosheets for superior capacitive energy storage. *Carbon,* **2021,** *171,* 819-828.
[http://dx.doi.org/10.1016/j.carbon.2020.09.057]

[11] Bajpai, A.; Kumbhakar, P.; Tiwary, C.S.; Biswas, K. Conducting graphene synthesis from electronic waste. *ACS Sustain. Chem. Eng.,* **2021,** *9*(42), 14090-14100.
[http://dx.doi.org/10.1021/acssuschemeng.1c03817]

[12] Li, M.; Lu, J.; Chen, Z.; Amine, K. 30 Years of lithium-ion batteries. *Adv. Mater.,* **2018,** *30*(33), 1800561.
[http://dx.doi.org/10.1002/adma.201800561] [PMID: 29904941]

[13] Punetha, V.D.; Rana, S.; Yoo, H.J.; Chaurasia, A.; McLeskey, J.T., Jr; Ramasamy, M.S.; Sahoo, N.G.; Cho, J.W. Functionalization of carbon nanomaterials for advanced polymer nanocomposites: A comparison study between CNT and graphene. *Prog. Polym. Sci.,* **2017,** *67,* 1-47.
[http://dx.doi.org/10.1016/j.progpolymsci.2016.12.010]

[14] Abbas, A.; Mariana, L.T.; Phan, A.N. Biomass-waste derived graphene quantum dots and their applications. *Carbon,* **2018,** *140,* 77-99.
[http://dx.doi.org/10.1016/j.carbon.2018.08.016]

[15] Ajuria, J.; Zarrabeitia, M.; Arnaiz, M.; Urra, O.; Rojo, T.; Goikolea, E. Graphene as vehicle for ultrafast lithium ion capacitor development based on recycled olive pit derived carbons. *J. Electrochem. Soc.,* **2019,** *166*(13), A2840-A2848.
[http://dx.doi.org/10.1149/2.0361913jes]

[16] Gómez-Urbano, J.L.; Moreno-Fernández, G.; Arnaiz, M.; Ajuria, J.; Rojo, T.; Carriazo, D. Graphene-coffee waste derived carbon composites as electrodes for optimized lithium ion capacitors. *Carbon,* **2020,** *162,* 273-282.
[http://dx.doi.org/10.1016/j.carbon.2020.02.052]

[17] Muramatsu, H.; Kim, Y.A.; Yang, K.S.; Cruz-Silva, R.; Toda, I.; Yamada, T.; Terrones, M.; Endo, M.; Hayashi, T.; Saitoh, H. Rice husk-derived graphene with nano-sized domains and clean edges. *Small,* **2014,** *10*(14), 2766-2770, 2740.
[http://dx.doi.org/10.1002/smll.201400017] [PMID: 24678046]

[18] Roy, I.; Sarkar, G.; Mondal, S.; Rana, D.; Bhattacharyya, A.; Saha, N.R.; Adhikari, A.; Khastgir, D.; Chattopadhyay, S.; Chattopadhyay, D. Synthesis and characterization of graphene from waste dry cell battery for electronic applications. *RSC Advances,* **2016,** *6*(13), 10557-10564.
[http://dx.doi.org/10.1039/C5RA21112C]

[19] Wang, H.; Xu, Z.; Kohandehghan, A.; Li, Z.; Cui, K.; Tan, X.; Stephenson, T.J.; King'ondu, C.K.; Holt, C.M.B.; Olsen, B.C.; Tak, J.K.; Harfield, D.; Anyia, A.O.; Mitlin, D. Interconnected carbon

nanosheets derived from hemp for ultrafast supercapacitors with high energy. *ACS Nano,* **2013**, *7*(6), 5131-5141.
[http://dx.doi.org/10.1021/nn400731g] [PMID: 23651213]

[20]   Zhao, H.; Zhao, T.S. Graphene sheets fabricated from disposable paper cups as a catalyst support material for fuel cells. *J. Mater. Chem. A Mater. Energy Sustain.,* **2013**, *1*(2), 183-187.
[http://dx.doi.org/10.1039/C2TA00018K]

[21]   Ruan, G.; Sun, Z.; Peng, Z.; Tour, J.M. Growth of graphene from food, insects, and waste. *ACS Nano,* **2011**, *5*(9), 7601-7607.
[http://dx.doi.org/10.1021/nn202625c] [PMID: 21800842]

[22]   Primo, A.; Atienzar, P.; Sanchez, E.; Delgado, J.M.; García, H. From biomass wastes to large-area, high-quality, N-doped graphene: Catalyst-free carbonization of chitosan coatings on arbitrary substrates. *Chem. Commun.,* **2012**, *48*(74), 9254-9256.
[http://dx.doi.org/10.1039/c2cc34978g] [PMID: 22875403]

[23]   Punetha, V.D.; Ha, Y.M.; Kim, Y.O.; Jung, Y.C.; Cho, J.W. Interaction of photothermal graphene networks with polymer chains and laser-driven photo-actuation behavior of shape memory polyurethane/epoxy/epoxy-functionalized graphene oxide nanocomposites. *Polymer,* **2019**, *181*, 121791.
[http://dx.doi.org/10.1016/j.polymer.2019.121791]

[24]   Pathak, M.; Tatrari, G.; Karakoti, M.; Pandey, S.; Sahu, P.S.; Saha, B.; Sahoo, N.G. Few layer graphene nanosheets from kinnow peel waste for high-performance supercapacitors: A comparative study with three different electrolytes. *J. Energy Storage,* **2022**, *55*, 105729.
[http://dx.doi.org/10.1016/j.est.2022.105729]

[25]   Wyss, K.M.; Chen, W.; Beckham, J.L.; Savas, P.E.; Tour, J.M. Holey and wrinkled flash graphene from mixed plastic waste. *ACS Nano,* **2022**, *16*(5), 7804-7815.
[http://dx.doi.org/10.1021/acsnano.2c00379] [PMID: 35471012]

[26]   Rashidi, L. 1 - Magnetic nanoparticles: Synthesis and characterization In: *Magnetic Nanoparticle-Based Hybrid Materials*; Woodhead Publishing, **2021**.
[http://dx.doi.org/10.1016/B978-0-12-823688-8.00035-1]

[27]   Kumar, N.; Biswas, K. Fabrication of novel cryomill for synthesis of high purity metallic nanoparticles. *Rev. Sci. Instrum.,* **2015**, *86*(8), 083903.
[http://dx.doi.org/10.1063/1.4929325] [PMID: 26329207]

[28]   Sharma, S.; Kalita, G.; Hirano, R.; Shinde, S.M.; Papon, R.; Ohtani, H.; Tanemura, M. Synthesis of graphene crystals from solid waste plastic by chemical vapor deposition. *Carbon,* **2014**, *72*, 66-73.
[http://dx.doi.org/10.1016/j.carbon.2014.01.051]

[29]   Wang, J.; Mu, X.; Sun, M.; Mu, T. Optoelectronic properties and applications of graphene-based hybrid nanomaterials and van der Waals heterostructures. *Appl. Mater. Today,* **2019**, *16*, 1-20.
[http://dx.doi.org/10.1016/j.apmt.2019.03.006]

[30]   Novoselov, K.S.; Geim, A.K.; Morozov, S.V.; Jiang, D.; Katsnelson, M.I.; Grigorieva, I.V.; Dubonos, S.V.; Firsov, A.A. Two-dimensional gas of massless Dirac fermions in graphene. *Nature,* **2005**, *438*(7065), 197-200.
[http://dx.doi.org/10.1038/nature04233] [PMID: 16281030]

[31]   Castro, E.V.; Ochoa, H.; Katsnelson, M.I.; Gorbachev, R.V.; Elias, D.C.; Novoselov, K.S.; Geim, A.K.; Guinea, F. Limits on charge carrier mobility in suspended graphene due to flexural phonons. *Phys. Rev. Lett.,* **2010**, *105*(26), 266601.
[http://dx.doi.org/10.1103/PhysRevLett.105.266601] [PMID: 21231692]

[32]   Du, X.; Skachko, I.; Barker, A.; Andrei, E.Y. Approaching ballistic transport in suspended graphene. *Nat. Nanotechnol.,* **2008**, *3*(8), 491-495.
[http://dx.doi.org/10.1038/nnano.2008.199] [PMID: 18685637]

[33] Berger, C.; Song, Z.; Li, X.; Wu, X.; Brown, N.; Naud, C.; Mayou, D.; Li, T.; Hass, J.; Marchenkov, A.N.; Conrad, E.H.; First, P.N.; de Heer, W.A. Electronic confinement and coherence in patterned epitaxial graphene. *Science,* **2006**, *312*(5777), 1191-1196.
[http://dx.doi.org/10.1126/science.1125925] [PMID: 16614173]

[34] Punetha, V.D.; Ha, Y.M.; Kim, Y.O.; Jung, Y.C.; Cho, J.W. Rapid remote actuation in shape memory hyperbranched polyurethane composites using cross-linked photothermal reduced graphene oxide networks. *Sens. Actuators B Chem.,* **2020**, *321*, 128468.
[http://dx.doi.org/10.1016/j.snb.2020.128468]

[35] Wang, J.; Song, J.; Mu, X.; Sun, M. Optoelectronic and photoelectric properties and applications of graphene-based nanostructures. *Mater. Today Phys.,* **2020**, *13*, 100196.
[http://dx.doi.org/10.1016/j.mtphys.2020.100196]

[36] Bao, Q.; Zhang, H.; Wang, Y.; Ni, Z.; Yan, Y.; Shen, Z.X.; Loh, K.P.; Tang, D.Y. Atomic-layer graphene as a saturable absorber for ultrafast pulsed lasers. *Adv. Funct. Mater.,* **2009**, *19*(19), 3077-3083.
[http://dx.doi.org/10.1002/adfm.200901007]

[37] Gusynin, V.P.; Sharapov, S.G.; Carbotte, J.P. Unusual microwave response of dirac quasiparticles in graphene. *Phys. Rev. Lett.,* **2006**, *96*(25), 256802.
[http://dx.doi.org/10.1103/PhysRevLett.96.256802] [PMID: 16907333]

[38] Sun, Z.; Hasan, T.; Torrisi, F.; Popa, D.; Privitera, G.; Wang, F.; Bonaccorso, F.; Basko, D.M.; Ferrari, A.C. Graphene mode-locked ultrafast laser. *ACS Nano,* **2010**, *4*(2), 803-810.
[http://dx.doi.org/10.1021/nn901703e] [PMID: 20099874]

[39] Punetha, V.D.; Dhali, S.; Rana, A.; Karki, N.; Tiwari, H.; Negi, P.; Basak, S.; Sahoo, N.G. Recent advancements in green synthesis of nanoparticles for improvement of bioactivities: A review. *Curr. Pharm. Biotechnol.,* **2022**, *23*(7), 904-919.
[http://dx.doi.org/10.2174/1389201022666210812115233] [PMID: 34387160]

[40] Yola, M.L. Development of novel nanocomposites based on graphene/graphene oxide and electrochemical sensor applications. *Curr. Anal. Chem.,* **2019**, *15*(2), 159-165.
[http://dx.doi.org/10.2174/1573411014666180320111246]

[41] Shao, Y.; Wang, J.; Wu, H.; Liu, J.; Aksay, I.A.; Lin, Y. Graphene based electrochemical sensors and biosensors: A review. *Electroanalysis,* **2010**, *22*(10), 1027-1036.
[http://dx.doi.org/10.1002/elan.200900571]

[42] Dong, X.; Cao, Y.; Wang, J.; Chan-Park, M.B.; Wang, L.; Huang, W.; Chen, P. Hybrid structure of zinc oxide nanorods and three dimensional graphene foam for supercapacitor and electrochemical sensor applications. *RSC Advances,* **2012**, *2*(10), 4364-4369.
[http://dx.doi.org/10.1039/c2ra01295b]

[43] Dong, X.; Ma, Y.; Zhu, G.; Huang, Y.; Wang, J.; Chan-Park, M.B.; Wang, L.; Huang, W.; Chen, P. Synthesis of graphene-carbon nanotube hybrid foam and its use as a novel three-dimensional electrode for electrochemical sensing. *J. Mater. Chem.,* **2012**, *22*(33), 17044-17048.
[http://dx.doi.org/10.1039/c2jm33286h]

[44] Olabi, A.G.; Abdelkareem, M.A.; Wilberforce, T.; Sayed, E.T. Application of graphene in energy storage device - A review. *Renew. Sustain. Energy Rev.,* **2021**, *135*, 110026.
[http://dx.doi.org/10.1016/j.rser.2020.110026]

[45] Scheuermann, G.M.; Rumi, L.; Steurer, P.; Bannwarth, W.; Mülhaupt, R. Palladium nanoparticles on graphite oxide and its functionalized graphene derivatives as highly active catalysts for the Suzuki-Miyaura coupling reaction. *J. Am. Chem. Soc.,* **2009**, *131*(23), 8262-8270.
[http://dx.doi.org/10.1021/ja901105a] [PMID: 19469566]

[46] Schedin, F.; Geim, A.K.; Morozov, S.V.; Hill, E.W.; Blake, P.; Katsnelson, M.I.; Novoselov, K.S. Detection of individual gas molecules adsorbed on graphene. *Nat. Mater.,* **2007**, *6*(9), 652-655.

[http://dx.doi.org/10.1038/nmat1967] [PMID: 17660825]

[47]   Radetić, M.M.; Jocić, D.M.; Jovančić, P.M.; Petrović, Z.L.; Thomas, H.F. Recycled wool-based nonwoven material as an oil sorbent. *Environ. Sci. Technol.*, **2003**, *37*(5), 1008-1012.
[http://dx.doi.org/10.1021/es0201303] [PMID: 12666933]

[48]   Li, A.; Sun, H-X.; Tan, D-Z.; Fan, W-J.; Wen, S-H.; Qing, X-J.; Li, G-X.; Li, S-Y.; Deng, W-Q. Superhydrophobic conjugated microporous polymers for separation and adsorption. *Energy Environ. Sci.*, **2011**, *4*(6), 2062-2065.
[http://dx.doi.org/10.1039/c1ee01092a]

[49]   Pathak, R.; Punetha, V. D.; Bhatt, S.; Punetha, M. Multifunctional role of carbon dot-based polymer nanocomposites in biomedical applications: A review. *J. Mater. Sci.*, **2023**, *58*(15), 6419-6443.
[http://dx.doi.org/10.1007/s10853-023-08408-4]

[50]   Toyoda, M.; Inagaki, M. Heavy oil sorption using exfoliated graphite. *Carbon*, **2000**, *38*(2), 199-210.
[http://dx.doi.org/10.1016/S0008-6223(99)00174-8]

[51]   Bhatt, S.; Punetha, V.D.; Pathak, R.; Punetha, M. Graphene in nanomedicine: A review on nano-bio factors and antibacterial activity. *Colloids Surf. B Biointerfaces*, **2023**, *226*, 113323.
[http://dx.doi.org/10.1016/j.colsurfb.2023.113323] [PMID: 37116377]

[52]   Liang, Y.; Wang, H.; Zhou, J.; Li, Y.; Wang, J.; Regier, T.; Dai, H. Covalent hybrid of spinel manganese-cobalt oxide and graphene as advanced oxygen reduction electrocatalysts. *J. Am. Chem. Soc.*, **2012**, *134*(7), 3517-3523.
[http://dx.doi.org/10.1021/ja210924t] [PMID: 22280461]

[53]   Shi, Y.; Zhu, W.; Shi, H.; Liao, F.; Fan, Z.; Shao, M. Mesocrystal PtRu supported on reduced graphene oxide as catalysts for methanol oxidation reaction. *J. Colloid Interface Sci.*, **2019**, *557*, 729-736.
[http://dx.doi.org/10.1016/j.jcis.2019.09.038] [PMID: 31563605]

[54]   Zhang, L.M.; Wang, Z.B.; Zhang, J.J.; Sui, X.L.; Zhao, L.; Gu, D.M. Honeycomb-like mesoporous nitrogen-doped carbon supported Pt catalyst for methanol electrooxidation. *Carbon*, **2015**, *93*, 1050-1058.
[http://dx.doi.org/10.1016/j.carbon.2015.06.022]

[55]   Barakat, N.; Ali Abdelkareem, M.; Abdelghani, E. Influence of Sn content, nanostructural morphology, and synthesis temperature on the electrochemical active area of Ni-Sn/C nanocomposite: Verification of methanol and urea electrooxidation. *Catalysts*, **2019**, *9*(4), 330.
[http://dx.doi.org/10.3390/catal9040330]

[56]   Barakat, N.A.M.; Moustafa, H.M.; Nassar, M.M.; Abdelkareem, M.A.; Mahmoud, M.S.; Almajid, A.A.; Khalil, K.A. Distinct influence for carbon nano-morphology on the activity and optimum metal loading of Ni/C composite used for ethanol oxidation. *Electrochim. Acta*, **2015**, *182*, 143-155.
[http://dx.doi.org/10.1016/j.electacta.2015.09.079]

[57]   Kou, R.; Shao, Y.; Wang, D.; Engelhard, M.H.; Kwak, J.H.; Wang, J.; Viswanathan, V.V.; Wang, C.; Lin, Y.; Wang, Y.; Aksay, I.A.; Liu, J. Enhanced activity and stability of Pt catalysts on functionalized graphene sheets for electrocatalytic oxygen reduction. *Electrochem. Commun.*, **2009**, *11*(5), 954-957.
[http://dx.doi.org/10.1016/j.elecom.2009.02.033]

[58]   Lim, D.H.; Wilcox, J. Mechanisms of the oxygen reduction reaction on defective graphene-supported Pt Nanoparticles from first-principles. *J. Phys. Chem. C*, **2012**, *116*(5), 3653-3660.
[http://dx.doi.org/10.1021/jp210796e]

[59]   Nakagawa, N.; Abdelkareem, M.A.; Sekimoto, K. Control of methanol transport and separation in a DMFC with a porous support. *J. Power Sources*, **2006**, *160*(1), 105-115.
[http://dx.doi.org/10.1016/j.jpowsour.2006.01.066]

[60]   Stankovich, S. Preparation and characterization of graphene oxide paper. *Nature*, **2007**, *44*, 457-460.

[61]   Tateishi, H.; Hatakeyama, K.; Ogata, C.; Gezuhara, K.; Kuroda, J.; Funatsu, A.; Koinuma, M.;

Taniguchi, T.; Hayami, S.; Matsumoto, Y. Graphene oxide fuel cell. *J. Electrochem. Soc.,* **2013**, *160*(11), F1175-F1178.
[http://dx.doi.org/10.1149/2.008311jes]

[62]   Ijaodola, O.; Ogungbemi, E.; Khatib, F.N.; Wilberforce, T.; Ramadan, M.; Hassan, Z.E.; Thompson, J.; Olabi, A.G. Evaluating the effect of metal bipolar plate coating on the performance of proton exchange membrane fuel cells. *Energies,* **2018**, *11*(11), 3203.
[http://dx.doi.org/10.3390/en11113203]

[63]   Kakati, B.K.; Ghosh, A.; Verma, A. Efficient composite bipolar plate reinforced with carbon fiber and graphene for proton exchange membrane fuel cell. *Int. J. Hyd. Energy,* **2013**, *38*(22), 9362-9369.
[http://dx.doi.org/10.1016/j.ijhydene.2012.11.075]

[64]   Logan, B.E.; Hamelers, B.; Rozendal, R.; Schröder, U.; Keller, J.; Freguia, S.; Aelterman, P.; Verstraete, W.; Rabaey, K. Microbial fuel cells: Methodology and technology. *Environ. Sci. Technol.,* **2006**, *40*(17), 5181-5192.
[http://dx.doi.org/10.1021/es0605016] [PMID: 16999087]

[65]   Sayed, E.T.; Nakagawa, N. Critical issues in the performance of yeast based microbial fuel cell. *J. Chem. Technol. Biotechnol.,* **2018**, *93*(6), 1588-1594.
[http://dx.doi.org/10.1002/jctb.5527]

[66]   Zhou, S.; Lin, M.; Zhuang, Z.; Liu, P.; Chen, Z. Biosynthetic graphene enhanced extracellular electron transfer for high performance anode in microbial fuel cell. *Chemosphere,* **2019**, *232*, 396-402.
[http://dx.doi.org/10.1016/j.chemosphere.2019.05.191] [PMID: 31158634]

[67]   Roslan, N.; Ya'acob, M.E.; Radzi, M.A.M.; Hashimoto, Y.; Jamaludin, D.; Chen, G. Dye Sensitized Solar Cell (DSSC) greenhouse shading: New insights for solar radiation manipulation. *Renew. Sustain. Energy Rev.,* **2018**, *92*, 171-186.
[http://dx.doi.org/10.1016/j.rser.2018.04.095]

[68]   Kumar, S.; Singh, P.K.; Punetha, V.D.; Singh, A.; Strzałkowski, K.; Singh, D.; Yahya, M.Z.A.; Savilov, S.V.; Dhapola, P.S.; Singh, M.K. In-situ N/O-heteroatom enriched micro-/mesoporous activated carbon derived from natural waste honeycomb and paper wasp hive and its application in quasi-solid-state supercapacitor. *J. Energy Storage,* **2023**, *72*, 108722.
[http://dx.doi.org/10.1016/j.est.2023.108722]

[69]   Liang, J.; Zhang, G.; Yang, J.; Sun, W.; Shi, M. TiO_2 hierarchical nanostructures: Hydrothermal fabrication and application in dye-sensitized solar cells. *AIP Adv.,* **2015**, *5*(1), 017141.
[http://dx.doi.org/10.1063/1.4906988]

[70]   Wang, X.; Zhi, L.; Müllen, K. Transparent, conductive graphene electrodes for dye-sensitized solar cells. *Nano Lett.,* **2008**, *8*(1), 323-327.
[http://dx.doi.org/10.1021/nl072838r] [PMID: 18069877]

[71]   Wang, H.; Leonard, S.L.; Hu, Y.H. Promoting effect of graphene on dye-sensitized solar cells. *Ind. Eng. Chem. Res.,* **2012**, *51*(32), 10613-10620.
[http://dx.doi.org/10.1021/ie300563h]

[72]   Sun, W.; Peng, T.; Liu, Y.; Yu, W.; Zhang, K.; Mehnane, H.F.; Bu, C.; Guo, S.; Zhao, X.Z. Layer-b--layer self-assembly of TiO_2 hierarchical nanosheets with exposed 001 facets as an effective bifunctional layer for dye-sensitized solar cells. *ACS Appl. Mater. Interfaces,* **2014**, *6*(12), 9144-9149.
[http://dx.doi.org/10.1021/am501233q] [PMID: 24881671]

[73]   Kilic, B.; Turkdogan, S. Fabrication of dye-sensitized solar cells using graphene sandwiched 3D-ZnO nanostructures based photoanode and Pt-free pyrite counter electrode. *Mater. Lett.,* **2017**, *193*, 195-198.
[http://dx.doi.org/10.1016/j.matlet.2017.01.128]

[74]   Sun, S.; Gao, L.; Liu, Y. Enhanced dye-sensitized solar cell using graphene-TiO_2 photoanode prepared by heterogeneous coagulation. *Appl. Phys. Lett.,* **2010**, *96*(8), 083113.
[http://dx.doi.org/10.1063/1.3318466]

[75]   Mathew, S.; Yella, A.; Gao, P.; Humphry-Baker, R.; Curchod, B.F.E.; Ashari-Astani, N.; Tavernelli, I.; Rothlisberger, U.; Nazeeruddin, M.K.; Grätzel, M. Dye-sensitized solar cells with 13% efficiency achieved through the molecular engineering of porphyrin sensitizers. *Nat. Chem.,* **2014**, *6*(3), 242-247.
        [http://dx.doi.org/10.1038/nchem.1861] [PMID: 24557140]

[76]   Neo, C.Y.; Ouyang, J. The production of organogels using graphene oxide as the gelator for use in high-performance quasi-solid state dye-sensitized solar cells. *Carbon,* **2013**, *54*, 48-57.
        [http://dx.doi.org/10.1016/j.carbon.2012.11.002]

[77]   Mishra, A.K.; Ramaprabhu, S. Functionalized graphene-based nanocomposites for supercapacitor application. *J. Phys. Chem. C,* **2011**, *115*(29), 14006-14013.
        [http://dx.doi.org/10.1021/jp201673e]

[78]   Kim, S.K.; Cho, K.H.; Kim, J.Y.; Byeon, G. Field study on operational performance and economics of lithium-polymer and lead-acid battery systems for consumer load management. *Renew. Sustain. Energy Rev.,* **2019**, *113*, 109234.
        [http://dx.doi.org/10.1016/j.rser.2019.06.041]

[79]   Li, Z.; Gadipelli, S.; Yang, Y.; He, G.; Guo, J.; Li, J.; Lu, Y.; Howard, C.A.; Brett, D.J.L.; Parkin, I.P.; Li, F.; Guo, Z. Exceptional supercapacitor performance from optimized oxidation of graphene-oxide. *Energy Storage Mater.,* **2019**, *17*, 12-21.
        [http://dx.doi.org/10.1016/j.ensm.2018.12.006]

[80]   Menachem, C.; Peled, E.; Burstein, L.; Rosenberg, Y. Characterization of modified NG7 graphite as an improved anode for lithium-ion batteries. *J. Power Sources,* **1997**, *68*(2), 277-282.
        [http://dx.doi.org/10.1016/S0378-7753(96)02629-8]

[81]   Gong, J.; Wu, H.; Yang, Q. Structural and electrochemical properties of disordered carbon prepared by the pyrolysis of poly(p-phenylene) below 1000°C for the anode of a lithium-ion battery. *Carbon,* **1999**, *37*(9), 1409-1416.
        [http://dx.doi.org/10.1016/S0008-6223(99)00002-0]

[82]   Xiao, J.; Mei, D.; Li, X.; Xu, W.; Wang, D.; Graff, G.L.; Bennett, W.D.; Nie, Z.; Saraf, L.V.; Aksay, I.A.; Liu, J.; Zhang, J.G. Hierarchically porous graphene as a lithium-air battery electrode. *Nano Lett.,* **2011**, *11*(11), 5071-5078.
        [http://dx.doi.org/10.1021/nl203332e] [PMID: 21985448]

[83]   Wang, Z.L.; Xu, D.; Xu, J.J.; Zhang, L.L.; Zhang, X.B. Graphene oxide gel-derived, free-standing, hierarchically porous carbon for high-capacity and high-rate rechargeable Li-O$_2$ batteries. *Adv. Funct. Mater.,* **2012**, *22*(17), 3699-3705.
        [http://dx.doi.org/10.1002/adfm.201200403]

[84]   Jeong, H.M.; Lee, J.W.; Shin, W.H.; Choi, Y.J.; Shin, H.J.; Kang, J.K.; Choi, J.W. Nitrogen-doped graphene for high-performance ultracapacitors and the importance of nitrogen-doped sites at basal planes. *Nano Lett.,* **2011**, *11*(6), 2472-2477.
        [http://dx.doi.org/10.1021/nl2009058] [PMID: 21595452]

[85]   Ahuja, P.; Ujjain, S. Graphene-based materials for flexible supercapacitors. In: *Self-standing Substrates, Materials and Applications*; Springer, **2019**; pp. 297-326.
        [http://dx.doi.org/10.1007/978-3-030-29522-6_10]

[86]   Choi, B.G.; Yang, M.; Hong, W.H.; Choi, J.W.; Huh, Y.S. 3D macroporous graphene frameworks for supercapacitors with high energy and power densities. *ACS Nano,* **2012**, *6*(5), 4020-4028.
        [http://dx.doi.org/10.1021/nn3003345] [PMID: 22524516]

CHAPTER 5

Two-Dimensional Boron Nitride From Synthesis to Energy Applications

Rakshit Pathak[1,*] and **Seyedeh Hanieh Ghiasi Limanjoobi[2]**

[1] *Centre of Excellence for Research, P.P. Savani University, Surat-394125, Gujarat, India*

[2] *Centre for Manufacturing and Materials (CMM) Coventry University, CV1 5FB, Coventry, United Kingdom*

Abstract: There is a great deal of interest in other 2D compounds due to the development and practicality of graphene in many fields. Due to its similar properties to graphene, the boron nitride (BN) nanosheet has become one of the most thoroughly studied and investigated nanomaterials in this area. The next wave of electrical and optoelectronic products can incorporate 2D-hBN and other 2D materials like graphene. In this chapter, our main aim is to summarize the 2D h-BN nanosheets and their synthesis process *via* two different mechanisms, *i.e.*, exfoliation techniques and the Chemical Vapour Deposition (CVD) process. The chapter also provides insights into the 2D h-BN's energy-associated properties and their applications in fabricating various energy storage devices, including batteries, supercapacitors, and solar and fuel cells.

Keywords: Batteries, Boron Nitride, Exfoliation, Chemical vapour deposition, Energy.

INTRODUCTION

The development of 2D nanomaterials opens up several opportunities for creating different contrivances at the micro- or nanoscale in several fields, such as energy devices to biomedical contrivances. Their magnificent characteristics, such as enormous surface, thermal and electrical stability, and high carrier mobility, have opened a new dimension in the material world [1 - 4]. In this context, several 2D nanostructures have shown their capabilities as potent materials in fabricating various energy storage and energy-producing devices. This material is also a catalyst in different biofuel, biodiesel, and other bioenergy-related reactions and processes [5 - 8]. Besides that, the material is also used in several biological applications such as antibacterial, antiviral, and antifungal activities [9 - 11].

---

[*] **Corresponding author Rakshit Pathak:** Centre of Excellence for Research, P.P. Savani University, Surat-394125, Gujarat, India, E-mail: rakshit.pathak@ppsu.ac.in

Vinay Deep Punetha (Ed.)

These properties make it very significant in several biomedical applications. Among all, the 2D material which have been much explored in the last two decades is graphene and its derivatives and Transition metal Dichalcogenides (TMDs) [12, 13]. A hexagonal honeycomb lattice of carbon atoms makes up this structure [14]. TMDs are a particular type of semiconductor with the structural formula MX_2, where M is a transition metal. Due to its durability and exceptional toughness compared to the other members of the TMDs family, MoS_2 is the most well-known material. Intriguing for essential research in high-end electronics, TMDCs display a singular combination of various characteristics such as bandgap, High spin-orbit coupling (SOC), and other physicochemical properties [15]. Besides that, other recently created 2D analogous, including transition metal carbides, MXenes are also getting an interest due to their surface chemistry. Some of these 2D structures, such as boron nitride, exhibit exceptional performance in energy-related applications. Boron nitride (BN) is a type of III-V compound (the materials having semiconductor properties and made of III and V group elements such as BN, InP, GaN, GaAs, *etc.*) with an amount of equal B and N is called boron nitride (BN). Although a rare natural occurrence has been observed, it is generally believed to be made of synthetic materials [16]. Numerous crystalline phases of this substance are known to exist, including cubic (c-BN), hexagonal (h-BN), wurtzite (w-BN), and rhombohedral (r-BN). The most stable substances are the cubic and hexagonal ones, which are made up of mixed sp^3/ sp^2-bonds (Fig. **1**). Compared to diamond, C-BN has a zinc-blend structure of boron and nitrogen atoms that are Td coordinated, while h-BN is layered, with each sp^2-bonded monolayer with a honeycomb structure made [17].

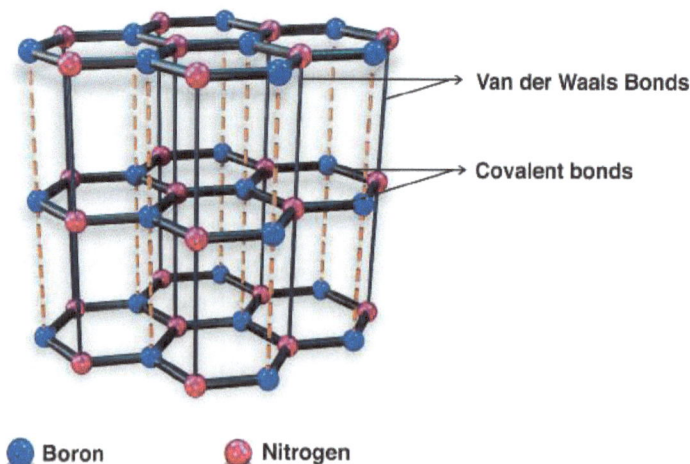

Fig. (1). Structure of hexagonal Boron Nitride and bonding between the atoms.

hBN is a significant new contender in the 2D materials group and possesses various fascinating characteristics. In recent times, there has been a significant rise in the study of 2D materials, and 2D has shown to be a flexible platform for various uses, especially in energy-related applications [18]. Due to their fascinating characteristics and exceptional physical occurrences, they are also called "white graphene/ inorganic graphite" since B and N atoms stand in for carbon at the planar sp^2, which forces pull neighboring hBN sheets together, the atomic structure of hBN exhibits geometric similarities to that of graphene. Some typical properties of Graphene and hBN are presented in Table **1**.

Table 1. Comparison of Physicochemical Properties of Graphene and Hexagonal Boron Nitride (hBN) [19 - 22].

Properties	Unit	Graphite/ Graphene	Hexagonal Boron Nitride
Hybridization	-	sp^2	sp^2
Morphology	-	Planner	Planner
Lattice Constant	nm	a= 0.25	a= 0.25
		c= 0.67	c= 0.67
Density	g/cm^3	2.3	2.09-2.23
Melting Point (MP.)	℃	3000	3560
Young's modulus	GPa	32-70	~1000
Thermal conductivity	Wm^{-1}K^{-1}	~550 (up to 5 Layers)	3080-3500
Electrical Conductivity	-	As Insulator	As Conductor
Band Gap	eV	~6	0

Although B (2.04) and N (3.04) have vastly different electro-negativities [23], these two elements combine to create highly polarised bonds in hBN, giving it an ionic characteristic and insulated environment [24]. Such characteristics create hBN as an insulating component with more excellent chemical steadiness and enhanced mechanical strength than graphite; as a result, it has been employed as a common lubricant. hBN has lately been shown to have additional unique properties and prospective uses thanks to the rapidly expanding field of 2D materials. Since hBN has a low electron delocalization and a large gap of around 6 eV [25]. In addition, hBN also has a plane surface that is free of electron traps or dangling bonds (it is an unsatisfied valence on an immobilized atom and chemically similar to a free radical). As a result, it is regarded as the best substrate for reducing the carrier's inelastic scattering. This enables hBN encapsulation to significantly lower the energy degeneracy in solid-state devices and enhance the impacted electrical properties of 2D materials lying on widely used amorphous substrates [26, 27]. Besides these properties, 2D hBN is also well-versed in several other attractive properties, making it an appropriate material for dielectrics. High thermal stability and mechanical strength also make it more applicable than other heteroatomic materials [28, 29]. Regarding optoelectronics, hBN is highly suitable for application in low-energy devices in UV and IR bands thanks to its inherent hyperbolicity and attractive photonic qualities. It also has exceptional qualities as a heat-dissipating material in thermal administration and energy-related equipment. Therefore, the present chapter, part of this book on two-dimensional materials, introduces hBN, a distinct two-dimensional substance comparable to graphene. The chapter primarily introduces the BNs in a fundamental manner already in the introduction part. Further, we will discuss the synthesis process in the next portion while focusing on the two main synthesis processes, *i.e.*, exfoliation and Chemical Vapor Deposition (CVD). The chapter also provides insights into the critical properties of hBN and associated applications in energy storage and related devices. The chapter will be wrapped up with a quick summary and outlook on how 2D-hBN-based materials and technology will develop in future energy-related disciplines.

SYNTHESIS/ GROWTH OF 2D-HBN

The structure and associated characteristics of hBN are highly responsible for its performance in several devices, and these characteristics are intensely subjective to the processes of synthesis and processing. Developing the most chosen 2D-hBN with the necessary structural behavior for specific uses is still difficult. To create h-BN, many processing techniques have been evaluated and reported, specifically producing materials with substantial lateral dimensions [30]. In general, there are two ways to create BN nanosheets: bottom-up, where B and N atoms create few-layered h-BN, and top-down, *i.e.*, exfoliation (Fig. **2**)

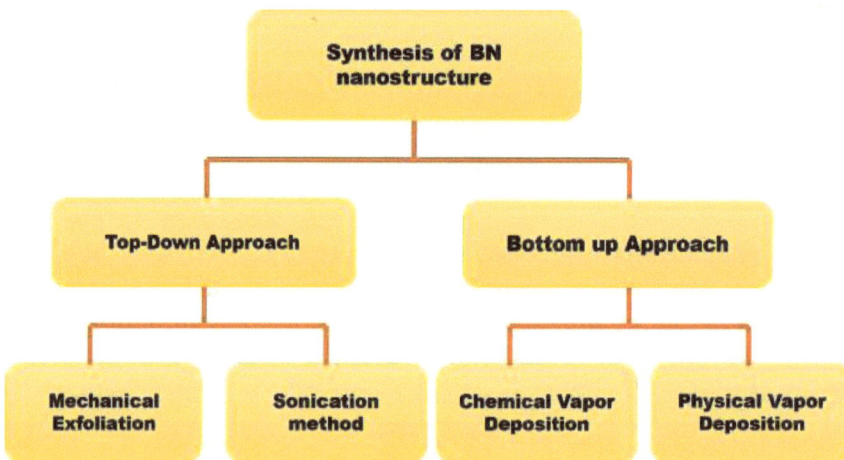

Fig. (2). Methods for h-BN nanostructure synthesis.

The route of synthesis decides the final physicochemical properties of synthesized hBNs. This might be accomplished in one of two ways: either by creating h-BN flakes between B- and N-containing materials (Bottom-up approach) or by disassembling the layered h-BN to create flakes (top-down approach) [31]. The previous techniques are frequently used to create 2D h-BN, which is where surface charge homogeneity is crucial (for example, in photonics or solar cells). To create big homogenous area h-BN films, chemical vapor deposition (CVD) has shown to be a capable approach. Besides these techniques, several synthesis routes have been used for the synthesis of hBN, which are very similar to the graphene synthesis route. A summary of different synthesis processes is given in Table **2**. The detailed methods of this synthesis route can be understood by the literature cited in the table's reference section.

Table 2. Top-down and bottom-up approach for hBN synthesis.

Process	Advantage	Disadvantage	Refs.
Top-Down Approach			
Exfoliation	Cost-effective Better Exfoliation Different medium Functionalization Size control possible	Defects in the structure. Issue of unwanted reactions.	[32, 33]
Ball milling	Provides a wet and dry atmosphere. Surface functionalization.	High defects Time taking Less reproducible	[34]
Solvothermal /Sonication	Energy efficient Reproducible Scalable Process Edge/Surface functionalization	Toxic by-products High temp.	[35]
Microwave	Fast process Reproducible	Issue of unwanted side reactions.	[36]
Bottom-up Approach			
Hydrothermal	Size and thickness can be Controlled Energy efficient process Selective functionalization	Toxic by-products 1:1 B:N ratio precursor needed	[37, 38]
Chemical Vapour Deposition (CVD)	Direct deposition on substrates Uniformity High conformity	High Cost Toxic by-products	[39, 40]

These are the several synthesis methods that are frequently used to produce 2D-BN. The most researched processing methods are mechanical and liquid exfoliation, CVD, and epitaxial. Among these, mechanical exfoliation offers a workable method for creating crystalline hBN structures, although the hBN layers produced are frequently irregularly dispersed, have small flake sizes, and have poor yields [41, 42]. However, liquid exfoliation is a more effective technique for producing a higher yield of hBN nanosheets at a much lower cost. Due to the arbitrary layers, the short size of the flake, and the potential for surface adulteration, this technique may be challenging to implement in large-scale practical manufacturing [43]. In regulating the large-scale generation of thin hBN, CVD and the epitaxial technique are superior to exfoliation. However, the need for elevated temperature frequently causes crystal structural flaws, which cause a significant variance in the characteristics of 2D-hBN and, consequently, in device performance. Assuming that single crystalline monolayer hBN nanosheets were created by Oshima in 1995 [44, 45]. Here, we discuss some of the basic synthesis mechanisms and approaches that have been established for the development of 2D-hBN nanostructures.

SYNTHESIS *VIA* EXFOLIATION TECHNIQUES

To reduce the Van der Waal interactions between the different layers of material, exfoliation is one of the most basic mechanisms in which extra external force is applied to break down the bond interaction between these layers. It is aided by the pressure created by the breakdown of numerous groups between the layers, which overcomes the van der Waals attractions. The production of alternative materials and the exploration into other sorts of them were spurred by graphene exfoliation [46, 47]. The significant benefit of the method is that it can produce hBNs nanosheets on a big scale. The sonication and chemical methods are the two most frequently used procedures for exfoliating different layered materials. During sonication, shear forces and cavitation act on the bulk material, causing exfoliation, while in the chemical method, high-temperature and high-vacuum conditions are not necessarily needed when processing bulk hBN compounds; only wet chemical dispensation is required [48].

The exfoliation technique produces 2D hBNs sheets from the bulk and may be categorized as a top-down process. Due to the inherent properties of certain materials, in-plane covalent connections between numerous fundamental elements are preserved. However, it is highly doable to directly build 2D nanosheets from the bulk crystals' atomic layers; hence exfoliation can be created. Graphene and h-BN are examples of layered materials that may be exfoliated in various ways, both chemically and mechanically [48]. Some of the common approaches and methods are incorporated in the below subsection.

Chemical Exfoliation

Chemical exfoliation is based on changing the 2D nanomaterial precursor's chemical structure through the proper reaction pathways. For instance, the

Hummers process has synthesized and exfoliated reduced graphene oxide (r-GO) from graphite. Other metal-based 2D materials have also been successfully generated using various chemical exfoliation processes that have been well-documented [49 - 51]. Despite multiple reports on developing various h-BN chemical exfoliation processes, methods for exfoliating h-BN to make 2D-hBN nanosheets have not been as effective. In 2013, Li *et al.* developed an innovative and efficient approach for creating h-BN nanosheets by exfoliating h-BN with NaOH and KOH molten salts.

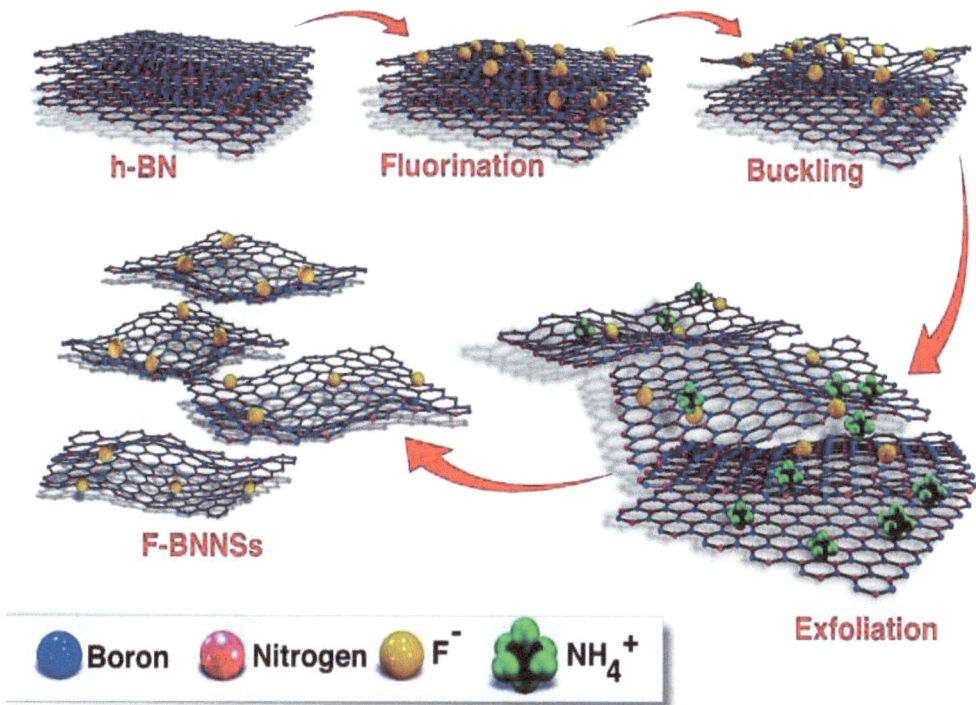

Fig. (3). Schematic presentation of exfoliation of F-2D-hBN.

Additionally, BN nano-scrolls are made. The finished compounds can be dispersed in solvents such as water and ethanol, creating stable dispersions [52]. Similarly, Du *et al.* used an upgraded Hummers' approach to exfoliate h-BN nanosheets by following the graphene oxide exfoliation process. They synthesized the 1g of h-BN sheet from 65mg of bulk BNs. Results showed that this approach is much more suitable than other methods and concluded that this cost-effective method produced different layers of monolayer BN nanosheets [53]. In a study, Lin *et al.* revealed that amine groups were used to functionalize and exfoliate h-BN. They created fluorinated 2D-hBN sheets by effectively exfoliating bulk-sized

BNs, as shown in Fig. (**3**). This synthesis method was found to be surface-based and efficient in the result; it was also claimed that the synthesized 2D nanosheets exhibit ferromagnetic characteristics and can be applied to different electronic devices [54, 55]. Moreover, Bhimanapati *et al.* have developed a simple and scalable method for the exfoliation and functionalization of h-BN utilizing a chemical approach. Using a phosphoric and sulfuric acid combination, bulk hBN was functionalized quickly and cheaply. The layered structure of hBN was exfoliated by this functionalization, with the sheets depending on the hBN powder utilized [56].

Exfoliation *via* Intercalation

Many researchers also endeavored to exfoliate h-BN using an intercalation approach. Three decades back, Doll *et al.* (1989) succeeded in the intercalation of BN sheets using alkali metal [57]. By intercalating potassium into h-BN, Lin *et al.* described a method for creating single-layer BN quantum dots (QDs) with a dimension of around 10 nm. These BN QDs, instead of mono-layered BN sheets, turn blue-green because of preparation-related flaws. Its optical behavior significantly increases the utility of a material that has already attracted much interest. It is further demonstrated that the QDs are suitable for bio-imaging and non-toxic to biological cells [55]. In another study, Zhao *et al.* (2018) experimented with the electrochemical intercalation of lithium ions into van der Waals (vdW) heterostructures by inserting graphene between h-BN. They applied the Hall effect to screen the electrochemical reaction [58].

Exfoliation *via* Liquid-phase Sonication

Liquid-phase sonication provides a novel method for exfoliating h-BN that needs to be discussed here. For h-BN liquid phase exfoliation, Han *et al.* (2008) revealed that a liquid phase sonication technique could successfully exfoliate h-BN sheets from the bulk crystalline structure of BN. Results also concluded that viable conditions for identifying the layers and stacking alignment are the moiré patterns and the fringe patterns [59]. After that, Zhi *et al.* used dimethyl formamide (DMF) and dynamic ultrasonication and centrifugation to exfoliate h-BN, yielding large-scale 2D sheets of h-BN. The method was reproducible and provided a high yield of h-BN [60]. Warner *et al.* also attempted to do liquid phase exfoliation in an ultrasonication bath similarly [61]. It has also been discovered that water, in addition to organic solvents, is a potent solvent for the ultrasonication of BN exfoliation. In this context, Lin *et al.* demonstrated that the hydrolysis process in distilled water could produce fresh and clean aqueous dispersals of BN material. It established the sonication-assisted approach may promote the development of exfoliated h-BN nanosheets based on ammonia tests and spectroscopic data [62]. Based on shared characteristics with inorganic graphene, Stengl *et al.* used a high-intensity ultrasonic exfoliation process to get a yield of up to 100% in a short period. This procedure involves synthesizing graphene-equivalent materials, such as h-BN, WS$_2$, *etc.*, in the appropriate solvents, followed by sonication to create a potent cavitation field [63]. Yuan *et al.* proposed a low-temperature thermal expansion-aided ultrasonic scalable exfoliation technique for producing bulky BN. This method uses a sonication procedure to exfoliate from bulk BN after first damaging the interlayer contacts between adjacent layers [64].

Mechanical Exfoliation

The weak contacts between material layers are broken down in this manner by the pulling energy, leaving the robust sp^2-based in-plane structure. The process itself is also critical and fascinating. Numerous bulk materials exfoliated similarly to graphene. However, due to structural defects, mechanical cleavage can result in meager yields and reduce the synthesis of good-quality h-BN. 3-D bulk materials exfoliation has been studied using a variety of techniques. In this manner, Alem *et al.* (2009) described a method for fabricating single and multilayer h- BNNSs using tape. They present a method that combines mechanical exfoliation and RIE to create monolayer and multilayer suspended h-BN sheets. The atoms are resolved using an ultrahigh-resolution transmission electron microscope, and the chemical makeup of each atom in the sample is determined using intensity profiles for reconstructed phase pictures. Even or odd h-BN multilayers exhibit significantly different reconstructed phase pictures, with uniquely detected and described notable flaws and edge restorations.

In addition to mechanical exfoliation methods, Ball milling has been touted by scientists in recent years as one of the most excellent methods for creating nanomaterials, particularly for exfoliating nanomaterials. Li and their teammates reported a low-energy ball milling approach that achieves high-yield exfoliation of bulk BNs by utilizing benzyl benzoate to provide modest shear stress. They concluded that this process decreased the thickness and produced defects in the dew layers of BN nanosheets [65]. Jinrui Ma and colleagues used a wet ball milling approach to show the exfoliation of a few-layered h-BN by surface modification. It was discovered that the BN produced after an 8-hour milling process had the most significant improvement in heat conductivity. In this instance, the thermal behavior of BN-reinforced epoxy is influenced by both the dispersion and BN size. Additionally, it was discovered that the modified BN improves the mechanical characteristics of epoxy composites more than the original BN [66]. Furthermore, Yang *et al.* showed how to make BN nanoplatelets *via* a ball-milling method assisted by amino acids [67].

Thermal Exfoliation

Thermal exfoliation is another method besides mechanical cleavage for the exfoliation of bulk BN structures. In a simple and scalable thermal oxidation method, Cui *et al.* explained the thermal exfoliation of h-BN. Using this method, they discovered that heating h-BN in the oxygen environment results in mass augmentation. After the reaction, it was stirred in DI water, which caused the hydrolyzation, and then exfoliated to produce hydroxylated h-BN [68]. Ko *et al.* created 2D h-BN nanostructures through a straightforward thermal exfoliation process. This standard procedure involved adding h-BN powder to an alumina crucible, putting it in a thermal furnace with air within, and then cooling it to room temperature. The material was removed, thoroughly cleaned with water, collected, and then filtered and centrifuged to remove bulky impurities before being dried to produce clean h-BN nanosheets [69]. Thermal exfoliation was used by Zhi *et al.* to create a polyaniline (PANI)/hydroxylated h-BN composite structure. *in situ* polymerization was used to create the hybrid PANI-BN, and after washing and drying, it was effective in producing the desired hybrid nanomaterial [60]. Acharya *et al.* (2021) recently showed how to manufacture BN nanosheets from bulk h-BN utilizing an innovative bi-thermal exfoliation method. In this technique, the dispersion was removed from the pressure cooker and let to cool before being frozen for 10-12 hours. Any remaining ice was then defrosted, and the partly exfoliated material's aqueous dispersion was then brought back to the cooker for pressure-heated heating. This exfoliation approach identified repeated heating and freezing as one bi-thermal cycle. As a result, after five more of these cycles, exfoliated 2D h-BN nanosheets could be effectively recovered by filtration [70]. Besides that, other exfoliation methods are also possible for the h-BN nanosheet preparation. Table **3** provides a brief overview of some exfoliation methods.

Table 3. Different exfoliation methods used for hBN nanosheet fabrication.

Exfoliation Method	Process	Yield	Key Observation	Refs.
Weak Sonication	By ionic liquids processing.	~50	One-step process Low cost	[71]
Thermal	Processing at a very high temp.	~65	Aqueous-suspended h-BN nanosheets.	[68]
Chemical	Use of Molten salts (NaOH/KOH); A critical step in the bulk BN exfoliation, intercalation of H_2 and MnO_2 nanoparticles	~0.191	Single step process Lower cost Meager yield	[62]

(Table 3) cont.....

Exfoliation Method	Process	Yield	Key Observation	Refs.
Green approach	Two-step process; K^+ and Zn^{2+} ions are intercalated with the help of sonication, and the compound -OH works in the h-BN nanosheet synthesis.	~16	Low Yield Few layers of hBN sheets obtained.	[72]
Ultrasound sonication	Exfoliation at a higher intensity.	100	Very high yield High purity with a minimum unwanted functional group.	[63]
Ball milling	Mechanical exfoliation in the presence of NaOH; milling impurity restricted by benzyl benzoate	~18	Large size sheets	[73]
Vortex fluid exfoliation	In the spinning tube, layers developed, with the shearing layer parallel to the rotation axis.	~5	Few layers of hBN nanosheet production.	[74]
Gas exfoliation	The cryogenic N_2 gasification and expansion of BN caused by the high temperature caused bulk BN to exfoliate.	~20	Short time reaction No chemical reagent	[75]
Liquid-based exfoliation	Applying supramolecular polymer, *i.e.*, polypropylene glycol	~83	High yield Simple process	[76]
Ultrasonic exfoliation	An aided low-temperature technique.	~26	Pristine hBN nanosheets.	[77]
Ball-milling technique	Using Lewis acid-base interactions to add a long alkyl chain amine to the molecule (defective h-BN).	~40	h-BN nanosheets with defects.	[54]

CHEMICAL VAPOR DEPOSITION METHOD

The process of condensing vapor-phase components into solid-phase material is known as chemical vapor deposition (CVD). This process modifies various materials' corrosion resistance and optical, electrical, and mechanical properties. Films made of composite materials and fabrics infused with various nanoparticles are produced using CVD [78]. Early in the 1970s, when the CVD technique started to gain popularity for producing thin films, researchers first used it to synthesize h-BN thin films. The fabrication of h-BN films on diverse substrates using the CVD technique was described in many publications that were later published [2, 79]. In the condition of ultra-high vacuum (UHV), single-layer BN was first produced by CVD in 1995. The production of atomic-layer-thin BN nanosheets did not receive much attention until recently, most likely because of certain complexities. The most efficient way to produce huge BN sheets with adjustable atomic thickness has been determined to be CVD. When nanosheets are utilized as the substrate for nanoscale electronics, this is very crucial. The quality of the BNNS is dramatically influenced by several important CVD factors.

The effects of different precursors, substrates, and substrate treatments are examined in this study since they are significant and may be changed or tweaked independently of the physical characteristics of the CVD system. The different CVD approaches are compiled here in terms of the precursor and substrate by taking the example of some highly cited literature.

Precursors for h-BN Nanosheets Synthesis *via* CVD

The different precursors for BN synthesis are listed in Table **4**. The primary precursor was the isostructural benzene analog borazine. When borazine is exposed to moisture in the air, it reacts and can polymerize, forming solid poly borazine [80]. As a result, its usage as a direct precursor for synthesis is constrained since it must be kept in a cold, inert environment for storage. In order to develop BN nanosheets, ammonia borane has frequently been employed as the CVD precursor. This substance is manageable, comparatively non-toxic, and air-stable. Some of the critical precursors used to synthesize BN nanosheets are covered in Table **4**.

Table 4. Impact of different precursor used for BN nanosheets synthesis *via* CVD process.

Precursor	Growth Parameters/Substrate	Results	Refs.
Ammonia Borane	The temperature range for NBN growth is 700 - 1100 °C.	BN sheets are obtained with single to few layers and in an area up to centimeters in size.	[81, 82]
Borazine	Conventional CVD process under the temp. range 700- 1100 °C.	Single and few-layered BN nanosheets with domains around 0.3 mm.	[30, 45]
BCl_3, NH_3, N_2, and H_2	The growth temperature was 1000 °C.	2D BN nanosheets with a thickness of around 10nm.	[83]
Trimethylamine borane	Growth temp. 1050°C for 5–20 min.	Monolayer BN sheets on Cu substrate with defects.	[84]
Decaborane and ammonia	Growth temp. 1000°C for 5 min.	2D BN nanosheets with a thickness of around 2nm.	[2]
Borazane	Cu base as a substrate and thermal annealing at 1020 °C.	Mono-few layers of BN sheets were obtained with very pristine quality and large size.	[30]
Borazane	Pt as a substrate and thermal annealing at 1100 °C.	hBN synthesized with the vast surface, and the thickness could be adjusted by varying the borazine concentration.	[20]
Borazine	Fe as a substrate	Monolayer h-BN obtained with very high carrier mobility.	[33]

hBN Nanosheets Synthesis with Different Substrates

The epitaxy-based methods could yield monolayers of hBN, which are typically very subtle towards the substrates, making them rather labor-intensive and challenging to integrate into large-scale practical manufacturing. Hence, new simple procedures must be devised to fulfill the needs and ideally be compatible with Si-based methods. In the CVD process for hBN synthesis, Cu, Ni, or Pt is used as a substrate [30]. The substrates recrystallize with bigger particles due to annealing at high temperatures, ensuring that the coverage of synthesized hBN ranges from tens of microns to several centimeters. In a hetero-epitaxial Co film, 2D hBN nanosheets might be produced. However, there are significant coverage issues. Lu *et al.* created monolayer hBN grains, roughly two orders bigger than those previously synthesized, using the Cu-Ni alloy as the substrate [50]. In a technique described by Qin and their co-workers, microwave plasma CVD (MPCVD) was used to generate triangular hBN nanosheets on Si substrates without the need for a catalyst [85]. Numerous attempts have been made to develop few-layer h-BN nanosheets, but it still needs to be challenging to fabricate wafer-size h-BN with regulated layers and good crystalline quality.

Low-energy electron microscopy (LEEM) was utilized by Orfero *et al.* (2013) to examine the structure, growth trajectory, and layers of thin h-BN films. Under low pressure, CVD is used to grow the h-BN films on hetero epitaxial Co. Their research has revealed that the development of monolayer films with two triangular BN domains are oppositely orientated and correspond to the Co lattice. Thicker domains only exist in patches, probably beginning between domain borders, suggesting that the development of h-BN is self-limiting at a monolayer [86]. Kim *et al.* (2015) synthesized the h-BN using Fe as a substrate and synthesized multilayer h-BN sheets using borazine as a precursor. Multilayer of BN obtained with very strong cathode luminescence, mechanical strength, and high carrier mobility ($\sim$24000 cm^{2}V^{-1}s^{-1}) [33]. By employing borazine (NH_3-BH_3) as the only precursor and no catalyst, Chen *et al.*, 2017 synthesized a porous sp^2 BN film directly formed on c-plane Al_2O_3 substrates by low-pressure CVD. The BN sheets produced under 20 m-bar were discovered to be porous and to include hollow nano-scaled BN of different sizes, with outer radii ranging from 7.20 nm to 13.32 nm. With a band gap of around 5.8 eV, all of the produced porous BN films displayed strong UV absorption edges around 210 nm [4].

OTHER NOVEL SYNTHESIS METHODS

The epitaxial and exfoliation are the two most frequent methods used to synthesize BN, as we discussed in the above section; besides that, other synthesis approaches are also reported. Some standard methods involve physical vapor deposition, Surface segregation-based method, and so on. A detailed review of these synthesis methods is not done here, as the chapter focuses on the exfoliation and CVD routes of synthesis. However, to better understand the possible synthesis routes for BN production, it is worth discussing these methods in a short time.

An effective method for growing films is physical vapor deposition (PVD), which can bypass the intricate interplay of growth factors. In light of this, Sutter *et al.* (2013) used radio frequency (RF) magnetron sputtering to deposit mono- and few-layer hBN on a B target in an ultrahigh-vacuum with N2/Ar environment [87]. The first-time surface segregation was utilized to create 2D-hBN by Xu *et al.* (2011). Surface segregation is a practical method for fabricating large area 2D layered materials. Using heat treatment, a few-layer hBN nanosheet was grown using an electropolished Fe-Cl-Ni alloy doped with B and N. Layer by layer, triangular hBN domains were seen on the alloy surface [88]. To lower the growth temperature and construct well-stoichiometric 2D h-BN sheets, Glavin *et al.* (2014) used the pulsed laser deposition (PLD) technology to create ultra-thin, highly crystalline hBN from an amorphous BN target at 700°C [89]. Another type of synthesis method, the pyrolysis method, has been created by Zhu *et al.* to create hBN nanosheets. In a nutshell, various boric acid and urea molar ratios were combined, dissolved in ultrapure water, and then the solvent was evaporated to allow for recrystallization. Pyrolyzing the material at high temperatures produced highly crystalline hBN with an extraordinarily high surface area, which could be adjusted by choosing various solvent species and adjusting the ratio [75, 90].

APPLICATION OF 2D H-BN NANOSHEETS

The h-BN nanosheets, which have similar properties to graphene, are also similar to graphene in terms of their application, primarily related to the energy storage device. Here, we mainly covered the specific uses of h-BN as a component for fabricating several energy-related devices such as batteries, supercapacitors, fuel cells, solar cells, *etc.* Table **5** provides a description of some energy storage devices fabricated using 2D h-BN nanostructure.

2D h-BN in Batteries

Electrode materials, separators, and electrolytes are the broad categories into which 2D h-BN can be applied to fabricate BN-based Li-ion batteries (BN-LIBs). It is well known that by including the charge/discharge ratio in the battery, the main shortcomings of traditional anodic electrodes may be solved [106]. H-BN is frequently used with other materials to fulfill its potential. The interlayer expansion brought on by Li-ion on the electrodes is minimized by the collective impact of the 2D h-BN/r-GO interface produced by van der Waals forces. In 2D h-BN/silicene heterostructures, the emergence of intermediate complexes at the interface lowers the bandgap, Li+ binding forces, and Li+ barrier are all lowered by the emergence of intermediate complexes at the interface [91].

Another important use of h-BN nanosheets in LIBs is electrode separation. A separator is required between the electrodes to avoid short circuits and regulate ion migration throughout the charge-discharge series. A LIB's short circuit frequently results from unchecked battery warming. The operability can, therefore, be enhanced and safety issues reduced by using a thermally stable separator. Common separators like polypropylene (PP) and polyethylene (PE) have uneven Li^+ diffusion flux and poor wettability against liquid electrolytes). As a separator, Li^+ is distributed uniformly in 2D h-BN with local surface polarities, and it has been demonstrated that this increases the electrolyte's ionic conductivity [107, 108]. 2D h-BN inhibits the development of Li-dendrites. Short circuits in LIB can be brought on by the Li- that results from the unchecked development of Li on the anodes penetrating through the separator. In order to enhance the mechanical stability and manufacturing robustness of the electrolyte, 2D h-BN first increases the ionic conductivity and Li transference number. In general, gel polymer or ionic electrolytes (GPE/GIE) are employed in place of liquid electrolytes to strengthen the mechanical stability of electrolytes. They are reinforced by inorganic fillers like TiO_2 or Al_2O_3 [18].

Table 5. 2D h-BN-based energy related devices and their performances.

Application	Substrate/ Medium/ CO-components	Device Properties	h-BN Role	Refs.
Lithium Ion Batteries	h-BN with r-GO	Current density 100 mAh/g; Stability up to 200 cycles	Electrode	[91]
	LTO with rGO/h-BN	capacity 100 mAh/g; Ultrafast charge rate of >10℃.	Electrode	[92]
	h-BN/silicene	Storage capacity around 1000 mAh/ g	Electrode	[93]
	h-BN/ PE-BN/PVDF	Low thermal shrinkage, fast electrolyte uptake	Separator	[94]
	2D h-BN/LMO graphene	Stability ~100; temp at 150℃.	Separator	[95]
Mg metal-based batteries (MMBs)	F-doped h-BN	Enhanced Conductivity	Electrode	[96]
Zinc Flow Batteries (ZFBs)	BN nanosheets	500 cycles; Efficiency up to 80%	Membrane	[97]

(Table 5) cont.....

Application	Substrate/ Medium/ CO-components	Device Properties	h-BN Role	Refs.
Supercapacitor	h-BN with rGO	Obtained capacitance around 140 F/g (Higher than the bare rGO); Exceptional cyclic Stability	Electrode	[98]
	h-BN with rGO	Capacitance up to 900 F/g; greater stability	Electrode	[99]
	$ZnCo_2S_4$ (ZCS)/2D h-BN with CNT	Capacitance around 800; higher stability	Electrode	[100]
	h-BN-doped PVA-H_2SO_4	Specific capacitance of 125 F/g, 99.2% capacitance	Electrolyte	[97]
Fuel Cell	Nafion/h-BN	Reduces the methanol crossover	Membrane	[101]
	Fe_2O_3/ BN-doped rGO	Circuit Potential 250 mV	Cathode Catalyst	[102]
	hBN coating on ZrO_2 powders	Active against H_2 diffusion	Hydrogen environmental barrier coating	[103]
Solar Cell	MoS_2/WSe_2/h-BN	Higher stability; 74% improvement in power conversion efficiency	Passivation	[104]
	Si /h-BN/graphene	Significant in the open circuit voltage and reduced interface recombination	The interface design of graphene/Si heterojunction	[105]

Abbreviation Used in the Table: LMO: Li-manganese oxide; LTO: Lithium titanium oxide; PE: Polyethylene; PVDF: Polyvinylidene fluoride.

Besides, Li-ion batteries, other alkali-based batteries can be benefitted *via* h-BN nanosheets, such as those based on magnesium (Mg-based batteries, *i.e.*, MMBs). Mg is a competitive option for usage in the secondary battery industry due to its availability in nature and electrochemical characteristics. Due to the highly polarized nature of Mg^{2+} ions, inserting or extracting Mg into the cathode is one significant barrier to commercializing these batteries. There have been reports of 2D h-BN nanomaterials with fluorine functionalities working effectively as MMB cathodes. Venkateswarlu *et al.* (2020) investigated graphene-based BN nanosheets for enhanced electroanalytical performance for a rechargeable magnesium storage system. Compared to graphene and other composites, the study found that the G/BN-20% composite has a discharge capability of 105 mAhg^{-1} within 100 cycles. The synergistic interaction of graphene and BN nanosheets is responsible for the abundant power of G/BN-20% [96].

2D-hBN sheets can also fabricate Zinc Flow Batteries (ZFBs). Due to their high energy density, low cost, and hazardless nature, ZFBs are frequently regarded as trustworthy energy storage solutions. Despite these benefits, these batteries have yet to be produced on a large scale or commercially. The uneven temperature on the electrode and the membrane's subpar mechanical properties significantly impact this process [109]. Several efforts have been made to use 2D h-BN to overcome these problems. Hu and their co-workers recently used an alkaline zinc-iron flow battery. Based on their results, 2D h-BN strengthens the membrane, alters the morphology of zinc, and functions as a heat transporter while facing the anode. The voltage-time plots show that the naked membrane was stable for up to 40 h before fluctuating due to micro-short circuiting and the development of zinc needle-like dendrites that penetrated the membrane. The electrolyte may cross over as a result of the dendritic growth's ability to harm the membrane. Self-discharge occurs as a result, and the concentration of active material decreases [97].

2D h-BN in Supercapacitor (SCs)

Supercapacitors (SCs) are the devices that store and supply energy by reversal of ions at the interfaces. Owing to their rapid charge-discharge rates, extended life cycle, high power, and high energy density, they are different from traditional capacitors [78]. In recent years, 2D h-BN has been acquainted with potential inorganic materials to boost the supercapacitor act. To improve the electrochemical behavior, the 2D h-BN alterations can produce free electrons that are accessible or increase the mobility of charge carriers. Moreover, the partial polarity of the B-N links makes the 2D h-BN lattice's charge separation easier. In fact, because of the partly ionic nature of the B-N bonds, 2D h-BN exhibits a pseudo-capacitive behavior [110]. BN can be modulated or uplift the super capacitance properties by changing the two distinct main components, i. e. electrode and electrolyte.

As Electrode Material

The most critical components of a supercapacitor assembly are the electrodes. Due to this reason, the performance of the electrodes has been the focus of most research on 2D h-BN-based materials. In one of the first efforts, Gao *et al.* used a simple liquid-phase exfoliation process to create a 2D h-BN/r-GO nanocomposite electrode. The constructed supercapacitor showed outstanding cyclic stability (105.5% capacitance retention after 1000 cycles), high specific capacitance (140 Fg-1 at 2 Ag-1), and good rate performance (71.5 Fg-1 at 50 Ag-1). These outstanding electrochemical results suggest that the BN/RGO nanocomposites have many potential uses in supercapacitors [98]. As a high-performance electrode, h-BN nanosheets have also been used in a quaternary composite of $ZnCo_2S_4$ (ZCS)/2D h-BN/CNT/Polypyrrole (PPY) [100].

As an Electrolyte Component

By incorporating h-BN nanosheets in SC as an electrolyte, it is possible to improve the different properties of SCs. For instance, Hu *et al.* (2017) used h-BN to develop electrolytes for SC applications. The h-BN-doped poly (vinyl alcohol)-sulfuric acid ($PVA-H_2SO_4$) gel polymer electrolyte (GPE) is first fabricated through a facile freeze-thaw technique [111]. As per the studies, it is also assumed that the development of the ionic routes contributes to the decreased charge transfer resistance of Supercapacitors containing 2D h-BN. Due to its solid ionic conductivity, h-BN has the potential to be used as a separator in different SCs.

In Fuel Cell

2D-hBN nanosheets can also be used for the fabrication of fuel cells. It is another energy-related application of this material. It may be utilized in fuel cells as a barrier coating, cathode catalyst, and membrane. One of the most common fuel cells, the direct methanol fuel cell (DMFC) [112], for instance, encounters several technical difficulties, including the slow oxidation and reduction of oxygen and the passage of methanol through the polymer electrolyte membrane (PEM). The development of a mixed potential at the cathode caused by the methanol crossover lowers the efficiency of DMFCs. Further difficulties with methanol crossover include unwanted oxidation, a decrease in the cathode's active sites for oxygen reduction, and CO toxicity [113]. Despite its exceptional thermal and chemical stability and proton conductivity, nafion (sulfonated tetrafluoroethylene-based fluoropolymer-copolymer)—the most often employed PEM, is inferred from the methanol crossover. Using 2D h-BN as reinforcement is one method of attempting to reduce the methanol crossover in Nafion. High proton conductivity results from polarised covalent bonds between B and N that cause valence electron build-up near nitrogen and an unequal dispersion of the electron cloud [114]. For this purpose, 2D h-BN can be functionalized *via* -OH groups, which will speed up the side-chain interactions with -mercaptopropyl tri methoxy silane and form continuous ionic channels in the PEM. After adding h-BN, the proton conductivity of bare Nafion increased from 135 to 214 $mS.cm^{-1}$. Adding 2D h-BN to the membrane simultaneously reduces the methanol crossover [101]. The results suggest that h-BN can be used as a membrane material in different fuel cells.

Although 2D h-BN can take part in catalytic processes, there needs to be more research or data published related to its use as a fuel cell electrode. However, it is now also being applied as a cathode in different fuel cells. Mahalingam *et al.*

(2020) developed a cathode catalyst of a microbial fuel cell (MFC) using 2D h-BN as a main component. Fe_2O_3 and BN-doped rGO nanosheets were synthesized using a surfactant-free hydrothermal process. The idea is to add 2D h-BN and Fe_2O_3 to r-GO nanosheets to increase surface area for easy reactant transport and accelerate electron transport to graphene [102]. In addition to the membrane and cathode catalyst, 2D h-BNs are employed as a covering for the hydrogen environment, which prevents material embrittlement and hydrogen diffusion [115]. Retaining hydrogen fuel is essential in addition to avoiding embrittlement. The deposition technique and resulting microstructure substantially impact the 2D h-BN barriers' efficacy. For example, ion beam-assisted deposition can form (002) planes that are orthogonal to the substrate yet cannot obstruct H_2 diffusion. Nevertheless, barriers produced by radiofrequency magnetron sputtering have a significantly reduced H_2 permeability [103, 116].

2D h-BN in Solar Cell

With a significant prospective to provide green, dependable, accessible, and economic strength to any nation, solar technology is regarded as the mainstream renewable technology. However, several problems in solar technology still need to be handled in terms of circuit design from the perspective of materials engineering [110]. Due to their distinctive features, 2D nanomaterials have led solar cell engineering and fabrication. In recent years, researchers have employed BN nanosheets in various solar cells to construct the interface or passivate the surface of active materials. hBN is a leading contender for solar cell application due to its unique properties, which include a lack of dangling bonds, a significant bandgap, and chemical and thermal stability [117]. In order to reduce the electrical loss, van der Waals heterojunction solar cell, 2D hBN, was employed as a passivation layer on top of the cell. This technique is supposed to improve the MoS_2/WSe_2 solar cell's photovoltaic performance. The study concluded that a significant rise in both short-circuit current and open-circuit voltage led to a 74% improvement in power conversion efficiency. The study also concluded that the h-BN top layer increased the tested 2D solar cell's long-term stability under ambient settings [104]. As the lack of a lattice-matched and broad bandgap passivation layer was a barrier to the development of InP, using 2D h-BN as a passive layer may be a significant achievement. 2D h-BN has also been used to passivate the anode aside from heterostructure solar cells (TiO_2 nanoparticles) [118]. Interface engineering for heterojunction structures is another use of 2D h-BN in solar cells. A solar cell with a Schottky junction made of graphene/Si has gained popularity because it may be used in low-cost photovoltaic systems. However, their short Schottky barrier causes charge carriers to recombine at the interface, significantly reducing their efficiency. Placing a 2D substance as an insulator between graphene and Si is one efficient technique to solve this problem.

Thus, h-BN nanostructures may be the most attractive choice for solar cell fabrication. Research on graphene/h-BN/Si solar cell platforms demonstrated that 2D h-BN might obstruct both the transport of holes (a minor carrier) from n-Si to graphene and the diffusion of electrons (the primary carrier). Due to the h-BN's characteristics and ideal band alignment with Si, it may act as an efficient layer that blocks the electrons, transports the holes, significantly boosts the open circuit voltage, and reduces interface recombination [105].

Similar to graphene [119], the development of scalable and economical synthesis routes for 2D-BN may unlock the door to a broad spectrum of applications. As graphene has shown significant contribution in the development of next-generation photothermal actuators [120], and energy storage solutions [121], 2D-BN's exceptional properties and versatile nature make it an ideal candidate for driving innovation across industries.

CONCLUSION

The use of 2D h-BN in energy storage and production systems shows the material's enormous promise for the sector. In batteries or other applications where it is necessary to separate conductive elements, 2D h-electrical BN's insulation avoids close contact between electrodes. By reducing heat accumulation, the enhanced heat dissipation offered by 2D h-thermally BN's conductive channels significantly improves control over heat management and safety problems. The current chapter provided an overview of the two primary processes—exfoliation and CVD—used to create 2D h-BN nanosheets. These two approaches are the most used ones. This chapter also covered all the unique characteristics linked to energy-related applications. This chapter effectively and concisely explains h-BN and will undoubtedly aid researchers in deciding on the synthesis method to use and creating various energy-related devices employing h-BN as a viable nanomaterial.

REFERENCES

[1]   Punetha, V.D.; Rana, S.; Yoo, H.J.; Chaurasia, A.; McLeskey, J.T., Jr; Ramasamy, M.S.; Sahoo, N.G.; Cho, J.W. Functionalization of carbon nanomaterials for advanced polymer nanocomposites: A comparison study between CNT and graphene. *Prog. Polym. Sci.,* **2017**, *67*, 1-47.
       [http://dx.doi.org/10.1016/j.progpolymsci.2016.12.010]

[2]   Bhatt, S.; Pathak, R.; Punetha, V.D.; Punetha, M. Recent advances and mechanism of antimicrobial efficacy of graphene-based materials: A review. *J. Mater. Sci.,* **2023**, *58*(19), 7839-7867.
       [http://dx.doi.org/10.1007/s10853-023-08534-z] [PMID: 37200572]

[3]   Pathak, R.; Punetha, V.D.; Bhatt, S.; Punetha, M. Multifunctional role of carbon dot-based polymer nanocomposites in biomedical applications: A review. *J. Mater. Sci.,* **2023**, *58*(15), 6419-6443.
       [http://dx.doi.org/10.1007/s10853-023-08408-4] [PMID: 37065681]

[4]   Sahoo, N.G.; Esteves, R.J.; Punetha, V.D.; Pestov, D.; Arachchige, I.U.; McLeskey, J.T., Jr Schottky diodes from 2D germanane. *Appl. Phys. Lett.,* **2016**, *109*(2), 023507.

[http://dx.doi.org/10.1063/1.4955463]

[5]   Pathak, R.; Mohsin, M.; Mehta, S.P.S. An Assessment of *in vitro* Antioxidant Potential of *Camelina sativa L.* Seed Oil and Estimation of Tocopherol Content using HPTLC Method. *J. Sci. Res.,* **2021,** *13*(2), 589-600.
[http://dx.doi.org/10.3329/jsr.v13i2.49783]

[6]   Inbaoli, A.; Sujith, K.C.S.; Jayaraj, S. Two-Dimensional (2D) layered materials as emerging nanocatalysts in the production of biodiesel. *Adv. Nanocataly. Biod. Prod.,* **2022,** (Sep), 185-197.
[http://dx.doi.org/10.1201/9781003120858-8]

[7]   Pathak, R.; Guleria, K.; Kumari, A.; Mehta, S.P.S. Deacidification of Camelina sativa L. seed oil by Physisorption method and characterization of produced biodiesel. *J. Appl. Nat. Sci.,* **2021,** *13*(1), 287-294.
[http://dx.doi.org/10.31018/jans.v13i1.2555]

[8]   Pathak, R.; Kumari, A.; Mohsin, M.; Bisht, G.; Bala, M. Phytochemical Assessment and *In vitro* Antioxidant potential of *Camelina sativa* L. seed cake. *Asian J. Res. Chem,* **2020,** *13*(1), 38-43.
[http://dx.doi.org/10.5958/0974-4150.2020.00009.7]

[9]   Balkrishna, A.; Pathak, R.; Bhatt, S.; Arya, V. Molecular insights of plant phytochemicals against diabetic neuropathy. *Curr. Diabetes Rev.,* **2022,** *2022,* 15733998196666220825124510.
[http://dx.doi.org/10.2174/15733998196666220825124510] [PMID: 36028963]

[10]   Bhatt, S.; Faridi, N.; Raj, S.P.M.; Pathak, D.; Agarwal, A. Cloning, expression and specificity evaluation of type III effector, Rip4, from *Ralstonia solanacearum. Ecol. Environ. Conserv.,* **2021,** *27,* S390-S397.

[11]   Bhatt, S.; Punetha, V.D.; Pathak, R.; Punetha, M. Two-dimensional carbon nanomaterial-based biosensors: Micromachines for advancing the medical diagnosis.*Recent Advances in Graphene Nanophotonics*; Springer: Cham, **2023,** pp. 181-225.
[http://dx.doi.org/10.1007/978-3-031-28942-2_9]

[12]   Punetha, V.D.; Dhali, S.; Rana, A.; Karki, N.; Tiwari, H.; Negi, P.; Basak, S.; Sahoo, N.G. Recent advancements in green synthesis of nanoparticles for improvement of bioactivities: A review. *Curr. Pharm. Biotechnol.,* **2022,** *23*(7), 904-919.
[http://dx.doi.org/10.2174/1389201022666210812115233] [PMID: 34387160]

[13]   Aryal, U.K.; Ahmadpour, M.; Turkovic, V.; Rubahn, H.G.; Di Carlo, A.; Madsen, M. 2D materials for organic and perovskite photovoltaics. *Nano Energy,* **2022,** *94,* 106833.
[http://dx.doi.org/10.1016/j.nanoen.2021.106833]

[14]   Zhen, Z.; Zhu, H. Structure and properties of graphene. In: *Graphene: Fabrication, Characterizations, Properties and Applications*; Academic Press, **2018**; pp. (Jan)1-12.
[http://dx.doi.org/10.1016/B978-0-12-812651-6.00001-X]

[15]   Manzeli, S.; Ovchinnikov, D.; Pasquier, D.; Yazyev, O. v.; Kis, A. 2D transition metal dichalcogenides. *Nat. Rev. Mater.,* **2017,** *2*(8), 1-15.
[http://dx.doi.org/10.1038/natrevmats.2017.33]

[16]   Dobrzhinetskaya, L.F.; Wirth, R.; Yang, J.; Green, H.W.; Hutcheon, I.D.; Weber, P.K.; Grew, E.S. Qingsongite, natural cubic boron nitride: The first boron mineral from the Earth's mantle. *Am. Mineral.,* **2014,** *99*(4), 764-772.
[http://dx.doi.org/10.2138/am.2014.4714]

[17]   Lopes, J.M.J. Synthesis of hexagonal boron nitride: From bulk crystals to atomically thin films. *Prog. Cryst. Growth Charact. Mater.,* **2021,** *67*(2), 100522.
[http://dx.doi.org/10.1016/j.pcrysgrow.2021.100522]

[18]   Gong, Y.; Xu, Z.Q.; Li, D.; Zhang, J.; Aharonovich, I.; Zhang, Y. Two-dimensional hexagonal boron nitride for building next-generation energy-efficient devices. *ACS Energy Lett.,* **2021,** *6*(3), 985-996.
[http://dx.doi.org/10.1021/acsenergylett.0c02427]

[19]   Yuan, C.; Li, J.; Lindsay, L.; Cherns, D. Modulating the thermal conductivity in hexagonal boron nitride *via* controlled boron isotope concentration. *Commun. Phys.,* **2019**, *2*(1), 1-8.
[http://dx.doi.org/10.1038/s42005-019-0145-5]

[20]   Kim, K.K.; Kim, S.M.; Lee, Y.H. A new horizon for hexagonal boron nitride film. *J. Korean Phys. Soc.,* **2014**, *64*(10), 1605-1616.
[http://dx.doi.org/10.3938/jkps.64.1605]

[21]   Yang, Y.; Peng, Y.; Saleem, M.F.; Chen, Z.; Sun, W. Hexagonal boron nitride on III–V compounds: A review of the synthesis and applications. *Materials,* **2022**, *15*(13), 4396.
[http://dx.doi.org/10.3390/ma15134396] [PMID: 35806522]

[22]   Smith, A.T.; LaChance, A.M.; Zeng, S.; Liu, B.; Sun, L. Synthesis, properties, and applications of graphene oxide/reduced graphene oxide and their nanocomposites. *Nano Mater. Sci.,* **2019**, *1*(1), 31-47.
[http://dx.doi.org/10.1016/j.nanoms.2019.02.004]

[23]   Kim, J.H.; Cho, H.; Pham, T.V.; Hwang, J.H.; Ahn, S.; Jang, S.G.; Lee, H.; Park, C.; Kim, C.S.; Kim, M.J. Dual growth mode of boron nitride nanotubes in high temperature pressure laser ablation. *Sci. Rep.,* **2019**, *9*(1), 15674.
[http://dx.doi.org/10.1038/s41598-019-52247-w] [PMID: 31666654]

[24]   Izyumskaya, N.; Demchenko, D.O.; Das, S.; Özgür, Ü.; Avrutin, V.; Morkoç, H. Recent development of boron nitride towards electronic applications. *Adv. Electron. Mater.,* **2017**, *3*(5), 1600485.
[http://dx.doi.org/10.1002/aelm.201600485]

[25]   Cassabois, G.; Valvin, P.; Gil, B. Hexagonal boron nitride is an indirect bandgap semiconductor. *Nat. Phot.,* **2016**, *10*(4), 262-266.
[http://dx.doi.org/10.1038/nphoton.2015.277]

[26]   Dean, C. R.; Young, A.F.; Meric, I.; Lee, C.; Wang, L.; Sorgenfrei, S.; Watanabe, K.; Taniguchi, T.; Kim, P.; Shepard, K.L.; Hone, J. Boron nitride substrates for high-quality graphene electronics. *Nat. Nanotechnol.,* **2010**, *5*(10), 722-726.
[http://dx.doi.org/10.1038/nnano.2010.172]

[27]   Auwärter, W. Hexagonal boron nitride monolayers on metal supports: Versatile templates for atoms, molecules and nanostructures. *Surf. Sci. Rep.,* **2019**, *74*(1), 1-95.
[http://dx.doi.org/10.1016/j.surfrep.2018.10.001]

[28]   Zhou, H.; Zhu, J.; Liu, Z.; Yan, Z.; Fan, X.; Lin, J.; Wang, G.; Yan, Q.; Yu, T.; Ajayan, P.M.; Tour, J.M. High thermal conductivity of suspended few-layer hexagonal boron nitride sheets. *Nano Res.,* **2014**, *7*(8), 1232-1240.
[http://dx.doi.org/10.1007/s12274-014-0486-z]

[29]   Kumar, R.; Rajasekaran, G.; Parashar, A. Optimised cut-off function for Tersoff-like potentials for a BN nanosheet: A molecular dynamics study. *Nanotechnology,* **2016**, *27*(8), 085706.
[http://dx.doi.org/10.1088/0957-4484/27/8/085706] [PMID: 26820110]

[30]   Liu, H.; You, C.Y.; Li, J.; Galligan, P.R.; You, J.; Liu, Z.; Cai, Y.; Luo, Z. Synthesis of hexagonal boron nitrides by chemical vapor deposition and their use as single photon emitters. *Nano Mater. Sci.,* **2021**, *3*(3), 291-312.
[http://dx.doi.org/10.1016/j.nanoms.2021.03.002]

[31]   Emanet, M.; Sen, Ö.; Taşkin, I.Ç.; Çulha, M. Synthesis, Functionalization, and Bioapplications of Two-Dimensional Boron Nitride Nanomaterials. *Front. Bioeng. Biotechnol.,* **2019**, *7*, 363.
[http://dx.doi.org/10.3389/fbioe.2019.00363] [PMID: 31921797]

[32]   Ye, H.; Lu, T.; Xu, C.; Han, B.; Meng, N.; Xu, L. Liquid-phase exfoliation of hexagonal boron nitride into boron nitride nanosheets in common organic solvents with hyperbranched polyethylene as stabilizer. *Macromol. Chem. Phys.,* **2018**, *219*(6), 1700482.
[http://dx.doi.org/10.1002/macp.201700482]

[33]   Kim, J.; Kwon, S.; Cho, D.H.; Kang, B.; Kwon, H.; Kim, Y.; Park, S.O.; Jung, G.Y.; Shin, E.; Kim, W.G.; Lee, H.; Ryu, G.H.; Choi, M.; Kim, T.H.; Oh, J.; Park, S.; Kwak, S.K.; Yoon, S.W.; Byun, D.; Lee, Z.; Lee, C. Direct exfoliation and dispersion of two-dimensional materials in pure water *via* temperature control. *Nat. Commun.,* **2015,** *6*(1), 8294.
        [http://dx.doi.org/10.1038/ncomms9294] [PMID: 26369895]

[34]   Deepika, A.; Li, L.H.; Glushenkov, A.M.; Hait, S.K.; Hodgson, P.; Chen, Y. High-efficient production of boron nitride nanosheets *via* an optimized ball milling process for lubrication in oil. *Sci. Rep.,* **2014,** *4*(1), 7288-7288.
        [http://dx.doi.org/10.1038/srep07288] [PMID: 25470295]

[35]   Tian, Z.; Chen, K.; Sun, S.; Zhang, J.; Cui, W.; Xie, Z.; Liu, G. Crystalline boron nitride nanosheets by sonication-assisted hydrothermal exfoliation. *J. Adv. Ceram.,* **2019,** *8*(1), 72-78.
        [http://dx.doi.org/10.1007/s40145-018-0293-1]

[36]   Mahdizadeh, A.; Farhadi, S.; Zabardasti, A. Microwave-assisted rapid synthesis of graphene-analogue hexagonal boron nitride (h-BN) nanosheets and their application for the ultrafast and selective adsorption of cationic dyes from aqueous solutions. *RSC Advances,* **2017,** *7*(85), 53984-53995.
        [http://dx.doi.org/10.1039/C7RA11248C]

[37]   Chen, Y.; Xu, X.; Li, C.; Bendavid, A.; Westerhausen, M.T.; Bradac, C.; Toth, M.; Aharonovich, I.; Tran, T.T. Bottom-up synthesis of hexagonal boron nitride nanoparticles with intensity-stabilized quantum emitters. *Small,* **2021,** *17*(17), 2008062.
        [http://dx.doi.org/10.1002/smll.202008062] [PMID: 33733581]

[38]   Kainthola, A.; Bijalwan, K.; Negi, S.; Sharma, H.; Dwivedi, C. Hydrothermal synthesis of highly stable boron nitride nanoparticles. *Mater. Today Proc.,* **2020,** *28*, 138-140.
        [http://dx.doi.org/10.1016/j.matpr.2020.01.452]

[39]   Jing, X.; Puglisi, F.; Akinwande, D.; Lanza, M. Chemical vapor deposition of hexagonal boron nitride on metal-coated wafers and transfer-free fabrication of resistive switching devices. *2d Mater.,* **2019,** *63*, 035021.
        [http://dx.doi.org/10.1088/2053-1583/ab1783]

[40]   Ren, X.; Dong, J.; Yang, P.; Li, J.; Lu, G.; Wu, T.; Wang, H.; Guo, W.; Zhang, Z.; Ding, F.; Jin, C. Grain boundaries in chemical-vapor-deposited atomically thin hexagonal boron nitride. *Phys. Rev. Mater.,* **2019,** *3*(1), 014004.
        [http://dx.doi.org/10.1103/PhysRevMaterials.3.014004]

[41]   Xu, Z.; Khanaki, A.; Tian, H.; Zheng, R.; Suja, M.; Zheng, J.G.; Liu, J. Direct growth of hexagonal boron nitride/graphene heterostructures on cobalt foil substrates by plasma-assisted molecular beam epitaxy. *Appl. Phys. Lett.,* **2016,** *109*(4), 043110.
        [http://dx.doi.org/10.1063/1.4960165]

[42]   Khalid, M.F.; Riaz, I.; Jalil, R.; Mahmood, U.; Mir, R.R.; Sohail, H.A. Dielectric properties of multi-layers hexagonal boron nitride. *Mater. Sci. Appl.,* **2020,** *11*(6), 339-346.
        [http://dx.doi.org/10.4236/msa.2020.116023]

[43]   Yu, C.; Zhang, J.; Tian, W.; Fan, X.; Yao, Y. Polymer composites based on hexagonal boron nitride and their application in thermally conductive composites. *RSC Advances,* **2018,** *8*(39), 21948-21967.
        [http://dx.doi.org/10.1039/C8RA02685H] [PMID: 35541702]

[44]   Xu, M.; Liang, T.; Shi, M.; Chen, H. Graphene-like two-dimensional materials. *Chem. Rev.,* **2013,** *113*(5), 3766-3798.
        [http://dx.doi.org/10.1021/cr300263a] [PMID: 23286380]

[45]   Jang, S. K.; Youn, J.; Song, Y. J.; Lee, S. Synthesis and characterization of hexagonal boron nitride as a gate dielectric. *Sci. Reports,* **2016,** *6*(1), 1-9.
        [http://dx.doi.org/10.1038/srep30449]

[46]   Alam, S.; Asaduzzaman Chowdhury, M.; Shahid, A.; Alam, R.; Rahim, A. Synthesis of emerging two-

dimensional (2D) materials - Advances, challenges and prospects. *FlatChem*, **2021**, *30*, 100305.
[http://dx.doi.org/10.1016/j.flatc.2021.100305]

[47]   Gao, E.; Lin, S.Z.; Qin, Z.; Buehler, M.J.; Feng, X.Q.; Xu, Z. Mechanical exfoliation of two-dimensional materials. *J. Mech. Phys. Solids*, **2018**, *115*, 248-262.
[http://dx.doi.org/10.1016/j.jmps.2018.03.014]

[48]   Gautam, C.; Chelliah, S. Methods of hexagonal boron nitride exfoliation and its functionalization: covalent and non-covalent approaches. *RSC Advances*, **2021**, *11*(50), 31284-31327.
[http://dx.doi.org/10.1039/D1RA05727H] [PMID: 35496870]

[49]   Hummers, W.S., Jr; Offeman, R.E. Preparation of graphitic oxide. *J. Am. Chem. Soc.*, **1958**, *80*(6), 1339.
[http://dx.doi.org/10.1021/ja01539a017]

[50]   Lu, G.; Wu, T.; Yuan, Q.; Wang, H.; Wang, H.; Ding, F.; Xie, X.; Jiang, M. Synthesis of large single-crystal hexagonal boron nitride grains on Cu-Ni alloy. *Nat. Commun.*, **2015**, *6*(1), 1-7.
[http://dx.doi.org/10.1038/ncomms7160]

[51]   Sohail, M.; Saleem, M.; Ullah, S.; Saeed, N.; Afridi, A.; Khan, M.; Arif, M. Modified and improved Hummer's synthesis of graphene oxide for capacitors applications. *Mod. Electron. Mater.*, **2017**, *3*(3), 110-116.
[http://dx.doi.org/10.1016/j.moem.2017.07.002]

[52]   Li, X.; Hao, X.; Zhao, M.; Wu, Y.; Yang, J.; Tian, Y.; Qian, G. Exfoliation of hexagonal boron nitride by molten hydroxides. *Adv. Mater.*, **2013**, *25*(15), 2200-2204.
[http://dx.doi.org/10.1002/adma.201204031] [PMID: 23436746]

[53]   Du, M.; Wu, Y.; Hao, X. A facile chemical exfoliation method to obtain large size boron nitride nanosheets. *CrystEngComm*, **2013**, *15*(9), 1782-1786.
[http://dx.doi.org/10.1039/c2ce26446c]

[54]   Gautam, C.; Chelliah, S. Methods of hexagonal boron nitride exfoliation and its functionalization: covalent and non-covalent approaches. *RSC Adv.*, **2021**, *11*(50), 31284-31327.
[http://dx.doi.org/10.1039/D1RA05727H]

[55]   Lin, L.; Xu, Y.; Zhang, S.; Ross, I.M.; Ong, A.C.M.; Allwood, D.A. Fabrication and luminescence of monolayered boron nitride quantum dots. *Small*, **2014**, *10*(1), 60-65.
[http://dx.doi.org/10.1002/smll.201301001] [PMID: 23839969]

[56]   Bhimanapati, G.R.; Kozuch, D.; Robinson, J.A. Large-scale synthesis and functionalization of hexagonal boron nitride nanosheets. *Nanoscale*, **2014**, *6*(20), 11671-11675.
[http://dx.doi.org/10.1039/C4NR01816H] [PMID: 25163394]

[57]   Doll, G.L.; Speck, J.S.; Dresselhaus, G.; Dresselhaus, M.S.; Nakamura, K.; Tanuma, S.I. Intercalation of hexagonal boron nitride with potassium. *J. Appl. Phys.*, **1989**, *66*(6), 2554-2558.
[http://dx.doi.org/10.1063/1.344219]

[58]   Zhao, S.Y.F.; Elbaz, G.A.; Bediako, D.K.; Yu, C.; Efetov, D.K.; Guo, Y.; Ravichandran, J.; Min, K.A.; Hong, S.; Taniguchi, T.; Watanabe, K.; Brus, L.E.; Roy, X.; Kim, P. Controlled electrochemical intercalation of graphene/ *h*- BN van der waals heterostructures. *Nano Lett.*, **2018**, *18*(1), 460-466.
[http://dx.doi.org/10.1021/acs.nanolett.7b04396] [PMID: 29268017]

[59]   Han, W.Q.; Wu, L.; Zhu, Y.; Watanabe, K.; Taniguchi, T. Structure of chemically derived mono- and few-atomic-layer boron nitride sheets. *Appl. Phys. Lett.*, **2008**, *93*(22), 223103.
[http://dx.doi.org/10.1063/1.3041639]

[60]   Zhi, C.; Bando, Y.; Tang, C.; Kuwahara, H.; Golberg, D. Large-scale fabrication of boron nitride nanosheets and their utilization in polymeric composites with improved thermal and mechanical properties. *Adv. Mater.*, **2009**, *21*(28), 2889-2893.
[http://dx.doi.org/10.1002/adma.200900323]

[61]   Warner, J.H.; Rümmeli, M.H.; Bachmatiuk, A.; Büchner, B. Atomic resolution imaging and

topography of boron nitride sheets produced by chemical exfoliation. *ACS Nano,* **2010**, *4*(3), 1299-1304.
[http://dx.doi.org/10.1021/nn901648q] [PMID: 20148574]

[62]   Lin, Y.; Williams, T.V.; Connell, J.W. Soluble, exfoliated hexagonal boron nitride nanosheets. *J. Phys. Chem. Lett.,* **2010**, *1*(1), 277-283.
[http://dx.doi.org/10.1021/jz9002108]

[63]   Štengl, V.; Henych, J.; Slušná, M.; Ecorchard, P. Ultrasound exfoliation of inorganic analogues of graphene. *Nanoscale Res. Lett.,* **2014**, *9*(1), 167.
[http://dx.doi.org/10.1186/1556-276X-9-167] [PMID: 24708572]

[64]   Chen, Y.; Liang, H.; Abbas, Q.; Liu, J.; Shi, J.; Xia, X.; Zhang, H.; Du, G. Growth and characterization of porous sp 2-BN films with hollow spheres under hydrogen etching effect *via* borazane thermal CVD. In: *Applied Surface Science*; Elsevier, **2017**.

[65]   Bai, Y.; Zhang, J.; Wang, Y.; Cao, Z.; An, L.; Zhang, B.; Yu, Y.; Zhang, J.; Wang, C. Ball milling of hexagonal boron nitride microflakes in ammonia fluoride solution gives fluorinated nanosheets that serve as effective water-dispersible lubricant additives. *ACS Appl. Nano Mater.,* **2019**, *2*(5), 3187-3195.
[http://dx.doi.org/10.1021/acsanm.9b00502]

[66]   Ma, J.; Luo, N.; Xie, Z.; Chen, F.; Fu, Q. Preparation of modified hexagonal boron nitride by ball-milling and enhanced thermal conductivity of epoxy resin. *Mater. Res. Exp.,* **2019**, *6*(10), 1050d8.
[http://dx.doi.org/10.1088/2053-1591/ab432a]

[67]   Yang, N.; Ji, H.; Jiang, X.; Qu, X.; Zhang, X.; Zhang, Y.; Liu, B. Preparation of boron nitride nanoplatelets *via* amino acid assisted ball milling: Towards thermal conductivity application. *Nanomaterials,* **2020**, *10*(9), 1652.
[http://dx.doi.org/10.3390/nano10091652]

[68]   Cui, Z.; Oyer, A.J.; Glover, A.J.; Schniepp, H.C.; Adamson, D.H. Large scale thermal exfoliation and functionalization of boron nitride. *Small,* **2014**, *10*(12), 2352-2355.
[http://dx.doi.org/10.1002/smll.201303236] [PMID: 24578306]

[69]   Ko, W.Y.; Chen, C.Y.; Chen, W.H.; Lin, K.J. Fabrication of hexagonal boron nitride nanosheets by using a simple thermal exfoliation process. *J. Chin. Chem. Soc.,* **2016**, *63*(3), 303-307.
[http://dx.doi.org/10.1002/jccs.201500335]

[70]   Acharya, L.; Babu, P.; Behera, A.; Pattnaik, S.P.; Parida, K. Novel synthesis of boron nitride nanosheets from hexagonal boron nitride by modified aqueous phase bi-thermal exfoliation method. *Mater. Today Proc.,* **2021**, *35*, 239-242.
[http://dx.doi.org/10.1016/j.matpr.2020.05.328]

[71]   Morishita, T.; Okamoto, H.; Katagiri, Y.; Matsushita, M.; Fukumori, K. A high-yield ionic liquid-promoted synthesis of boron nitride nanosheets by direct exfoliation. *Chem. Commun.,* **2015**, *51*(60), 12068-12071.
[http://dx.doi.org/10.1039/C5CC04077A] [PMID: 26121635]

[72]   Gonzalez Ortiz, D.; Pochat-Bohatier, C.; Cambedouzou, J.; Bechelany, M.; Miele, P. Exfoliation of Hexagonal Boron Nitride (h-BN) in Liquide Phase by Ion Intercalation. *Nanomaterials,* **2018**, *8*(9), 716.
[http://dx.doi.org/10.3390/nano8090716]

[73]   Lee, D.; Lee, B.; Park, K.H.; Ryu, H.J.; Jeon, S.; Hong, S.H. Scalable exfoliation process for highly soluble boron nitride nanoplatelets by hydroxide-assisted ball milling. *Nano Lett.,* **2015**, *15*(2), 1238-1244.
[http://dx.doi.org/10.1021/nl504397h] [PMID: 25622114]

[74]   Chen, X.; Dobson, J.F.; Raston, C.L. Vortex fluidic exfoliation of graphite and boron nitride. *Chem. Commun.,* **2012**, *48*(31), 3703-3705.
[http://dx.doi.org/10.1039/c2cc17611d] [PMID: 22314550]

[75]   Zhu, W.; Gao, X.; Li, Q.; Li, H.; Chao, Y.; Li, M.; Mahurin, S.M.; Li, H.; Zhu, H.; Dai, S. Controlled gas exfoliation of boron nitride into few-layered nanosheets. *Angew. Chem. Int. Ed.,* **2016**, *55*(36), 10766-10770.
[http://dx.doi.org/10.1002/anie.201605515] [PMID: 27444210]

[76]   Muhabie, A.A.; Cheng, C-C.; Huang, J-J.; Liao, Z-S.; Huang, S-Y.; Chiu, C-W.; Lee, D-J.; Li, H.; Zhu, H.; Dai, S. Non-covalently functionalized boron nitride mediated by a highly self-assembled supramolecular polymer. *Chem. Mater.,* **2017**, *29*(19), 8513-8520.
[http://dx.doi.org/10.1021/acs.chemmater.7b03426]

[77]   Yuan, F.; Jiao, W.; Yang, F.; Liu, W.; Liu, J.; Xu, Z.; Wang, R. Scalable exfoliation for large-size boron nitride nanosheets by low temperature thermal expansion-assisted ultrasonic exfoliation. *J. Mater. Chem. C Mater. Opt. Electron. Devices,* **2017**, *5*(25), 6359-6368.
[http://dx.doi.org/10.1039/C7TC01692A]

[78]   Tahir, M.B.; Rafique, M.; Rafique, M.S.; Nawaz, T.; Rizwan, M.; Tanveer, M. Photocatalytic nanomaterials for degradation of organic pollutants and heavy metals. In: *Nanotechnology and Photocatalysis for Environmental Applications*; Elsevier, 2020; pp. 119-138.
[http://dx.doi.org/10.1016/B978-0-12-821192-2.00008-5]

[79]   Takahashi, T.; Itoh, H.; Takeuchi, A. Chemical vapor deposition of hexagonal boron nitride thick film on iron. *J. Cryst. Growth,* **1979**, *47*(2), 245-250.
[http://dx.doi.org/10.1016/0022-0248(79)90248-3]

[80]   Li, J.; Bernard, S.; Salles, V.; Gervais, C.; Miele, P. Preparation of polyborazylene-derived bulk boron nitride with tunable properties by warm-pressing and pressureless pyrolysis. *Chem. Mater.,* **2010**, *22*(6), 2010-2019.
[http://dx.doi.org/10.1021/cm902972p]

[81]   Babenko, V.; Lane, G.; Koos, A. A.; Murdock, A. T.; So, K.; Britton, J.; Meysami, S. S.; Moffat, J.; Grobert, N. Time dependent decomposition of ammonia borane for the controlled production of 2D hexagonal boron nitride. *Sci. Reports,* **2017**, *7*(1), 1-12.
[http://dx.doi.org/10.1038/s41598-017-14663-8]

[82]   Koepke, J.C.; Wood, J.D.; Chen, Y.; Schmucker, S.W.; Liu, X.; Chang, N.N.; Nienhaus, L.; Do, J.W.; Carrion, E.A.; Hewaparakrama, J.; Rangarajan, A.; Datye, I.; Mehta, R.; Haasch, R.T.; Gruebele, M.; Girolami, G.S.; Pop, E.; Lyding, J.W. Role of pressure in the growth of hexagonal boron nitride thin films from ammonia-borane. *Chem. Mater.,* **2016**, *28*(12), 4169-4179.
[http://dx.doi.org/10.1021/acs.chemmater.6b00396]

[83]   Zhang, C.; Hao, X.; Wu, Y.; Du, M. Synthesis of vertically aligned boron nitride nanosheets using CVD method. *Mater. Res. Bull.,* **2012**, *47*(9), 2277-2281.
[http://dx.doi.org/10.1016/j.materresbull.2012.05.042]

[84]   Tay, R.Y.; Li, H.; Tsang, S.H.; Zhu, M.; Loeblein, M.; Jing, L.; Leong, F.N.; Teo, E.H.T. Trimethylamine borane: A new single-source precursor for monolayer h-BN single crystals and h-BCN thin films. *Chem. Mater.,* **2016**, *28*(7), 2180-2190.
[http://dx.doi.org/10.1021/acs.chemmater.6b00114]

[85]   Qin, L.; Yu, J.; Li, M.; Liu, F.; Bai, X. Catalyst-free growth of mono- and few-atomic-layer boron nitride sheets by chemical vapor deposition. *Nanotechnology,* **2011**, *22*(21), 215602.
[http://dx.doi.org/10.1088/0957-4484/22/21/215602] [PMID: 21451227]

[86]   Orofeo, C.M.; Suzuki, S.; Kageshima, H.; Hibino, H. Growth and low-energy electron microscopy characterization of monolayer hexagonal boron nitride on epitaxial cobalt. *Nano Res.,* **2013**, *6*(5), 335-347.
[http://dx.doi.org/10.1007/s12274-013-0310-1]

[87]   Sutter, P.; Lahiri, J.; Zahl, P.; Wang, B.; Sutter, E. Scalable synthesis of uniform few-layer hexagonal boron nitride dielectric films. *Nano Lett.,* **2013**, *13*(1), 276-281.
[http://dx.doi.org/10.1021/nl304080y] [PMID: 23244762]

[88]   Xu, M.; Fujita, D.; Chen, H.; Hanagata, N. Formation of monolayer and few-layer hexagonal boron nitride nanosheets *via* surface segregation. *Nanoscale,* **2011**, *3*(7), 2854-2858.
[http://dx.doi.org/10.1039/c1nr10294j] [PMID: 21611645]

[89]   Glavin, N.R.; Jespersen, M.L.; Check, M.H.; Hu, J.; Hilton, A.M.; Fisher, T.S.; Voevodin, A.A. Synthesis of few-layer, large area hexagonal-boron nitride by pulsed laser deposition. *Thin Solid Films,* **2014**, *572*, 245-250.
[http://dx.doi.org/10.1016/j.tsf.2014.07.059]

[90]   Zhu, W.; Dai, B.; Wu, P.; Chao, Y.; Xiong, J.; Xun, S.; Li, H.; Li, H. Graphene-analogue hexagonal bn supported with tungsten-based ionic liquid for oxidative desulfurization of fuels. *ACS Sustain. Chem. Eng.,* **2015**, *3*(1), 186-194.
[http://dx.doi.org/10.1021/sc5006928]

[91]   Li, H.; Tay, R.Y.; Tsang, S.H.; Liu, W.; Teo, E.H.T. Reduced graphene oxide/boron nitride composite film as a novel binder-free anode for lithium ion batteries with enhanced performances. *Electrochim. Acta,* **2015**, *166*, 197-205.
[http://dx.doi.org/10.1016/j.electacta.2015.03.109]

[92]   Ergen, O. Hexagonal boron nitride incorporation to achieve high performance $Li_4Ti_5O_{12}$ electrodes. *AIP Adv.,* **2020**, *10*(4), 045040.
[http://dx.doi.org/10.1063/5.0004376]

[93]   Wang, T.; Zhang, S.; Yin, L.; Li, C.; Xia, C.; An, Y.; Wei, S. Silicene/boron nitride heterostructure for the design of highly efficient anode materials in lithium-ion battery. *J. Phys. Condens. Matter,* **2020**, *32*(35), 355502.
[http://dx.doi.org/10.1088/1361-648X/ab8c8c] [PMID: 32325446]

[94]   Waqas, M.; Ali, S.; Lv, W.; Chen, D.; Boateng, B.; He, W. High-performance PE-BN/PVDF-HFP bilayer separator for lithium-ion batteries. *Adv. Mater. Interfaces,* **2019**, *6*(1), 1801330.
[http://dx.doi.org/10.1002/admi.201801330]

[95]   de Moraes, A.C.M.; Hyun, W.J.; Luu, N.S.; Lim, J.M.; Park, K.Y.; Hersam, M.C. Phase-inversion polymer composite separators based on hexagonal boron nitride nanosheets for high-temperature lithium-ion batteries. *ACS Appl. Mater. Interfaces,* **2020**, *12*(7), 8107-8114.
[http://dx.doi.org/10.1021/acsami.9b18134] [PMID: 31973532]

[96]   Venkateswarlu, G.; Madhu, D.; Rani, J.V. Graphene supported boron nitride nanosheets as advanced electroanalytical performance for rechargeable magnesium storage system. *ChemistrySelect,* **2020**, *5*(7), 2247-2254.
[http://dx.doi.org/10.1002/slct.201904872]

[97]   Hu, J.; Yue, M.; Zhang, H.; Yuan, Z.; Li, X. A Boron nitride nanosheets composite membrane for a long-life zinc-based flow battery. *Angew. Chem. Int. Ed.,* **2020**, *59*(17), 6715-6719.
[http://dx.doi.org/10.1002/anie.201914819] [PMID: 32022372]

[98]   Gao, T.; Gong, L.; Wang, Z.; Yang, Z.; Pan, W.; He, L.; Zhang, J.; Ou, E.; Xiong, Y.; Xu, W. Boron nitride/reduced graphene oxide nanocomposites as supercapacitors electrodes. *Mater. Lett.,* **2015**, *159*, 54-57.
[http://dx.doi.org/10.1016/j.matlet.2015.06.072]

[99]   Saha, S.; Jana, M.; Samanta, P.; Murmu, N.C.; Kim, N.H.; Kuila, T.; Lee, J.H. Investigation of band structure and electrochemical properties of h-BN/rGO composites for asymmetric supercapacitor applications. *Mater. Chem. Phys.,* **2017**, *190*, 153-165.
[http://dx.doi.org/10.1016/j.matchemphys.2017.01.025]

[100]   Maity, C.K.; Goswami, N.; Verma, K.; Sahoo, S.; Nayak, G.C. A facile synthesis of boron nitride supported zinc cobalt sulfide nano hybrid as high-performance pseudocapacitive electrode material for asymmetric supercapacitors. *J. Energy Storage,* **2020**, *32*, 101993.
[http://dx.doi.org/10.1016/j.est.2020.101993]

[101] Parthiban, V.; Sahu, A.K. Performance enhancement of direct methanol fuel cells using a methanol barrier boron nitride-Nafion hybrid membrane. *New J. Chem.,* **2020**, *44*(18), 7338-7349.
[http://dx.doi.org/10.1039/D0NJ00433B]

[102] Mahalingam, S.; Ayyaru, S.; Ahn, Y.H. Enhanced cathode performance of Fe_2O_3, boron nitride-doped rGO nanosheets for microbial fuel cell applications. *Sustain. Energy Fuels,* **2020**, *4*(3), 1454-1468.
[http://dx.doi.org/10.1039/C9SE01243E]

[103] Bull, S.K.; Champ, T.A.; Raj, S.V.; O'Brien, R.C.; Musgrave, C.B.; Weimer, A.W. Atomic layer deposited boron nitride nanoscale films act as high temperature hydrogen barriers. *Appl. Surf. Sci.,* **2021**, *565*, 150428.
[http://dx.doi.org/10.1016/j.apsusc.2021.150428]

[104] Cho, A.J.; Kwon, J.Y. Hexagonal boron nitride for surface passivation of two-dimensional van der waals heterojunction solar cells. *ACS Appl. Mater. Interfaces,* **2019**, *11*(43), 39765-39771.
[http://dx.doi.org/10.1021/acsami.9b11219] [PMID: 31577117]

[105] Meng, J.H.; Liu, X.; Zhang, X.W.; Zhang, Y.; Wang, H.L.; Yin, Z.G.; Zhang, Y.Z.; Liu, H.; You, J.B.; Yan, H. Interface engineering for highly efficient graphene-on-silicon Schottky junction solar cells by introducing a hexagonal boron nitride interlayer. *Nano Energy,* **2016**, *28*, 44-50.
[http://dx.doi.org/10.1016/j.nanoen.2016.08.028]

[106] Han, D.; Zhang, J.; Weng, Z.; Kong, D.; Tao, Y.; Ding, F.; Ruan, D.; Yang, Q.H. Two-dimensional materials for lithium/sodium-ion capacitors. *Mater. Today Energy,* **2019**, *11*, 30-45.
[http://dx.doi.org/10.1016/j.mtener.2018.10.013]

[107] Yu, L.; Miao, J.; Jin, Y.; Lin, J.Y.S. A comparative study on polypropylene separators coated with different inorganic materials for lithium-ion batteries. *Front. Chem. Sci. Eng.,* **2017**, *11*(3), 346-352.
[http://dx.doi.org/10.1007/s11705-017-1648-9]

[108] Rahman, M.M.; Mateti, S.; Cai, Q.; Sultana, I.; Fan, Y.; Wang, X.; Hou, C.; Chen, Y. High temperature and high rate lithium-ion batteries with boron nitride nanotubes coated polypropylene separators. *Energy Storage Mater.,* **2019**, *19*, 352-359.
[http://dx.doi.org/10.1016/j.ensm.2019.03.027]

[109] Xue, T.; Fan, H.J. From aqueous Zn-ion battery to $Zn-MnO_2$ flow battery: A brief story. *J. Energy Chem.,* **2021**, *54*, 194-201.
[http://dx.doi.org/10.1016/j.jechem.2020.05.056]

[110] Angizi, S.; Alem, S. A. A.; Pakdel, A. Towards integration of two-dimensional hexagonal Boron Nitride (2D h-BN) in energy conversion and storage devices. *Energies,* **2022**, *15*(3), 1162.
[http://dx.doi.org/10.3390/en15031162]

[111] Hu, J.; Xie, K.; Liu, X.; Guo, S.; Shen, C.; Liu, X.; Li, X.; Wang, J.; Wei, B. Dramatically enhanced ion conductivity of gel polymer electrolyte for supercapacitor *via* h-BN nanosheets doping. *Electrochim. Acta,* **2017**, *227*, 455-461.
[http://dx.doi.org/10.1016/j.electacta.2017.01.045]

[112] Zuo, Y.; Sheng, W.; Tao, W.; Li, Z. Direct methanol fuel cells system–A review of dual-role electrocatalysts for oxygen reduction and methanol oxidation. *J. Mater. Sci. Technol.,* **2022**, *114*, 29-41.
[http://dx.doi.org/10.1016/j.jmst.2021.10.031]

[113] Gouda, M. H.; Tamer, T. M.; Konsowa, A. H.; Farag, H. A.; Eldin, M. S. M. Organic-inorganic novel green cation exchange membranes for direct methanol fuel cells. *Energies,* **2021**, *14*(15), 4686.
[http://dx.doi.org/10.3390/en14154686]

[114] Kregar, A.; Frühwirt, P.; Ritzberger, D.; Jakubek, S.; Katrašnik, T.; Gescheidt, G. Sensitivity based order reduction of a chemical membrane degradation model for low-temperature proton exchange membrane fuel cells. *Energies,* **2020**, *13*(21), 5611.
[http://dx.doi.org/10.3390/en13215611]

[115] Mahato, N.; Banerjee, A.; Gupta, A.; Omar, S.; Balani, K. Progress in material selection for solid oxide fuel cell technology: A review. In: *Progress in Materials Science*; Elsevier, **2018**. [http://dx.doi.org/10.1016/j.pmatsci.2015.01.001]

[116] Sahoo, N.G.; Sandeep, M. A process of Manufacturing. Indian Patent 352780, 2016.

[117] Punetha, V.D.; Ha, Y.M.; Kim, Y.O.; Jung, Y.C.; Cho, J.W. Rapid remote actuation in shape memory hyperbranched polyurethane composites using cross-linked photothermal reduced graphene oxide networks. *Sens. Actuators B Chem.*, **2020**, *321*, 128468. [http://dx.doi.org/10.1016/j.snb.2020.128468]

[118] Kumar, S.; Singh, P.K.; Punetha, V.D.; Singh, A.; Strzałkowski, K.; Singh, D.; Yahya, M.Z.A.; Savilov, S.V.; Dhapola, P.S.; Singh, M.K. In-situ N/O-heteroatom enriched micro-/mesoporous activated carbon derived from natural waste honeycomb and paper wasp hive and its application in quasi-solid-state supercapacitor. *J. Energy Storage,* **2023**, *72*, 108722. [http://dx.doi.org/10.1016/j.est.2023.108722]

CHAPTER 6

Functionalization Strategies and Applications of Two-Dimensional Boron Nitride

Mayank Punetha[1,*], Abbas Zaarifi[2], Anton Kuzmin[3,4] and **Sadafara Pillai[5]**

[1] *Centre of Excellence for Research, P.P. Savani University, Surat-394125, Gujarat, India*

[2] *Department of Physics, Yasouj University, Yasouj, 75918-74934, Iran*

[3] *Scientific Laboratory "Advanced Composite Materials and Technologies," Plekhanov Russian University of Economics, Stremyanny Ln, 36, 117997, Moscow, Russia*

[4] *Department of Mechanization of Agricultural Products Processing, National Research Mordovia State, University, Bolshevistskaya st., 68, 430005, Saransk, Republic of Mordovia, Russia*

[5] *School of Science, P.P. Savani University, Surat-394125, Gujarat, India*

Abstract: Recently, 2D Boron Nitride (BN) and its derivatives have emerged as materials of great interest due to their intriguing structure, similar to graphene, and possessing remarkable physical, chemical, and optoelectronic properties. BN has shown great applications in various fields, including electronics, energy storage and conversion, advanced composites, lubricants, and many more. Moreover, the hybrid materials of 2D BN with graphene and other nanomaterials have evolved as excellent dielectric substrates widely used in electronic devices. However, the extensive application of this material is severely restricted for various reasons. The book chapter elaborates different 2D BN nanostructures with a focused view on their striking applications. The mechanistic aspects of surface revamping through covalent functionalization have been discussed for the readers' comprehensive overview and a concise discussion on the challenges associated with this. The book chapter reviews the application of BN in electronics, biomedical applications, and smart composites in depth. This book chapter will provide a comprehensive outlook to the readers in understanding the recent and significant epistemological evidence.

Keywords: BN heterostructures, Composites, Drug delivery, Electronic, Functionalization, Microbial, Nanomaterial, Nano-ribbons, 2D Boron Nitride.

INTRODUCTION: HISTORY OF 2D BNS

Two-dimensional (2D) materials have become an intriguing class of materials with special characteristics and fundamental differences from their bulk analogs. 2D nanomaterials attracted the significant attention of the worldwide research

* Corresponding author Mayank Punetha: Centre of Excellence for Research, P.P. Savani University, Surat-394125, Gujarat, India; E-mail: Mayank.punetha@ppsu.ac.in

community and has been considered as a frontrunner in shaping the fourth industrial revolution [1]. Now after almost two decades since this discovery, various new members have been added to this clan of 2D materials because of its characteristic bandgap, great thermal stability, and exceptional dielectric features; 2D BN nanosheets have gained significant attention from the research community [2]. The first synthesis of BN was reported back in 1959 by heating boron (B) and nitrogen (N) in an electric arc furnace [3]. The BN sheets have shown tremendous potential for their use as a substrate material in graphene-based electronics due to their high thermal stability [4, 5]. The first successful graphene transfer onto a BN substrate was announced in 2010 by the University of Manchester researchers. In addition, properties such as the high thermal stability of BN sheets offer the scope of broader applications such as super flat and insulating surfaces for graphene that prevent charge carrier dispersion [6, 7]. The interatomic distance between 'B' and 'N' is 1.45 Å, while inter planer distance between sheets in bulk BN is near 3.33 Å (Fig. **1**). The 2D BN has an average Young's modulus of 0.865TPa with a fracture strength of 70.5GPa. Unlike graphene, the strength of bulk BN and 2D BN remains relatively consistent and does not exhibit a direct correlation between resilience and the number of layers [4, 5].

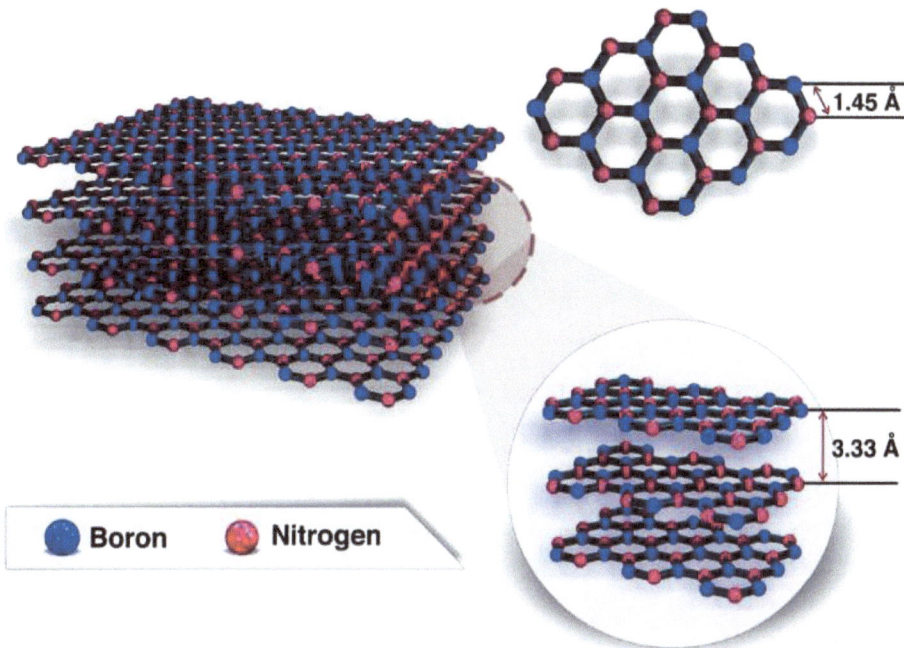

Fig. (1). Two-dimensional structure of BN, showing flat 2D sheet, stacked sheets, and zoomed view of lattice.

The synthesis of BN nanosheets can be achieved by employing diverse techniques, such as high-temperature annealing, chemical vapor deposition (CVD) [8], and liquid-phase exfoliation (LFE) [9]. In recent years, multiple groups have reported these methods with high-quality 2D BN synthesis [4]. The LFE has been the most popular method for creating BN nanosheets which helps synthesize BN of desired structural parameters. Recent research works are focused more on investigations related to the outstanding mechanical electrical, and optical properties of this material [10, 11] that can be elaborated for similar studies conducted on graphene-related materials [12 - 15]. The excellent mechanical superlatives, such as high Young's modulus, strength, and unique optoelectronic properties, make it a promising material for fabricating high-quality, flexible, and wearable electronics [10].

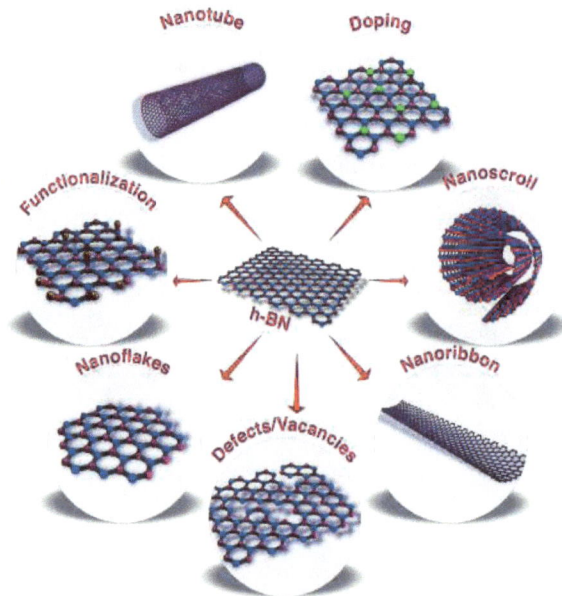

Fig. (2). Nano-structured derivatives of 2D BNs.

Moreover, properties such as excellent thermal conductivity and exceptional stability at high temperatures suggest their potential use in advanced materials for thermal control in high-end electronic devices [16]. A wide range of nano BN-based materials have been discovered as of now, which include 2D BN sheets, Doped 2D BN sheets [17], functionalized 2D BN sheets [18], BN nano-scrolls, BN nanoflakes and BN nano-ribbons (Fig. **2**). However, their widespread application is limited by the limited scope of surface revamping. Significant progress has been made by the research community in recent times in revamping BN nano surfaces to exploit their properties. Covalent or non-covalent pathways can update the surface of the 2D BN with possibilities of optimization in their properties. The advancement in the structural revamping of 2D BN has positioned it well to play a significant part in advancing future technologies. The following chapter has been dedicated to science and challenges in surface revamping of the 2D BNs [19 - 22].

FUNCTIONALIZATION OF 2D BNS AND ASSOCIATED CHALLENGES

To enter into the details of the functionalization of 2D BN, it is vital to understand the basic structure that makes the revamping so essential and challenging. A set of comparative arguments between graphene and 2D BN can make this discussion more perceptible [19]. BN and graphene exhibit 2D structures with significant differences in reactivity toward various functional groups. These differences are mainly due to the difference in the fundamental atomic structures of 'C,' 'B,' and 'N .' In case of graphene, the functionalization is more pronounced and feasible, while in BN 2D structures, it is complex and rare [19]. The surface revamping in the case of graphene is also supported by extensive established literature for mechanistic detailing of C, while the mechanism for B and N has yet to be thoroughly explored. In addition, in the case of graphene, the functionalization is favored by certain factors, such as bond formation between 'C' atoms; zero difference in the electronegativity limits the scope of selectivity of the attacking group [21].

Moreover, the presence of hydrogen atoms on the graphene surface mandates electrophilic substitution on the surface. Contrary to this, 2D 'BN' structures are the alternative arrangement of 'B' and 'N' atoms with significant differences in their electronegativity. This brings selectivity to play an important role. In addition, the absence of a leaving group restricts electrophilic attack on the surface and stimulates reaction to proceed through the addition mechanism. However, the attacking group does not need to attack at two adjacent positions, as resonance can separate the charges formed due to the attack of a reagent to different positions [22]. The heterolytic bond fission between 'B' and 'N' atoms simultaneously develops charges on them. This restricts the reaction from proceeding in the forward direction, as it will leave the charged species on the basal plane, making the overall change thermodynamically non-spontaneous. To balance the charge, subsequent and simultaneous attack has to take place. Being one of the three most electronegative elements of the periodic table after F and O, Nitrogen atoms bear a negative charge when bonded to B in the 'BN' structures. Due to these reasons, 'B' seems susceptible to the attack of nucleophiles, while electrophilic species prefer to attack the electronegative 'N' atoms. There are limited functional groups successfully introduced on the 'BN' 2D surface due to these structural reasons [21]. The most studied functionalized derivatives of 2D

'BN' consist of functional groups containing –OH (hydroxy), -R (alkoxy), and –NH$_2$ (amine). However, further revamping can be done with the help of these groups [7, 8, 17, 18, 22]

The functionalization of 2D 'BN' alters many of its physiochemical as well as optoelectronic properties of it and is frequently used in material optimization for various applications. One of the hurdles in the widespread use of 2D 'BN' is its poor dispersibility, which hampers its wide applications [6, 9, 22]. In addition, extensive use in composite fabrication to exploit its exceptional stability is hindered due to the tendency to aggregate in the polymer matrix. The most widely functionalized moiety on the 2D 'BN' surface is the –OH and –NH$_2$ group. The structure provides a scope for further subsequent revamping through its reactivity and offers excellent opportunities to disperse these materials in various matrix systems [22].

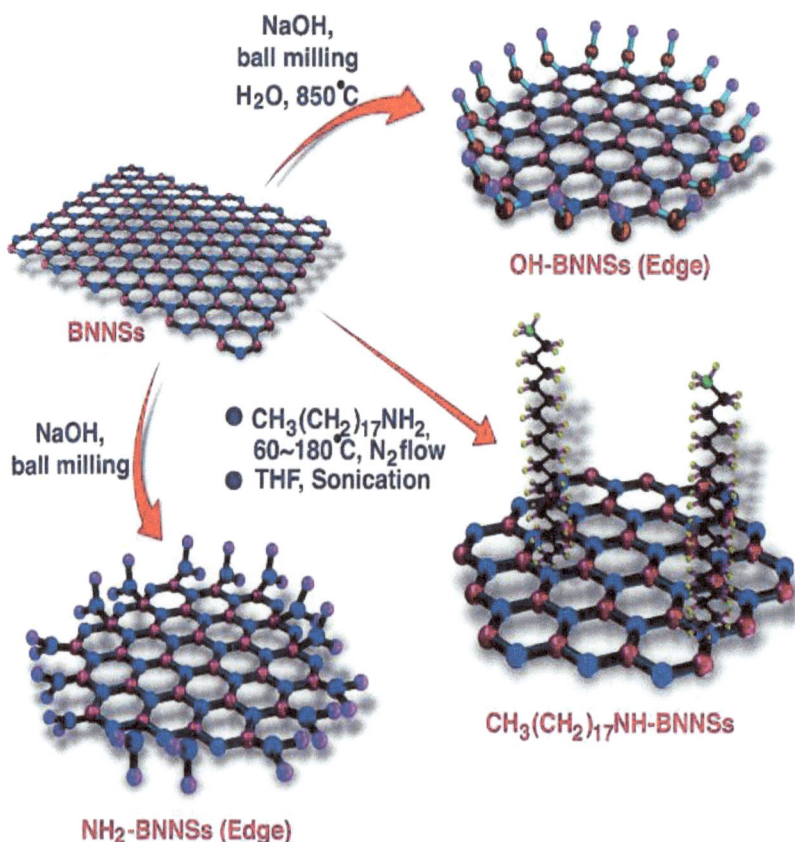

Fig. (3). Single-step functionalization of 2D BN.

Hydroxyl (–OH) Functionalized 2D 'BN'

The chemical strategy to functionalize the 'BN' derivatives depends a lot on the structural features of the precursor. If sheets are stacked, the first step is to exfoliate them to provide a larger surface area for attacking species with more reaction centers. In some cases, the –OH groups can be introduced on 2D 'BN' surface by direct sonication in aqueous media, while in some cases, active -OH donating reagents are required [23]. Other methods widely use acid peroxide mixtures such as sulphuric acid (H_2SO_4 and H_2O_2). Lin *et al.* synthesized –OH functionalized 'BN' derivatives by sonicating the aqueous solution of 'BN' stacks. The functionalized material exhibited enhanced dispersibility in the aqueous media. Coleman *et al.* synthesized –alkyl functionalized 2D 'BN' by using t-butoxy groups and hydrolyzed the product of this reaction with H_2SO_4/H_2O_2 to break the ethereal linkage [23]. Lee *et al.* developed a method for the bulk synthesis of –OH functionalized 'BN' derivatives using hydroxide-assistant ball milling. This method used chemical and mechanical actions to provide a few layered –OH functionalized sheets. The reagent in this process furnishes the –OH, while the sheer force generated by the ball mills separates the stacked layers by applying mechanical pressure on the surfaces. Various single-step functionalizations of 2D BN have been illustrated in the Fig. (**3**).

Amino (–NH$_2$) and Alkyl (-R) Functionalized 2D 'BN'

As ball mill assisted –OH functionalization, Lei and Chen *et al.* developed a method for ball mill administered –NH$_2$ functionalized 2D BNs using urea as the precursor. Due to the amino group's capacity to exhibit hydrogen bonding, the functionalized material exhibited great dispersion in water and enhanced solubility. Moreover, the dispersion of NH$_2$-f-2D 'BN' gave colloidal solutions in water. The –NH$_2$ group can further be used to extend the scope of covalent extensions on the 'BN' surface. The direct alkylation of 2D 'BN' structures is very difficult. However, the alkyl group can be substituted if a group for substitution can be installed on the surface, which can act as a nucleofuge as the alkyl group attacks. Sainsbury and his group developed an innovative pathway to introduce –alkyl groups on the 'BN' surface. They introduced dibromo carbene into the 'BN' bonds through incersion and substituted –Br with desired –alkyl group as it acts as an excellent leaving group [24].

Other Functional Groups

Due to their high reactivity, the –OH and –NH$_2$ groups can further be used to functionalize other polymeric chains *via* two-step functionalization methods (Fig. **4**). Such as –OH and -NH$_2$ groups can participate in the formation of ester or peptide linkages on reacting with –COOH bearing structures. Organic acids such as $CF_3CF_2CF_2COOH$ react readily with the $-OH$ functionalized 2D 'BNs'. The fluorine substitution of acids is used to enhance the acidic nature of the respective acids. When $-COOH$ group releases proton, which forms a negatively charged carboxylate ion. Stabilization of this carboxylate ion increases with an electron-withdrawing group like fluorine. Sainsbury and Zhi synthesized esterified 2D 'BNs' using 2D 'BN' and substituting carboxylic acid as a precursor. Huang *et al.* reported polyhedral oligo silsesquioxane (POSS) functionalized 2D 'BNs' [25, 26].

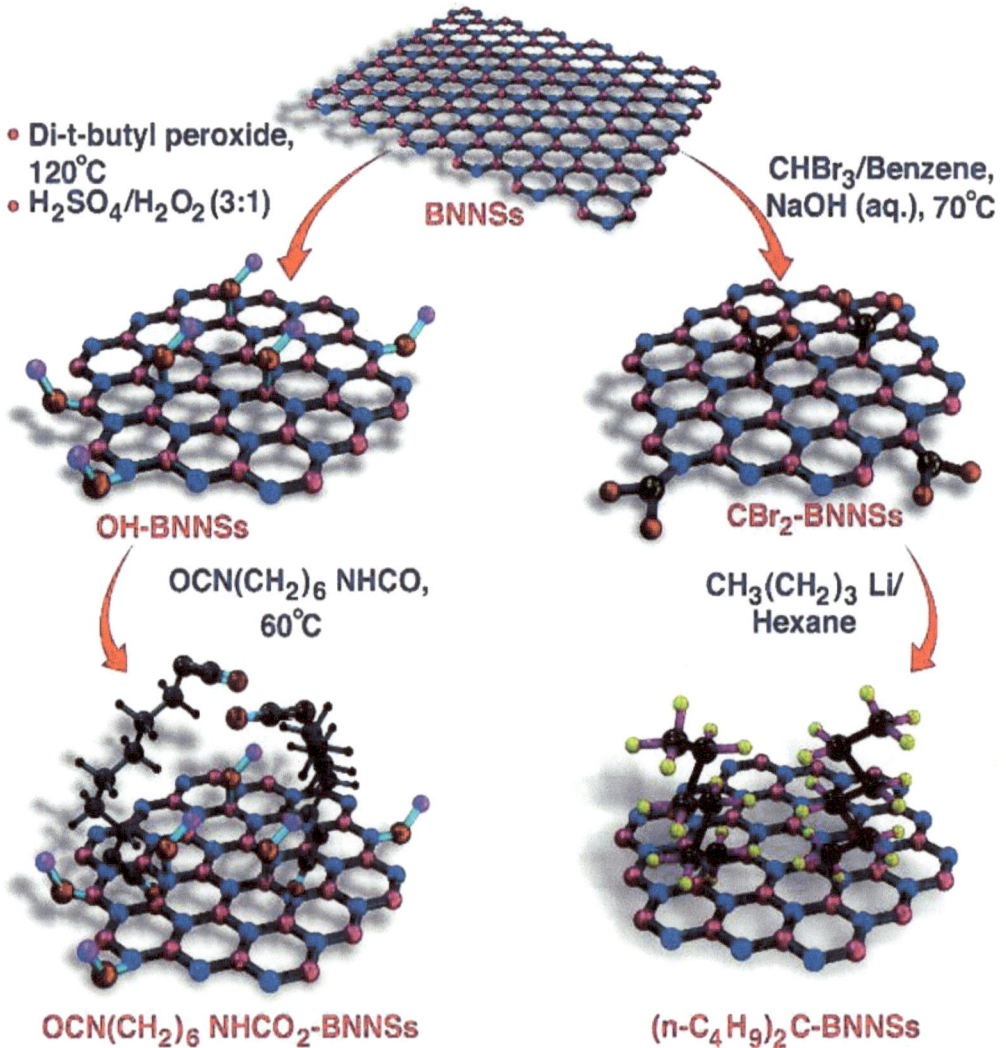

Fig. (4). Two-step functionalization of 2D BN.

APPLICATIONS OF 2D 'BN' IN ENERGY APPLICATIONS

2D 'BNs' have shown tremendous potential to revolutionize the existing scaffolds of electronics. A wide range of properties associated with this material, such as exceptional thermal stability and conduction, particularly its dielectric properties and a bandgap of ~5.49 eV, have shown many opportunities in their potential use in electronics. It is called graphene due to its great potential in revamping modern electronics. A broad array of epistemological evidence indicates that 'BN' derivatives could be utilized in numerous ways, including the innovation of electrolyte compositions with 2D BNs, use as separators, excellent substrates for doping, ion-conducting channels, fillers with high thermal resistance, and much more. The BN derivative-based electrolytes are found to increase oxidation resistance. Various ionic liquids and 'BN' based electrolyte compositions allow Li-ion batteries to operate efficiently over a wide temperature range and as high as 150 °C [17, 24, 25]. In addition, coating the 2D layer of 'BN' on anodes prevents the growth and erosion of the lithium electrode in various organic electrolytes. The protective coating of BN derivatives offers superior mechanical support to the sandwiched Li and shows smooth electrochemical activities with excellent performance. The protected electrodes exhibited stable and high cycling over fifty cycles, along with a high columbic efficiency of ~97% [27, 28].

Applications of 2D 'BN' in the Fabrication of Advanced Composite Materials

2D 'BNs' have found many other applications in fabricating various new innovative materials. Their exceptional mechanical strength, inertness toward different chemicals, ability to ear extreme pressure and temperature, and rapid heat evacuation, Thin h-BN is an applied technology that has been proven effective in protecting the underlying material from exposure to various elements by its chemical inertness and stability even at the high temperatures [29].

Certain factors affect the superiority of 2D 'BNs' as reinforcement in the fabrication of the composites. The ratio of the length of sheets and thickness plays an essential role in the interaction of polymer matrix and BN-based reinforcement, also known as aspect ratio. In addition, chemical inertness in 2D BNs provides exceptional dispersion of the fillers in the matrix [30]. The synergic action of these two properties is a chance to effectively stress-transfer between 2D sheets and the matrix. Moreover, the 2D BN sheets are aligned well in the matrix and exhibit effective reinforcement of properties. When an external load is applied to the reinforced composite, stress dispersion occurs from the matrix to the mechanically robust support, which increases the overall mechanical strength [29, 13]. The matrix and reinforcements are called two different phases, and the stress dispersion occurs from the mechanically weak phase to the mechanically robust phase. The boundary separating these phases is known as the phase boundary. The effective stress dispersion is directly proportional to the phase boundary and offers excellent load transfer [13]. Better distribution creates significant phase boundaries, which help in the effective stress dispersion between the phases. However, the chemical inertness of the 2D BN sheets prevents important phase boundaries due to poor interaction between the phases, but this can be worked by revamping the 2D BN surface with suitable functional groups. The functional groups enhance interaction between phases providing opportunities to create more significant phase boundaries resulting in the effective stress dispersion and overall enhancement of the mechanical properties. Moreover, stacking is also prevented by the mode of functionalization and allows the formation of more extensive phase boundaries [15].

In addition, the effective dispersion of the 2D BN sheets also offers compelling resonance between matrix properties and reinforcement. These properties include thermal, electrical, and optoelectronic properties. The thermal properties of 2D BN-reinforced composites are generally superior to that of pure polymer due to the better thermal resistance of BN sheets. Effective reinforcement affects properties such as thermal stability, thermal conductivity, and the composites', most importantly, glass transition temperature in the absence of free electrons [31]. As BN crystals do not have free electrons, the atomic vibration in the skeleton plays a crucial role in transferring heat or phonons. In polymers, in general, the polymer chains are distributed randomly, mainly following a random arrangement. One of the most crucial aspects of using BN as reinforcement in the polymer matrix comes from enhanced thermal stability. In general, polymers are highly susceptible to thermal degradation. This thermal degradation is an inherent characteristic of the polymers. When BN derivatives are used as reinforcement, they significantly reduce the degradation process. The mechanism involves the role of BN sheets as a barrier to the oxygen molecules inside the polymer matrix. In the absence of oxygen, the most spontaneous oxidation phenomenon does not occur, and oxide formation is slowed, resulting in enhanced thermal stability. Simultaneously, the newly formed gases could not escape from the matrix due to the barrier caused by the BN sheets. Overall, this reduces the oxygen concentration; by not letting oxygen gas enter; and, second, not allowing oxidized fragments of carbon chains to move out.

In addition, the impermeability of a 2D skeleton plays a crucial role in enhancing the thermal degradation process. It has been observed that the 2D sheets are impermeable to gases. This enhances the path gas needs to travel to come in or move out, resulting in decreased thermal degradation. From the discussion, a higher aspect ratio is required to enhance thermal stability. The higher the surface area of BN sheets, the greater the barrier for gases to move and the greater the strength [32]. Another exciting application of BN comes from its great utility in fabricating composites for power electronics. Dielectric polymers are generally used for such applications as they exhibit better performance at higher voltage.

Biomedical Applications of 2D BN

The crucial parameters to develop effective drug carriers require fulfilling certain conditions before use. These include effective loading of the drug, low toxicity in the biological atmosphere, identification of the target, and effective loading [33]. Owing to these properties, 2D BN has recently attracted significant interest as a drug delivery agent and an anticancer drug such as doxorubicin [7] for its anticancer potential. One of the most remarkable properties of BN is that developing it with porous morphology strengthens its potential as an effective drug-delivery carrier [34 - 37, 42]. The porous form of BN can be synthesized by using various physical methods and materials. Among the most essential materials are different activated carbon forms, SiO_2, zeolites, graphene and derivatives, and many more. However, using BN structures as an effective drug-delivery tool faces some key challenges. The associated challenges can be categorized into two major segments. (1) The issues about the hydrophobicity and (2) The biocompatibility of this material [38 - 43].

It is established that 2D BN structures exhibit severe hydrophobicity; this water-repelling nature of BN restricts its broader application in the biological atmosphere where H_2O acts as an active natural solvent. The poor dispersion of BN sheets in water affects the state of dispersion and causes agglomeration. Addressing these challenges is a prerequisite before materializing this substance's full potential and scope in active drug delivery. Similar to graphene-based delivery systems, 2D BNs exhibit tremendous solubility transformation scope *via* surface functionalization. Functionalization of 2D BN surface with moieties having an affinity for the water can address these challenges effectively.

In comparison to the solubility of the graphene derivatives, BN derivatives have shown some promising results. In general, it is a well-known fact that surface functionalization with guest molecules such as –OH, -COOH, and -NH$_2$ enhances the solubility in water. In particular, the amine-functionalized BN sheets exhibit the remarkable potential to show promising results with enhanced solubility of 32mg/mL from 2mg/mL [26].

Biocompatibility is the second key issue in using boron nitride in biological applications. Its greater applicability in various domains, such as coating agents, drug delivery vehicles, and antimicrobial material, depends on the point of being nontoxic or injurious and capable of escaping immunological rejections [45 - 47]. Several epistemological pieces of evidence exist that these materials are relatively safe for such uses. The biocompatibility of these materials vastly depends upon the surface morphology and size of the sheets. Mateti *et al.* first investigated the possible relation between the dimension of the nanosheets and their biocompatibility [48]. To synthesize the materials of different morphologies, they opted for ball milling in the presence of argon and ammonia gas. The process was focused on creating two classes on BN sheets with higher (100 nm) thickness and lower (3 nm) thickness. The biocompatibility and its relation with the size and morphology were evaluated using the developed material against the human osteosarcoma cells. The results suggest an association between biocompatibility and the size of BN sheets. The results conclude that as the size of the BN sheets increases, their biocompatibility also increases, and suggested precautions are required in using the BN sheets of small dimensions. The group further investigated the mechanistic aspects of BN-induced toxicity at the low thickness and high surface area. It was concluded that the higher surface area of the sheets controls the generation of reactive species. The higher surface area of the sheets makes the surface reactive, primarily due to the presence of boron atoms on its surface [49]. A summary of the application of BN derivatives has been given in Table **1**.

Table 1. Applications of 2D boron nitride as a drug delivery agent.

S. NO.	Boron Nitride /Derivative	Application	Refs.
1.	2DBN/CpG oligonucleotides	DD Agent	[35]
2.	2DBN /CpG oligonucleotides	DD Agent	[36]
3.	2DBN -Dox	DD Agent	[33]
4.	2DBN –Folic Acid/Dox	DD Agent	[34]
5.	2DBN nanoparticles	DD Agent	[40]
6.	2DBN	Sensing	[41]
7.	Porous-2DBN	DD Agent	[42]
8.	BN NPs	DD Agent	[43]
9.	Porous BN	Antimicrobial	[44]
10.	2D BN	Antioxidant	[45]
11.	BN derivative	Chemotherapy	[46]

*cpg – cytosine guanine linked with phosphate, Dox- Doxorubicin, Cyst- Cysteine, NPs- Nanoparticles, DD- drug delivery

The wonder 2D BN sheets have also found significant applications in various other applications. Some of the conclusions made by researchers need to be more consistent. Ikram *et al.*, in their research work, reported that the BN Nanosheets could exhibit antimicrobial activity against *E. coli* and *S. aureus*, particularly more against the former [50]. However, Parra *et al.* suggested the BN surface to be harmless against the bacterial cultures and exploited this property to eradicate bacteria-induced bio-decomposition (bio-corrosion) of the copper substrate by E. *coli*. His study suggests that the bacteria viability on the surface of boron nitride increases by 18%, though the sheets do not degrade and remain intact. This phenomenon invited the application of BN sheets in the antimicrobial coating, as the structure stands intact against the bacterial strains. The same group investigated the applications of the BN coating in the successful bio-corrosion prevention of copper sheets. In addition, several other derivatives of BN sheets have been investigated for antimicrobial performance [51]. Huang *et al.* studied the antibacterial behavior of the doped BN sheets. His group introduced host Zr atoms on the BN surface by doping and found that the doped host entities enhanced the antimicrobial performance of the sheets. The antibacterial performance of the 2D BN sheets dramatically depends on several parameters such as surface area, shape and surface energy, and extent of doping. For the Zr-doped material, the antimicrobial inhibition was directly proportional to the extent of surface coverage by the host atoms [52]. Moreover, the applications of 2D BNs have been slowly but steadily taking place in several other areas. A summary of significant applications has been summarized in Table **2**.

Table 2. Applications of boron nitride in bioengineering.

S. NO.	Boron Nitride /Derivative	Application	Refs.
1.	2DBN nanoparticles/PPF	Bio-engineering	[53]
2.	2DBN s/AKM scaffolds	Bio-engineering	[54]
3.	2DBN /GESM	Bio-engineering	[55]
4.	Hydroxyl-2DBN /PVA	Bio-engineering	[56]
5.	2DBN/SiH$_4$	Bio-engineering	[57]
6.	2DBN nanoparticles	Skin formulations	[58]
7.	2DBN	Biomedicine	[59]
8.	2D BN	Lubricant	[60]
9.	2DBN	Lubricant	[61]
10.	2DBN	Semiconductor	[62]
11.	TiO$_2$-f-2DBN	Nanofiber	[63]
12.	Porous 2D BN	Water Purification	[64]
13.	2DBN	Catalysis	[65]
14.	2DBN	H$_2$ generation	[66]
15.	2DBN	Water purification	[67]
16.	2D BN	Water Purification	[68]

(Table 2) cont.....

S. NO.	Boron Nitride /Derivative	Application	Refs.
17.	TiO$_2$-f-2DBN	Photo-catalysis	[69]

CONCLUSION

Because of their excellent physicochemical characteristics, 2D-BNs are regarded as one of the most promising candidates for a variety of applications. The most significant progress can be seen in developing new and innovative drug and gene delivery methods, advanced bioengineering materials, pharmaceutics, and many more. Surface functionalization offers unique opportunities to innovate applications of 2D BN in several other applications. However, functionalization and synthesis of BN sheets remain critical to exploit the maximum of its potential. The synthesis remains challenging as several quality parameters, such as crystallinity, shape, size, and purity, depend on the precursors and synthesis mode. Large-scale commercial production of 2D BNs is still facing several challenges, as it requires high temperatures, making the synthesis cost ineffective and environmentally harmful. However, few attempts have been made to synthesize the 2D BNs at low temperatures, but significant progress has yet to be made. The other challenges include limited scope and improvement in its surface revamping. The hydrophobic nature and chemically inert surface of BN sheets restrict its wide use in several applications such as drug delivery, gene delivery, composite fabrication, *etc.* The theoretical work using quantum calculations can be of great significance in ascertaining chemical behavior and properties of its surface. The theoretical analysis and study of detailed reaction mechanisms, reaction kinetics, and energetics of involved chemical processes can open a wide range of opportunities for synthetic chemists to deliver new wonder BN derivatives, which can be used in addressing several other challenges to humanity.

ACKNOWLEDGEMENTS

We would like to acknowledge Ms. Himani Pant, ChemX VV for her valuable editorial support in the preparation of this document.

REFERENCES

[1]   Punetha, V.D.; Rana, S.; Yoo, H.J.; Chaurasia, A.; McLeskey, J.T., Jr; Ramasamy, M.S.; Sahoo, N.G.; Cho, J.W. Functionalization of carbon nanomaterials for advanced polymer nanocomposites: A comparison study between CNT and graphene. *Prog. Polym. Sci.,* **2017**, *67*, 1-47.
        [http://dx.doi.org/10.1016/j.progpolymsci.2016.12.010]

[2]   Arenal, R.; Lopez-Bezanilla, A. Boron nitride materials: An overview from 0D to 3D (nano)structures. *Wiley Interdiscip. Rev. Comput. Mol. Sci.,* **2015**, *5*(4), 299-309.
        [http://dx.doi.org/10.1002/wcms.1219]

[3]   Kakarla, A.B.; Kong, I. *In vitro* and *In vivo* Cytotoxicity of Boron Nitride Nanotubes: A Systematic Review. *Nanomaterials.,* **2022**, *12*(12), 2069.

[http://dx.doi.org/10.3390/nano12122069]

[4]   Li, S.; Li, H.; Qu, L. Preparation of hydroxylated boron nitride-modified epoxy acrylate emulsion and its application in waterborne anticorrosion coating. *Int J Electrochem Sci,* **2023**, *18*(6), 100091.
[http://dx.doi.org/10.1016/j.ijoes.2023.100091]

[5]   Jo, I.; Pettes, M.T.; Kim, J.; Watanabe, K.; Taniguchi, T.; Yao, Z.; Shi, L. Thermal conductivity and phonon transport in suspended few-layer hexagonal boron nitride. *Nano Lett.,* **2013**, *13*(2), 550-554.
[http://dx.doi.org/10.1021/nl304060g] [PMID: 23346863]

[6]   Antidormi, A.; Colombo, L.; Roche, S. Emerging properties of non-crystalline phases of graphene and boron nitride based materials. *Nano Mater. Sci.,* **2022**, *4*(1), 10-17.
[http://dx.doi.org/10.1016/j.nanoms.2021.03.003]

[7]   Ares, P.; Novoselov, K.S. Recent advances in graphene and other 2D materials. *Nano Mater. Sci.,* **2022**, *4*(1), 3-9.
[http://dx.doi.org/10.1016/j.nanoms.2021.05.002]

[8]   Lourie, O.R.; Jones, C.R.; Bartlett, B.M.; Gibbons, P.C.; Ruoff, R.S.; Buhro, W.E. CVD growth of boron nitride nanotubes. *Chem. Mater.,* **2000**, *12*(7), 1808-1810.
[http://dx.doi.org/10.1021/cm000157q]

[9]   Shi, Z.; Wang, L.; Jiang, Y.; Liu, Q.; Huang, C. Vapor–liquid-solid growth of large-area multilayer hexagonal boron nitride on dielectric substrates. *Nat. Commun.,* **2015**, *6*, 6160.
[PMID: 32051410]

[10]   Weissmantel, S.; Reisse, G.; Keiper, B.; Schulze, S. Microstructure and mechanical properties of pulsed laser deposited boron nitride films. *Diamond Rela. Mater.,* **1999**, *8*(2-5), 377-381.
[http://dx.doi.org/10.1016/S0925-9635(98)00394-X]

[11]   Kumar, A.; Venkatappa Rao, T.; Ray Chowdhury, S.; Ramana Reddy, S.V.S.; Ray, B.C. Compatibility confirmation and refinement of thermal and mechanical properties of poly (lactic acid)/poly (ethylene-co -glycidyl methacrylate) blend reinforced by hexagonal boron nitride. *React. Funct. Polym.,* **2017**, *117*, 1-9.
[http://dx.doi.org/10.1016/j.reactfunctpolym.2017.05.005]

[12]   Punetha, V.D.; Ha, Y.M.; Kim, Y.O.; Jung, Y.C.; Cho, J.W. Interaction of photothermal graphene networks with polymer chains and laser-driven photo-actuation behavior of shape memory polyurethane/epoxy/epoxy-functionalized graphene oxide nanocomposites. *Polymer,* **2019**, *181*, 121791.
[http://dx.doi.org/10.1016/j.polymer.2019.121791]

[13]   Punetha, V.D.; Ha, Y.M.; Kim, Y.O.; Jung, Y.C.; Cho, J.W. Rapid remote actuation in shape memory hyperbranched polyurethane composites using cross-linked photothermal reduced graphene oxide networks. *Sens. Actuators B Chem.,* **2020**, *321*, 128468.
[http://dx.doi.org/10.1016/j.snb.2020.128468]

[14]   Bhatt, S.; Punetha, V.D.; Pathak, R.; Punetha, M. Graphene in nanomedicine: A review on nano-bio factors and antibacterial activity. *Colloids Surf. B Biointerfaces,* **2023**, *226*, 113323.
[http://dx.doi.org/10.1016/j.colsurfb.2023.113323] [PMID: 37116377]

[15]   Pathak, R.; Punetha, V.D.; Bhatt, S.; Punetha, M. Carbon nanotube-based biocompatible polymer nanocomposites as an emerging tool for biomedical applications. *Eur. Polym. J.,* **2023**, *196*, 112257.
[http://dx.doi.org/10.1016/j.eurpolymj.2023.112257]

[16]   Soares, G.P.; Guerini, S. Structural and electronic properties of impurities on boron nitride nanotube. *J. Mod. Phys.,* **2011**, *2*(8), 857-863.
[http://dx.doi.org/10.4236/jmp.2011.28102]

[17]   Huang, C.; Chen, C.; Zhang, M.; Lin, L.; Ye, X.; Lin, S.; Antonietti, M.; Wang, X. Carbon-doped BN nanosheets for metal-free photoredox catalysis. *Nat. Commun.,* **2015**, *6*(1), 7698.
[http://dx.doi.org/10.1038/ncomms8698] [PMID: 26159752]

[18]   Qian, Y.; Xu, Y.; Yan, Z.; Jin, Y.; Chen, X.; Yuan, W-E.; Fan, C. Boron nitride nanosheets functionalized channel scaffold favors microenvironment rebalance cocktail therapy for piezocatalytic neuronal repair. *Nano Energy,* **2021,** *83,* 105779.
[http://dx.doi.org/10.1016/j.nanoen.2021.105779]

[19]   Jedrzejczak-Silicka, M.; Trukawka, M.; Dudziak, M.; Piotrowska, K.; Mijowska, E. Hexagonal boron nitride functionalized with Au nanoparticles-Properties and potential biological applications. *Nanomaterials,* **2018,** *8*(8), 605.
[http://dx.doi.org/10.3390/nano8080605] [PMID: 30096857]

[20]   Chen, X.; Wu, P.; Rousseas, M.; Okawa, D.; Gartner, Z.; Zettl, A.; Bertozzi, C.R. Boron nitride nanotubes are noncytotoxic and can be functionalized for interaction with proteins and cells. *J. Am. Chem. Soc.,* **2009,** *131*(3), 890-891.
[http://dx.doi.org/10.1021/ja807334b] [PMID: 19119844]

[21]   Sainsbury, T.; Satti, A.; May, P.; Wang, Z.; McGovern, I.; Gun'ko, Y.K.; Coleman, J. Oxygen radical functionalization of boron nitride nanosheets. *J. Am. Chem. Soc.,* **2012,** *134*(45), 18758-18771.
[http://dx.doi.org/10.1021/ja3080665] [PMID: 23101481]

[22]   Lyalin, A.; Nakayama, A.; Uosaki, K.; Taketsugu, T.; Shah, P.B. Functionalization of monolayer h-BN by a metal support for the oxygen reduction reaction. *J. Phys. Chem. C,* **2013,** *117*(41), 21359-21370.
[http://dx.doi.org/10.1021/jp406751n]

[23]   Jing, L.; Li, H.; Tay, R.Y.; Sun, B.; Tsang, S.H.; Cometto, O.; Lin, J.; Teo, E.H.T.; Tok, A.I.Y. Biocompatible hydroxylated boron nitride nanosheets/poly(vinyl alcohol) interpenetrating hydrogels with enhanced mechanical and thermal responses. *ACS Nano,* **2017,** *11*(4), 3742-3751.
[http://dx.doi.org/10.1021/acsnano.6b08408] [PMID: 28345866]

[24]   Chen, C.; Wang, J.; Liu, D.; Yang, C.; Liu, Y.; Ruoff, R.S.; Lei, W. Functionalized boron nitride membranes with ultrafast solvent transport performance for molecular separation. *Nat. Commun.,* **2018,** *9*(1), 1902.
[http://dx.doi.org/10.1038/s41467-018-04294-6] [PMID: 29765025]

[25]   Sainsbury, T.; Satti, L.; Mayne, G.; Wang, A.; McGovern, A.; Gun'ko, J. Covalently functionalized hexagonal boron nitride nanosheets by nitrene addition. *Chem. Eur. J.,* **2012,** *18*(35), 10808-10812.
[http://dx.doi.org/10.1002/chem.201201734]

[26]   Zhang, K.; Feng, Y.; Wang, F.; Yang, Z.; Wang, J. Two dimensional hexagonal boron nitride (2D-hBN): synthesis, properties and applications. *J. Mater. Chem. C Mater. Opt. Electron. Devices,* **2017,** *5*(46), 11992-12022.
[http://dx.doi.org/10.1039/C7TC04300G]

[27]   Wang, J.; Hao, J.; Liu, D.; Qin, S.; Portehault, D.; Li, Y.; Chen, Y.; Lei, W. Porous boron carbon nitride nanosheets as efficient metal-free catalysts for the oxygen reduction reaction in both alkaline and acidic solutions. *ACS Energy Lett.,* **2017,** *2*(2), 306-312.
[http://dx.doi.org/10.1021/acsenergylett.6b00602] [PMID: 28217747]

[28]   Kim, K.B.; Jang, W.; Cho, J.Y.; Woo, S.B.; Jeon, D.H.; Ahn, J.H.; Hong, S.D.; Koo, H.Y.; Sung, T.H. Transparent and flexible piezoelectric sensor for detecting human movement with a boron nitride nanosheet (BNNS). *Nano Energy,* **2018,** *54,* 91-98.
[http://dx.doi.org/10.1016/j.nanoen.2018.09.056]

[29]   Zhi, C.; Bando, Y.; Tang, C.; Kuwahara, H.; Golberg, D. Large-scale fabrication of boron nitride nanosheets and their utilization in polymeric composites with improved thermal and mechanical properties. *Adv. Mater.,* **2009,** *21*(28), 2889-2893.
[http://dx.doi.org/10.1002/adma.200900323]

[30]   Liu, L.; Xiao, L.; Li, M.; Zhang, X.; Chang, Y.; Shang, L.; Ao, Y. Effect of hexagonal boron nitride on high-performance polyether ether ketone composites. *Colloid Polym. Sci.,* **2016,** *294*(1), 127-133.
[http://dx.doi.org/10.1007/s00396-015-3733-2]

[31]   Du Frane, W.L.; Cervantes, O.; Ellsworth, G.F.; Kuntz, J.D. Consolidation of cubic and hexagonal boron nitride composites. *Diamond Rela. Mater.,* **2016**, *62*, 30-41.
[http://dx.doi.org/10.1016/j.diamond.2015.12.003]

[32]   Zhang, Y.; Wang, Y-N.; Sun, X-T.; Chen, L.; Xu, Z-R. Boron nitride nanosheet/CuS nanocomposites as mimetic peroxidase for sensitive colorimetric detection of cholesterol. *Sens. Actuators B Chem.,* **2017**, *246*, 118-126.
[http://dx.doi.org/10.1016/j.snb.2017.02.059]

[33]   Emanet, M.; Şen, Ö.; Çulha, M.; Şahinöz, M. Evaluation of boron nitride nanotubes and hexagonal boron nitrides as nanocarriers for cancer drugs. *Nanomedicine,* **2017**, *12*(7), 797-810.
[http://dx.doi.org/10.2217/nnm-2016-0322] [PMID: 28322118]

[34]   Permyakova, E.S.; Sukhorukova, I.V.; Antipina, L.Y.; Konopatsky, A.S.; Kovalskii, A.M.; Matveev, A.T.; Lebedev, O.I.; Golberg, D.V.; Manakhov, A.M.; Shtansky, D.V. Synthesis and characterization of folate conjugated boron nitride nanocarriers for targeted drug delivery. *J. Phys. Chem. C,* **2017**, *121*(50), 28096-28105.
[http://dx.doi.org/10.1021/acs.jpcc.7b10841]

[35]   Zhang, H.; Yamazaki, T.; Zhi, C.; Hanagata, N.; Zhang, Y. Identification of a boron nitride nanosphere-binding peptide for the intracellular delivery of CpG oligodeoxynucleotides. *Nanoscale,* **2012**, *4*(20), 6343-6350.
[http://dx.doi.org/10.1039/c2nr31189e] [PMID: 22941279]

[36]   Hanagata, N.; Zhang, H.; Chen, ; Zhi, ; Yamazaki, Chitosan-coated boron nitride nanospheres enhance delivery of CpG oligodeoxynucleotides and induction of cytokines. *Int. J. Nanomedicine,* **2013**, *8*, 1783-1793.
[http://dx.doi.org/10.2147/IJN.S43251] [PMID: 23674892]

[37]   Li, X.; Wang, X.; Zhang, J.; Hanagata, N.; Wang, X.; Weng, Q.; Ito, A.; Bando, Y.; Golberg, D. Hollow boron nitride nanospheres as boron reservoir for prostate cancer treatment. *Nat. Commun.,* **2017**, *8*(1), 13936.
[http://dx.doi.org/10.1038/ncomms13936] [PMID: 28059072]

[38]   Kıvanç, M.; Barutca, B.; Koparal, A.T.; Göncü, Y.; Bostancı, S.H.; Ay, N. Effects of hexagonal boron nitride nanoparticles on antimicrobial and antibiofilm activities, cell viability. *Mater. Sci. Eng. C,* **2018**, *91*, 115-124.
[http://dx.doi.org/10.1016/j.msec.2018.05.028] [PMID: 30033238]

[39]   Gonzalez-Ortiz, D.; Salameh, C.; Bechelany, M.; Miele, P. Nanostructured boron nitride–based materials: synthesis and applications. *Mater Today Adv,* **2020**, *8*, 100107.
[http://dx.doi.org/10.1016/j.mtadv.2020.100107]

[40]   Tang, C.; Bando, Y.; Huang, Y.; Zhi, C.; Golberg, D. Synthetic routes and formation mechanisms of spherical boron nitride nanoparticles. *Adv. Funct. Mater.,* **2008**, *18*(22), 3653-3661.
[http://dx.doi.org/10.1002/adfm.200800493]

[41]   Nurunnabi, M.; Nafiujjaman, M.; Lee, S.J.; Park, I.K.; Huh, K.M.; Lee, Y. Preparation of ultra-thin hexagonal boron nitride nanoplates for cancer cell imaging and neurotransmitter sensing. *Chem. Commun.,* **2016**, *52*(36), 6146-6149.
[http://dx.doi.org/10.1039/C5CC10650H] [PMID: 27074347]

[42]   Weng, Q.; Wang, B.; Wang, X.; Hanagata, N.; Li, X.; Liu, D.; Wang, X.; Jiang, X.; Bando, Y.; Golberg, D. Highly water-soluble, porous, and biocompatible boron nitrides for anticancer drug delivery. *ACS Nano,* **2014**, *8*(6), 6123-6130.
[http://dx.doi.org/10.1021/nn5014808] [PMID: 24797563]

[43]   Sukhorukova, I.V.; Zhitnyak, I.Y.; Kovalskii, A.M.; Matveev, A.T.; Lebedev, O.I.; Li, X.; Gloushankova, N.A.; Golberg, D.; Shtansky, D.V. Boron nitride nanoparticles with a petal-like surface as anticancer drug-delivery systems. *ACS Appl. Mater. Interfaces,* **2015**, *7*(31), 17217-17225.
[http://dx.doi.org/10.1021/acsami.5b04101] [PMID: 26192448]

[44]   Song, Q.; Fang, Y.; Liu, Z.; Li, L.; Wang, Y.; Liang, J.; Huang, Y.; Lin, J.; Hu, L.; Zhang, J.; Tang, C. The performance of porous hexagonal BN in high adsorption capacity towards antibiotics pollutants from aqueous solution. *Chem. Eng. J.,* **2017,** *325,* 71-79.
[http://dx.doi.org/10.1016/j.cej.2017.05.057]

[45]   Li, Y.; Yang, M.; Xu, B.; Sun, Q.; Zhang, W.; Zhang, Y.; Meng, F. Synthesis, structure and antioxidant performance of boron nitride (hexagonal) layers coating on carbon nanotubes (multi-walled). *Appl. Surf. Sci.,* **2018,** *450,* 284-291.
[http://dx.doi.org/10.1016/j.apsusc.2018.04.205]

[46]   Scorei, R.I.; Popa, R., Jr Boron-containing compounds as preventive and chemotherapeutic agents for cancer. *Anticancer. Agents Med. Chem.,* **2010,** *10*(4), 346-351.
[http://dx.doi.org/10.2174/187152010791162289] [PMID: 19912103]

[47]   Mateti, S.; Wong, C.S.; Liu, Z.; Yang, W.; Li, Y.; Li, L.H.; Chen, Y. Biocompatibility of boron nitride nanosheets. *Nano Res.,* **2018,** *11*(1), 334-342.
[http://dx.doi.org/10.1007/s12274-017-1635-y]

[48]   Salvetti, A.; Rossi, L.; Iacopetti, P.; Li, X.; Nitti, S.; Pellegrino, T.; Mattoli, V.; Golberg, D.; Ciofani, G. *In vivo* biocompatibility of boron nitride nanotubes: Effects on stem cell biology and tissue regeneration in planarians. *Nanomedicine,* **2015,** *10*(12), 1911-1922.
[http://dx.doi.org/10.2217/nnm.15.46] [PMID: 25835434]

[49]   Merlo, A.; Mokkapati, V.R.S.S.; Pandit, S.; Mijakovic, I. Boron nitride nanomaterials: Biocompatibility and bio-applications. *Biomater. Sci.,* **2018,** *6*(9), 2298-2311.
[http://dx.doi.org/10.1039/C8BM00516H] [PMID: 30059084]

[50]   Ikram, M.; Hassan, J.; Imran, M.; Haider, J.; Ul-Hamid, A.; Shahzadi, I.; Ikram, M.; Raza, A.; Qumar, U.; Ali, S. 2D chemically exfoliated hexagonal boron nitride (hBN) nanosheets doped with Ni: synthesis, properties and catalytic application for the treatment of industrial wastewater. *Appl. Nanosci.,* **2020,** *10*(9), 3525-3528.
[http://dx.doi.org/10.1007/s13204-020-01439-2]

[51]   Parra, C.; Montero-Silva, F.; Henríquez, R.; Flores, M.; Garín, C.; Ramírez, C.; Moreno, M.; Correa, J.; Seeger, M.; Häberle, P. Suppressing bacterial interaction with copper surfaces through graphene and hexagonal-boron nitride coatings. *ACS Appl. Mater. Interfaces,* **2015,** *7*(12), 6430-6437.
[http://dx.doi.org/10.1021/acsami.5b01248] [PMID: 25774864]

[52]   Eichler, J.; Lesniak, C. Boron nitride (BN) and BN composites for high-temperature applications. *J. Eur. Ceram. Soc.,* **2008,** *28*(5), 1105-1109.
[http://dx.doi.org/10.1016/j.jeurceramsoc.2007.09.005]

[53]   Farshid, B.; Lalwani, G.; Shir Mohammadi, M.; Simonsen, J.; Sitharaman, B. Boron nitride nanotubes and nanoplatelets as reinforcing agents of polymeric matrices for bone tissue engineering. *J. Biomed. Mater. Res. B Appl. Biomater.,* **2017,** *105*(2), 406-419.
[http://dx.doi.org/10.1002/jbm.b.33565] [PMID: 26526153]

[54]   Shuai, C.; Han, Z.; Feng, P.; Gao, C.; Xiao, T.; Peng, S. Akermanite scaffolds reinforced with boron nitride nanosheets in bone tissue engineering. *J. Mater. Sci. Mater. Med.,* **2015,** *26*(5), 188.
[http://dx.doi.org/10.1007/s10856-015-5513-4] [PMID: 25917828]

[55]   Nagarajan, S.; Belaid, H.; Pochat-Bohatier, C.; Teyssier, C.; Iatsunskyi, I.; Coy, E.; Balme, S.; Cornu, D.; Miele, P.; Kalkura, N.S.; Cavaillès, V.; Bechelany, M. Design of boron nitride/gelatin electrospun nanofibers for bone tissue engineering. *ACS Appl. Mater. Interfaces,* **2017,** *9*(39), 33695-33706.
[http://dx.doi.org/10.1021/acsami.7b13199] [PMID: 28891632]

[56]   Lei, W.; Mochalin, V.N.; Liu, D.; Qin, S.; Gogotsi, Y.; Chen, Y. Boron nitride colloidal solutions, ultralight aerogels and freestanding membranes through one-step exfoliation and functionalization. *Nat. Commun.,* **2015,** *6*(1), 8849.
[http://dx.doi.org/10.1038/ncomms9849] [PMID: 26611437]

[57]   Al-Saadi, S.; Banerjee, P.C.; Anisur, M.R.; Raman, R.K.S.; Raza, A.; Qaiser, M. Hexagonal boron nitride impregnated silane composite coating for corrosion resistance of magnesium alloys for temporary bioimplant applications. *Metals,* **2017**, *7*(12), 518.
       [http://dx.doi.org/10.3390/met7120518]

[58]   Fiume, M.M.; Bergfeld, W.F.; Belsito, D.V.; Hill, R.A.; Klaassen, C.D.; Liebler, D.C.; Marks, J.G., Jr; Shank, R.C.; Slaga, T.J.; Snyder, P.W.; Andersen, F.A. Safety assessment of boron nitride as used in cosmetics. *Int. J. Toxicol.,* **2015**, *34*(3_suppl), 53S-60S.
       [http://dx.doi.org/10.1177/1091581815617793] [PMID: 26684796]

[59]   Lu, T.; Wang, L.; Jiang, Y.; liu, Q.; Huang, C. Hexagonal boron nitride nanoplates as emerging biological nanovectors and their potential applications in biomedicine. *J. Mater. Chem. B Mater. Biol. Med.,* **2016**, *4*(36), 6103-6110.
       [http://dx.doi.org/10.1039/C6TB01481J] [PMID: 32263498]

[60]   Turkoglu, M.; Sahin, I.; San, T.; Baykara, A. Evaluation of hexagonal boron nitride as a new tablet lubricant. *Pharm. Dev. Technol.,* **2005**, *10*(3), 381-388.
       [http://dx.doi.org/10.1081/PDT-65684] [PMID: 16176018]

[61]   Uğurlu, T.; Turkoğlu, M. Hexagonal boron nitride as a tablet lubricant and a comparison with conventional lubricants. *Int. J. Pharm.,* **2008**, *353*(1-2), 45-51.
       [http://dx.doi.org/10.1016/j.ijpharm.2007.11.018] [PMID: 18160235]

[62]   Cassabois, G.; Valvin, P.; Gil, B. Hexagonal boron nitride is an indirect bandgap semiconductor. *Nat. Photonics,* **2016**, *10*(4), 262-266.
       [http://dx.doi.org/10.1038/nphoton.2015.277]

[63]   Lin, L.; Jiang, W.; Nasr, M.; Bechelany, M.; Miele, P.; Wang, H.; Xu, P. Enhanced visible light photocatalysis by TiO_2–BN enabled electrospinning of nanofibers for pharmaceutical degradation and wastewater treatment. *Photochem. Photobiol. Sci.,* **2019**, *18*(12), 2921-2930.
       [http://dx.doi.org/10.1039/c9pp00304e] [PMID: 31691716]

[64]   Lei, W.; Portehault, D.; Liu, D.; Qin, S.; Chen, Y.; Chen, X. Porous boron nitride nanosheets for effective water cleaning. *Nat. Commun.,* **2013**, *4*(1), 1777.
       [http://dx.doi.org/10.1038/ncomms2818] [PMID: 23653189]

[65]   Fan, D.; Feng, J.; Liu, J.; Gao, T.; Ye, Z.; Chen, M.; Lv, X. Hexagonal boron nitride nanosheets exfoliated by sodium hypochlorite ball mill and their potential application in catalysis. *Ceram. Int.,* **2016**, *42*(6), 7155-7163.
       [http://dx.doi.org/10.1016/j.ceramint.2016.01.105]

[66]   Zhang, M.; Zhou, M.; Luo, Z.; Zhang, J.; Wang, S.; Wang, X. Molten salt assisted assembly growth of atomically thin boron carbon nitride nanosheets for photocatalytic H_2 evolution. *Chem. Commun.,* **2020**, *56*(17), 2558-2561.
       [http://dx.doi.org/10.1039/C9CC09524A] [PMID: 32010905]

[67]   Li, J.; Jin, P.; Dai, W.; Wang, C.; Li, R.; Wu, T.; Tang, C. Excellent performance for water purification achieved by activated porous boron nitride nanosheets. *Mater. Chem. Phys.,* **2017**, *196*, 186-193.
       [http://dx.doi.org/10.1016/j.matchemphys.2017.02.049]

[68]   Liu, F.; Yu, J.; Ji, X.; Qian, M.; Li, Y. Nanosheet-structured boron nitride spheres with a versatile adsorption capacity for water cleaning. *ACS Appl. Mater. Interfaces,* **2015**, *7*(3), 1824-1832.
       [http://dx.doi.org/10.1021/am507491z] [PMID: 25552343]

[69]   Liu, D.; Zhang, M.; Xie, W.; Sun, L.; Chen, Y.; Lei, W. Porous BN/TiO_2 hybrid nanosheets as highly efficient visible-light-driven photocatalysts. *Appl. Catal. B,* **2017**, *207*, 72-78.
       [http://dx.doi.org/10.1016/j.apcatb.2017.02.011]

CHAPTER 7

Two-Dimensional Germanene Synthesis, Functionalization, and Applications

Vinay Deep Punetha[1,*], **Gaurav Nath**[2], **Sadafara Pillai**[3] and **Golnaz Taghavi Pourian Azar**[4]

[1] *Centre of Excellence for Research, P.P. Savani University, Surat-394125, Gujarat, India*

[2] *Department of Materials and Geosciences, Technische Universität Darmstadt, Darmstadt, Germany*

[3] *Department of Chemistry, School of Science, P.P. Savani University, Surat-394125, Gujarat, India*

[4] *The Functional Materials and Chemistry Research Group, Research Centre for Manufacturing and Materials (CMM), Coventry University, Coventry, CV1 5FB, United Kingdom*

Abstract: The discovery of graphene stimulated the intense search for possibilities of other 2D analogs of it. These investigations resulted in many wonder materials, especially from elements of the 14th group of the periodic table. One of the most celebrated 2D structures of the 14th group after graphene is a germanium-based 2D structure known as germanene. Like graphene, germanene is also a single-atom-thick 2D structure. There are several similarities in the structures and properties of graphene and germanene; however, they are distinct in several other properties due to the difference in atomic size, effective nuclear charge, and band structures. One of the most defining phenomena in the structures of graphene and germanene is the buckled structure of the germanene derivative. The buckled structure allows unique orbital mixing and changes the hybridization mode among combining germanium atoms. On the one hand, carbon atoms in graphene exhibit a planer geometry with mesmerizing consistency of the sp^2-hybridized orbitals. On the other hand, germanium atoms tend to exhibit mixed sp^2 and sp^3 hybridizations. Germanene has gained more popularity due to ease in manipulating its band structure with possibilities to revamp the existing electronics. In addition, mixed hybridization offers the remarkable potential to use this material in various energy and catalytic applications. This chapter deals with various aspects of its chemistry and properties ranging from different methods of synthesis of germanene and its functionalized derivatives, band gap manipulation in these structures, and catalytic applications.

Keywords: Band structure, Catalytic, Germanene, Germanane, 2D materials.

---

* **Corresponding author Vinay Deep Punetha:** Centre of Excellence for Research, P.P. Savani University, Surat-394125, Gujarat, India; E-mail: Vinaydeep.punetha@ppsu.ac.in

Vinay Deep Punetha (Ed.)

INTRODUCTION

Generally, two-dimensional (2D) nanosheets are associated with a high degree of anisotropy, nanoscale thickness, and infinite length in other dimensions. These unique properties lead to novel ultrathin 2D nanomaterials with many budding applications such as energy storage, energy conversion, electronics, and many more [1]. The material's properties at the nanoscale differ from their respective 3D bulk forms. This led to a sharp escalation in research focused on developing various atomically thin two-dimensional (2D) materials. Germanene is one of the engineered 2D materials, which is an allotropic form of germanium. It took almost a decade to synthesize it successfully since the revolutionary discovery of graphene [2]. The idea of 2D structures of Ge was first investigated by Takeda *et al.* in 1994 [3]. The investigations were based on the local-density functional (LDF) method. His group investigated possibilities of graphite-like architectures with Si and Ge. His investigations concluded that carbon, unlike Si and Ge, prefers to exhibit a planer 2D structure, while Si and Ge prefer to exhibit a corrugated stage (Fig. **1**). After nearly two decades, at the end of the third quarter of 2014, G. Le Lay and his team announced the successful synthesis of a nearly flat 2D allotrope of germanium. They synthesized it with the help of molecular beam epitaxy upon a gold surface. This 2D architecture was simultaneously and independently claimed by two other groups in 2014 [4].

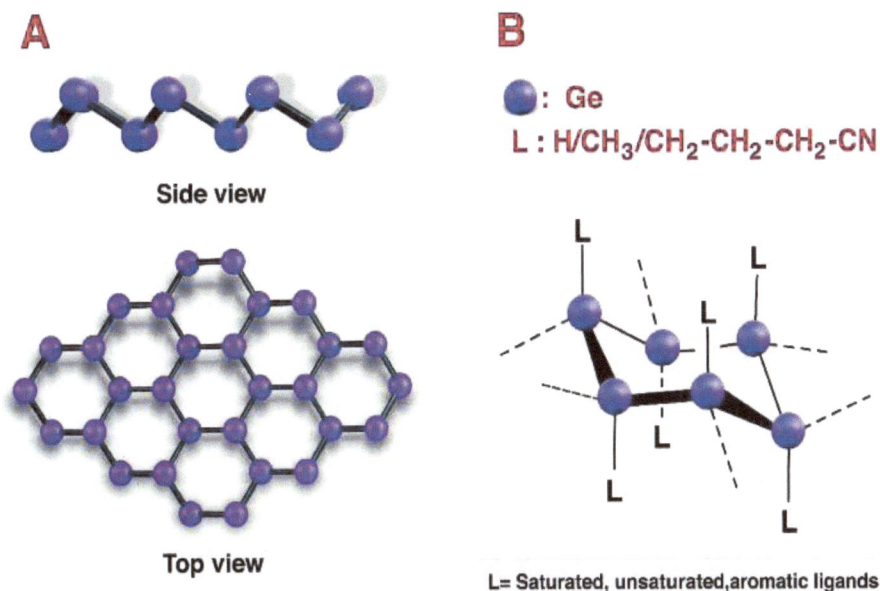

A

Side view

Top view

B

⬤ : Ge

L : $H/CH_3/CH_2\text{-}CH_2\text{-}CH_2\text{-}CN$

L
L
L
L
L
L

L= Saturated, unsaturated,aromatic ligands

Fig. (1). The 2D skeleton of germanene with (a) Side, top view, and (b) Functionalised groups shown with puckered structure.

The elemental monolayer of germanium was found to possess remarkable electronic properties, as concluded from various theoretical studies, and emphasized the revamping of existing electronics by using fundamental 2D structures in various electronic devices. Several electronic properties of germanene set it apart from graphene [1]. The most promising property is the scope of manipulation in energy required to excite the electron from its bound state to the conduction band. Graphene is generally considered a zero band gap material; contrary to this, the significant band gap in the germanene enables it to exhibit the mesmerizing phenomenon of the Quantum Hall effect (QHE). In addition, the band gap in germanene and its derivatives can be manipulated, as surface revamping alters the structure and energies of its molecular orbitals [2]. The hydrogenation of germanene gives it a hydrogenated form (germanane), which is buckled with a significant band gap change. This band gap can be regulated as it is a function of the extent of hydrogenation. This phenomenon can also be observed with other derivatives of germanene and offers the most incredible opportunity of tuning the band gap for suitable applications in electronics [2, 4].

Since the rise of germanene, various attempts have been made to revamp its surface to investigate and optimize the synthesis of its application-driven derivatives. The functionalization of germanene is fundamentally different from that of graphene due to inherent differences in their electronic structures. The chemical stability in environmental conditions of graphene offers the scope of revamping it through established synthetic routes. In addition, established reaction mechanisms in organic chemistry help formulate productive reaction schemes [2]. On the other hand, the lack of literature on the mechanisms of reactions involving Ge and its chemical instability poses a crucial challenge in the functionalization of germanene. In principle, the methodologies opted for the surface revamping of germanane can be categorized into the following four major groups. (i) Exfoliation of Zintl phase CaGe$_2$ with suitable reactants, (ii) solvothermal method, (iii) Ge–H activation, and (iv) alloy-induced Ge–H activation [5]. Each of these methods has its pros and cons. One key difference in this method is the extent of revamping and consistency in surface modifications.

We are attempting this book chapter to assemble the significant pieces of epistemological evidence; documented in recent years so that a complete picture of the recent findings can be plotted. This chapter covers a gamut of recent findings and observations, along with vigorous discussions on fundamental concepts such as synthesis methods, the scope of surface revamping, and the effect of surface revamping on the band structure. In addition to the germanene, its functionalized derivatives have been investigated for their scope in different applications ranging from modern electronics to energy storage and conversion devices.

STRUCTURE AND SYNTHESIS OF GERMANENE

The atomic structure of C and Ge plays a crucial role in the structural, physiochemical, and optoelectronic behavior of graphene and germanene. The atomic radius of carbon is approximately 170 pm, compared to 211 pm for germanium. Due to the larger atomic size, Ge exhibits reluctancy to form π-bonds resulting due to the lateral overlap of the 'p' orbitals, as the significant internuclear distance between combining atoms does not offer a significant overlap between participating orbitals, resulting in a poor concentration of sp^2 mixing [1, 6]. Contrary to this, the smaller size of carbon is suitable for the lateral overlap of the 'p' orbitals. This also reflects the hybridization of the C and Ge in their respective 2D structure. C prefers to exist in the sp^2 hybridized state, while Ge prefers the sp^3 mode, with no scope of forming π-bonds. This results in the fundamental differences in their structures; graphene exists in the hexagonal phase, while germanene is in the cubic phase with puckered structure (Fig. **1**). In addition, there are several differences in the bond parameters of graphene and germanene, such as the bond length Ge-Ge in germanene is 2.431Å. In contrast, it is 1.422Å for C-C in 2D flat graphene [7]. The lattice constants of germanene and graphene are 4.043Å and 2.463Å, respectively. The difference in the size of C and Ge is reflected in the lattice constant (LC), which is directly proportional to the atomic radius. The LC for germanene is calculated to be 4.043Å and 2.463Å for graphene. The increased lattice constant insinuates the presence of loosely bound electrons in the germanene in comparison to that of graphene [7]. Table **1** summarizes a comparative analysis of different properties of germanene and graphene.

Table 1. A comparative analysis of various physical properties of Graphene and Germanene.

S. No.	Parameter	Graphene	Germanene	Refs.
1.	Atomic size	170pm	211pm	[1, 2, 8 - 10]
2.	Bond length	1.422Å	2.431Å	[1, 2, 8 - 10]
3.	Mode of hybridization	sp^2	sp^2 & sp^3	[1, 2, 8 - 10]
4.	Structure	planer	puckered	[1, 2, 8 - 10]
5.	Lattice constant	2.463Å	4.043Å	[1, 2, 8 - 10]
6.	Elastic constant	328.02(Jm^{-2})	56.01(Jm^{-2})	[1, 2, 8 - 10]
7.	Buckling	No	Yes	[1, 2, 8 - 10]
8.	Electron-phonon coupling strength	Less	More	[1, 2, 8 - 10]

(Table 1) cont.....

S. No.	Parameter	Graphene	Germanene	Refs.
9.	Fermi velocity	3.39×10^5 cm²V⁻¹s⁻¹	6.09×10^5 cm²V⁻¹s⁻¹	[1, 2, 8 - 10]
10.	Mobility of electrons	Low	High	[1, 2, 8 - 10]
11.	Surface Reactivity	Low	High	[1, 2, 8 - 10]

In most methods, 2D monolayer germanene is fabricated by epitaxial growth [11 - 15]. Epitaxy is a process of developing a crystal layer on a crystal surface. The germanene is synthesized by heteroepitaxy, where Ge atoms are arranged on different metallic substrates. In some cases, semi-metals and non-metal substrates are also used in this synthesis. The substrate acts as a seed crystal during the process of epitaxy. Germanene crystals grow on the substrate by deposition of germanium atoms at high temperatures and ultrahigh vacuum [13]. The Ge atoms originate from a bulk of Ge; mostly metallic rods are used for this purpose [14]. The evaporation of atoms from the bulk is done with the help of an electron-beam evaporator of high energy. The atoms of Ge occupy desired orientations determined by the seed crystals. These methods are highly efficient in developing products with remarkable consistency. The details of the epitaxial growth method are given in Fig. (**1**).

Though there are some cons of this method, as the growth of germanene on the metal surface encounters stacking of metal atoms into the basal plane of germanene [15], the stacking of metal atoms induces alteration in the state of hybridization and strain resulting into the different electronic properties [16]. To preserve the intrinsic electronic properties of the germanene, other non-metallic substrates are used, including semimetallic Sb and non-metallic MoS_2 as the substrate. The germanene obtained by the epitaxial growth is found to be of long-range order and with the slightest structural defects [17 - 19]. However, restricted space and inconsistency in the crystal structure on a vertical scale, difficulty in transferring grown layers from the seed crystal, and the inability of mass production remain a challenge to be addressed. A summary of various substrates used for the growth of germanene is given in Table **2**, along with the year these methods were reported [20].

Table 2. Epitaxial growth of germanene on various substrates.

S. No.	Seed Crystal	Method	Year	Refs.
1.	Cu(111)	Epitaxial growth	2017	[5]
2.	Pt (111)	Epitaxial growth	2014	[6]
3.	Al(111)	Epitaxial growth	2015	[7]
4.	Al(111)	Epitaxial growth	2016	[11]

(Table 2) cont.....

S. No.	Seed Crystal	Method	Year	Refs.
5.	Al(111)	Epitaxial growth	2017	[12]
6.	Au(111)	Epitaxial growth	2016	[13]
7.	Au(111)	Epitaxial growth	2014	[14]
8.	Au(111)	Epitaxial growth	2017	[15]
9.	Au(111)	Epitaxial growth	2017	[16]
10.	Ag(111)	Epitaxial growth	2018	[17]
11.	Sb(111)	Epitaxial growth	2016	[18]
12.	MoS_2	Epitaxial growth	2016	[19]
13.	MoS_2	Epitaxial growth	2016	[20]

Various spectroscopic and other techniques confirm the 2D structure of the germanene. Scanning tunneling microscopy (STM) is used to evaluate the germanene surface at the atomic level. The selected area electron diffraction pattern helps determine unit cell type. In addition, crystal parameters and other information can be evaluated with the help of other techniques such as X-ray Diffraction (XRD), Raman, and FT-IR spectroscopy. The peaks obtained from Raman frequencies help determine various vibrational modes [19, 20].

SYNTHESIS OF GERMANANE

Hydrogenated germanene is known as germanane with abundant Ge-H bonds. Germanane synthesis mainly uses Zintle phase precursors such as α-$CaGe_2$, β-$CaGe_2$, or Zintle phase-like structures such as $EuGe_2$ [21, 22]. By definition, the Zintle phase is identified by the intermetallic crystal formation. For this phase to exist, a significant difference in the ionic characters of the participating metals is a prerequisite. These crystals generally form between 's' block metals and 'p' block metalloids or metals. In order to prepare germanane, one of the most prolific strategies is to synthesize zintle phase crystals of the Ca and Ge as it offers an extended covalent network of Ge-Ge bonds with Ca^{+2} cations intercalated between the network [22]. Removing Ca^{+2} ions from the network results in the extended 2D framework of germanium atoms. In general, the intercalated ions are removed by acid treatment (mostly HCl), in which Cl anions form compounds with the intercalated Ca cations and remove them from the lattice (Fig. **2**). The Zintle phase between Ca and Ge is achieved by reacting them at very high temperatures in stoichiometric ratios. The prepared Zintle phase compound of Ca and Ge is then immersed into HCl at a low temperature (-40°C) to avoid the Ge-Cl bond formation.

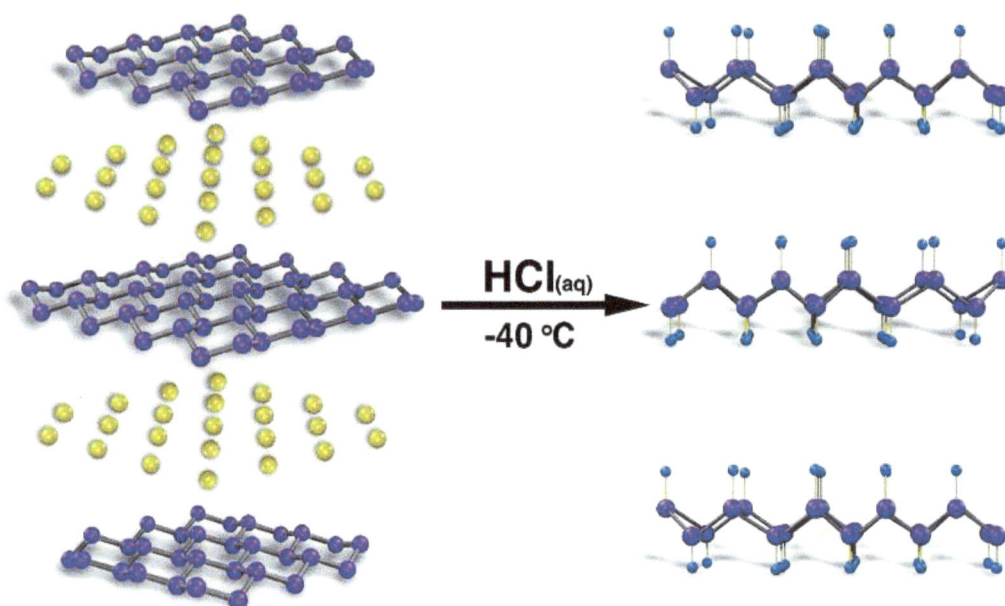

Fig. (2). Removal of calcium ions (in yellow) from single-phase.

The greater reactivity toward Ca^{+2} and Cl^- ions results in the removal of Ca^{+2} from the phase. The final product left after repeated washing can be separated. The acid-assisted wash allows the formation of Ge-H bonds and separates the interlayer distance from that of the zintle phase. It is observed that the formation of hydrogenated germanane increases the interlayer distance by 0.4 Å. Generally, the sheets obtained by deintercalation of Ca^{+2} ions are a few 2 to 5 mm long and 50-100 nanometres thick. The first Ge-G synthesis was done back in 2004, though it remained uncelebrated under the scarcity of rigorous structural elucidation. The hydrogenation of the germanene offers the extended scope of utilizing the germanene skeleton in various applications as it provides exceptional stability. It is noted that the Ge-H bond shows inertness at room temperature and can prevent significant oxidation for 5 to 6 months [23, 24]. Table **3** gives a summarized account of various precursors used in the synthesis of germanene derivatives.

Table 3. Synthesis of germanane with different initial precursors and methods.

S. No.	Precursor	Method	Year	Refs.
1	Epitaxial thin $CaGe_2$ films	Topochemical transformation	2000	[21]
2	Zintl-phase $CaGe_2$	Ion-exchange method	2013	[22]
3	Rhombohedral $CaGe_2$	Ion-exchange method	2018	[23]
4	Rhombohedral $CaGe_2$	Ion-exchange method	2016	[24]

(Table 3) cont.....

S. No.	Precursor	Method	Year	Refs.
5	Rhombohedral CaGe$_2$	Pb as a reactive flux	2007	[25]
6	EuGe$_2$	Ion-exchange method	2018	[23]
7	MBE grown CaGe$_2$	Ion-exchange method	2014	[26]

SYNTHESIS OF FUNCTIONALISED GERMANENE

As discussed in the previous segment, functionalization alters a wide range of properties in germanane and offers the scope of application-driven manipulation. Recently, a wide range of ligands have been functionalized onto the germanene skeleton. Most of the organic fragments were introduced by using haloalkanes. Jiang *et al*. [27] first synthesized the methyl-terminated germanane using CH$_3$I as an attacking reagent and water. The reaction successfully methylated the germanane keeping its basic 2D skeleton intact. However, this process suffered from minor drawbacks, such as partial oxidation of the surface and inconsistent methylation of the surface. In a revised method, his group used CH$_3$CN and different stoichiometric ratios of water and developed a more reliable method to methylate the germanane skeleton. They prepared a reaction mixture of CH$_3$I and H$_2$O (1:5) in CH$_3$CN as the solvent and found that methylation took place with greater consistency and without oxidizing germanane atoms on the surface. The methylation of germanane results in enhanced thermal stability photoluminescence [28].

Similar to the -CH$_3$ group, many other alkyl groups were functionalized on the germanane surface using the halide substituted reagents of the same. These include alkyl groups with saturation such as CH$_3$- CH$_2$-, CH$_3$- CH$_2$- CH$_2$-, CH$_3$- CH$_2$- CH$_2$- CH$_2$-, CH$_3$- CH$_2$- CH$_2$- CH$_2$- CH$_2$- CH$_2$- CH$_2$- ; alkyl groups with unsaturation or aromatic ring such as -CH$_2$CH=CH$_2$, 4-(CF$_3$)C$_6$H$_5$CH$_2$- and chains with the heteroatoms such as -CH$_2$OCH$_3$. The detailed mechanism of these reactions has yet to be worked out, with profound evidence gathered from the information on reaction pathways and participating intermediates [27 - 33].

Functionalization can extensively be exploited to tune the properties of germanane for different applications. The germanane skeleton can be modified suitably by optimizing the required band gap for electronic applications. Several factors affect the band gap; more importantly, ligand size and its electronic effects (electron pulling or donating) are vital in favorable functionalization. The effect of surface modification reflects in the various spectroscopic and crystal parameters. For instance, functionalized germanane exhibits changes in the lattice constants, XRD Patterns, and, more notably, the Raman spectra [33]. The FT-IR spectra exceptionally can be of great significance in the preliminary determination of the functionalization. Due to different stretching and wagging modes, germanane exhibits an absorption band for the Ge-H covalent bond. As H gets replaced, the ligand intensities of these absorptions decrease. A stretching mode can be seen in the range of ~ 2000 cm^{-1}, while waging modes can be seen in the fingerprint region, particularly at ~ 569, 508, and 474 cm^{-1} [33].

Moreover, FT-IR spectra can also be used to determine the presence of Ge-O moieties on the germanane skeleton due to its sensitivity toward the Ge-O band. In addition, XPS spectra could also be used to determine surface revamping. It helps determine the extent of functionalization, Ge's oxidation state, and the bonding's nature. Ge-H generally exhibits signals centered at 1217 eV, and this peak corresponds to Ge $2p_{3/2}$. The removal of H in a condition when alkyl or any substituent attacks significant changes can be observed as it affects Ge's electronic environment and oxidation state. Moreover, evidence obtained from the Raman spectra dramatically helps determine Ge-H formation and functionalization [27 - 33]. Table **4** summarizes various applications of functionalized germanene derivatives.

Table 4. Functionalization of germanene with various substrates.

Precursor	Reagent	Ligand	Observations	Refs.
CaGe$_2$	CH$_3$I/H$_2$O	-CH$_3$	Enhanced thermal stability.	[27]
CaGe$_2$	CaGe$_2$/CH$_3$I/H$_2$O 30: 10: 60	-CH$_3$	Photoluminescence and transport properties.	[27]
CaGe$_2$	CH$_3$I/H$_2$O In CH$_3$CN	-CH$_3$	Consistent functionalization.	[28]
CaGe$_2$	CaGe$_2$/CH$_3$I/H$_2$O at 1:30:10	-CH$_3$	Decomposition of rhodamine b.	[29]
CaGe$_2$	CH$_3$I/ CH$_3$CN	-CH$_3$	Thin-film device.	[30]
CaGe$_2$	4-(CF$_3$)BnBr	-CH$_2$BnCF$_3$ 4-(trifluoromethyl)benzyl bromide	Enhanced thermal stability.	[31]
CaGe$_2$	GeCH$_2$0CH$_3$	CH$_3$0CH$_2$X	Enhanced thermal stability and photoluminescence.	[32]
CaGe$_2$	GeCH$_2$CH=CH$_2$	CH$_2$=CHCH$_2$X	Enhanced thermal stability and photoluminescence.	[32]
CaGe$_2$	GeCH$_2$C$_6$H$_5$	C$_6$H$_5$CH$_2$X.	Enhanced thermal stability.	[32]
CaGe$_2$	Ge(CH$_2$)$_2$CH$_3$	CH$_3$ (CH$_2$)$_2$X	Enhanced thermal stability.	[32]

(Table 4) cont.....

Precursor	Reagent	Ligand	Observations	Refs.
$CaGe_2$	$Ge(CH_2)_3CH_3$	$CH_3(CH_2)_3X$	Enhanced Thermal stability.	[32]
$CaGe_2$	$Ge(CH_2)_6CH_3$	$CH_3(CH_2)_6X$	Enhance thermal stability and photoluminescence.	[32]
$CaGe_2$	GeBr	CCl_3Br	Enhance thermal stability and photoluminescence.	[33]

The crystalline Ge exhibits a signature peak at around 297 cm^{-1}, showing a slight blueshift to 302 cm^{-1} as the germanane synthesis takes place. XRD spectra enable some striking information on the transformation of Ge-H to Ge-CH$_3$. In the case of Ge-H, in general, five signature peaks are obtained for (002), (100), (101), (110), and (112) planes at ~ 16°, ~ 27°, ~ 28°, ~ 45° ~ 48°. Among all these peaks, the most significant changes are seen in the shift of the (002) peak from ~ 16° to ~ 10°. In addition to this, new peaks arise near ~ the 4° for the (001) plane, and the peak corresponding to the (112) plane at ~ 48° disappears.

ORIGIN OF OPTOELECTRONIC PROPERTIES IN GERMANENE DERIVATIVES

Germanene and its derivatives exhibit mesmerizing optoelectronic properties quite different from graphene. In the case of graphene, the atomic size of carbon allows lattice points to lie in one plane, as the small size of carbon atoms allows the formation of π-bonds by the lateral overlap of the p-orbitals. On the other hand, the atomic size of germanium atoms in the germanane does not allow a flat structure due to its larger size [1]. The larger size of Ge prevents the formation of the pi-bond. Consequently, the germanene structure gets buckled to manage a long array of Ge atoms. Due to the buckling, two sub-lattices in germanene are separated from each other with a distance of 0.64 Å. The separation is due to the formation of two sub-lattices. A unique phenomenon takes place in which the p$_z$ orbital partial mixes with s, p$_x$, and p$_y$ orbitals. This gives rise to the sp^2–sp^3 hybridization in germanene and dramatically affects its optoelectronic properties. In principle, the germanene and graphenes exhibit different properties due to the decreased sigma and pi bands and significant sp^2–sp^3 hybridization in germanene [3, 6]. The buckled structure of germanene is responsible for the narrow bandgap opening equivalent to ~23.9 meV [34]. The band gap opening on hydrogenation of germanene is shown in Fig. (**3**).

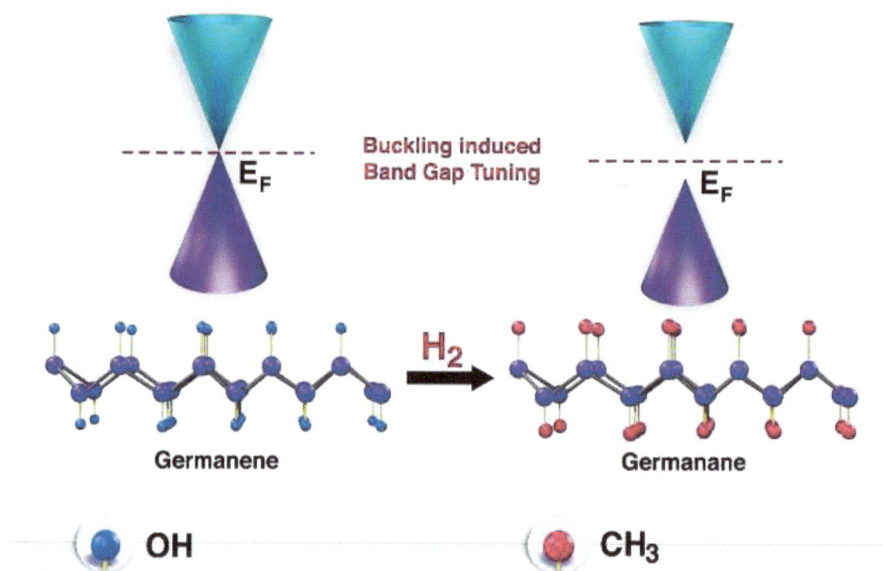

Fig. (3). Band gap manipulation in germanane with the help of functionalization.

Ye *et al.* reported band gap manipulation by absorption of alkali metals on the Germanene surface by performing DFT analysis [35]. Wang *et al.* concluded in their studies that small organic molecules alter the band gap in germanene [6]. Livache *et al.* recently investigated the optoelectronic properties of CH_3-Ge with a critical focus on the photoluminescence mechanism and transport properties. The photoluminescence in CH_3-Ge is observed for two reasons: (1) contribution from the band edge and (2) contribution from trap states with exciton dynamics ranging over 1–100 ns. Moreover, his group concluded p-type conduction in the sheets [36].

APPLICATION OF GERMANENE AND ITS STRUCTURAL DERIVATIVES

Germanene and its derivatives have shown tremendous potential in various applications. Researchers have evaluated potential application of germanene derivatives in a wide range of fields, including energy storage devices, gas sensing and drug delivery agents, antimicrobial agents, electronics, electrocatalysts, and many more (Fig. **4**). However, an in-depth discussion on all the topics is beyond the scope of this book chapter, though a generalized and more concise discussion has been made on energy, electrocatalysis, gas sensing, and biological applications.

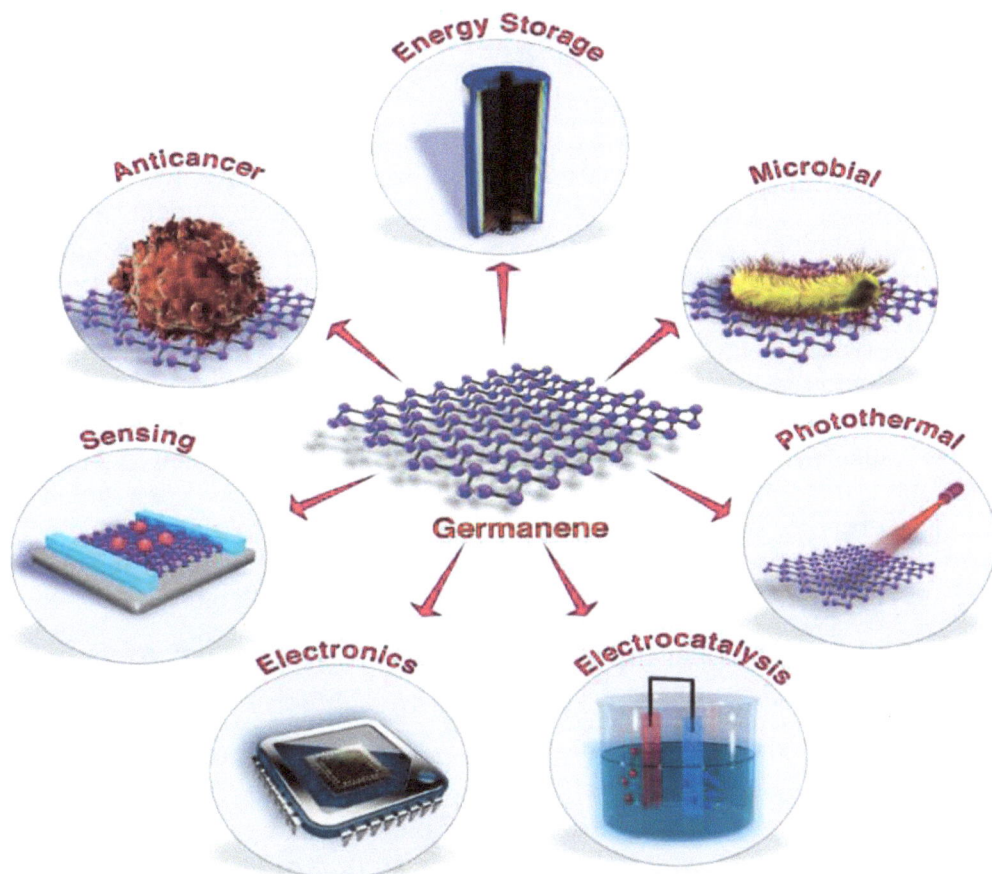

Fig. (4). Various applications of germanane and its derivatives.

Germanene Derivatives in Energy Applications:

Batteries with more energy and high power density are one of the most desired technological aspirations to manage the dependence on fossil fuels and other non-environment-friendly sources. In addition, the development of these can escalate the role of a more significant number of portable devices, especially in the next generation of electric vehicles. Serino *et al*. developed germanane-based battery electrodes for Li-ion batteries [37]. The developed anodes were capable of exhibiting energy storage in resonance with the theoretical maximum. The germanane electrodes showed a reversible capacity of 1107.99 mAh/g at high cycling rates. Moreover, the developed electrodes maintain this capacity for more than 100 cycles. Attempts have been made to explore the electrochemical performance of Ge-H and Ge-CH$_3$ in electrode development in coin cells. Wu *et al*. assessed the performance of Ge-H and Ge-CH$_3$-based hybrids with single-wall carbon nanotubes (SWCNTs) [38]. The metallic coin cells were developed with

SWCNT/Ge-H and SWCNT/Ge-CH$_3$ to produce an electrode (anode) and compared the performance with the metallic Li. The evidence from the electrochemical studies was very promising as the SWCNT/Ge-H electrode showed a tremendous capacity of 1032.1 mA h g^{-1} at a high current density of 2000 mA g^{-1}. Moreover, the electrodes exhibit a remaining capacity of 654 mA h g^{-1} even after 100 cycles at 500 mA g^{-1}. Loaiza, in his study, concluded that the layered morphology of germanane and derivatives affects the volume variations during the electrochemical process of lithiation; this allows Li$^+$ ions to diffuse rapidly [39]. In addition, several attempts have been made to assess the insertion of Na or Li ions on 2D germanane layers. Mortazavi *et al.* recognized the favorable and most stable binding sites and binding energies of single Na or Li ions on germanene [40]. His study came up with striking findings; the mechanistic detailing suggests that Na and Li ions occupy the vacant hexagonal spaces in the germanane with reduced electron density. The reduced electron density in Na and Li ions is observed due to the difference in the electronegativities. The Ge atoms try to drag the electron cloud from the Na and Li ions. Consequently, the electron decreases from the trapped ions and increases around the Ge atoms.

Germanene Derivatives in Electrocatalytic Applications

The unique surface of germanane has found several other exciting applications. Liu *et al.* explored the catalytic application of germanane for hydrogen production in the ammonia borane process [41]. They compared the catalytic performance of the Ge-H and Ge-CH$_3$ in the presence of visible light ($\lambda \geq 420$ nm). They also determined crucial catalytic parameters such as turnover number (TON) and turnover frequency (TOF) to assess the catalytic performance of both. He found that Ge-CH$_3$ exhibits superior catalytic performance compared to Ge-H. Moreover, the hydrogen selectivity was also superior for the Ge-CH$_3$. The TOF for Ge-H was found to be half that of Ge-CH$_3$. The Ge-CH$_3$ exhibited the TOF value of 18.159 mol(Hydrogen gas)$\times$min$^{-1}\times$mol(catalyst)$^{-1}$. The reusability of the Ge-CH$_3$ catalyst was also reported to be significantly higher than that of the Ge-H. Ni *et al.* developed metal-functionalized Ge-H (Mt-f-Ge-H) sheets to study the photocatalytic performance of the developed materials [42]. A wide range of metallic nano-functionalities were introduced on the Ge-H surface and applied as photocatalysts. The surface-mediated reduction was used as the synthetic route to introduce gold (Au), silver (Ag), copper (Cu), palladium (Pd), and platinum (Pt) nanoparticles onto the Ge-H surface. Oxidation of benzyl alcohol was successfully achieved into the benzaldehyde. In addition, the group worked on the reaction mechanism for the observed oxidation. The study results show a manifold increase in the conversion of benzyl alcohol to benzaldehyde when metal nanoparticles (MNPs) were decorated on the Ge-H surface. The study was parameterized based on conversion efficiency, selectivity, and yield. Initially, the conversion % shown by the MNPs was markedly less than 10% in all the cases. The MNPs functionalized Ge-H exhibited a tremendous rise in the conversion, more than 35% in all the cases and a maximum near 80% for Ag and Pt decorated Ge-H surfaces. In addition, selectivity for all the MNPs-f-Ge-H surfaces was reported to be more than 60%, with the highest selectivity for Pt-f-Ge-H. The experiment concluded some striking results that when Ge-H was used in combination with the nanoparticles, selectivity in all the cases increased to nearly 80%; however, a sharp decrease in the overall yield of the product was observed. They concluded that the reaction using a Ge-H surface might go through a radical mechanism on exposure to the light due to no difference in the electronegativity of the Ge-Ge bond. The Ge radical produced due to the homolytic bond cleavage might attack the benzyl alcohol and releases the hydrogen gas, producing the benzyl alcohol radical, which in further steps gets converted into benzaldehyde [42]. Liu *et al.* studied the degradation capacity of Ge-H on the rhodamine B (RhB) and compared the performance with the N-doped P25. RhB is a hazardous dye used in printing and dyeing in textiles, paper, paints, and leathers and poses an environmental threat due to its slow degradation. It was concluded that Ge-H exhibits a superior capacity for catalytic degradation. It takes only 4 minutes to degrade 80% of the RhB, while for the same time interval, only 70% degradation can be achieved by N-doped P25. The experiment was conducted under visible light irradiation. The reaction proceeded with the release of a considerable amount of H_2 gas [43]. In addition, the superiority of the degradation was confirmed by the intact Ge-H structure even after repeated use. In addition, several theoretical calculations were also performed to investigate the catalytic ability and stability of various germanane derivatives. In an exciting finding, Kenecny *et al.* reported successfully transferring the metal-carbonyl group to the benzyl group [44]. Various metal carbonyls of chromium, molybdenum, and tungsten were made to react with benzyl-decorated germanane, and found that the germanane surface offers the optimum surface for the interaction. Ng *et al.* synthesized various functionalized derivatives of germanane viz. methyl, propyl, hydroxypropyl, and 2-(methoxycarbonyl)ethyl and evaluated their potential as photo-electrocatalyst in the hydrogen evolution and photo-oxidation of H_2O. The electrocatalytic performance of various derivatives is evaluated based on overpotential, which is in inverse relation to each other. In addition, to further consolidate the results, a Tafel analysis was also performed by Ng *et al.* The overpotentials for Ge-CH$_3$, Ge-CH$_2$-CH$_2$-CH$_3$, Ge--CH$_2$CH$_2$CH$_2$OH were ~ 879 mV,~ 859 mV, and ~ 812 respectively, suggesting −CH$_2$CH$_2$CH$_2$OH to be the superior for the catalytic performance. The Tafel analysis concluded that the mechanistic route is dominated by electrochemical adsorption (Volmer process) and desorption followed by the Heyrovsky process [45].

Germanene Derivatives in Gas Sensing Applications

As discussed previously, the functionalization of germanane opens the band gap; this phenomenon can be utilized in developing gas-sensing materials. Nagrajan *et al.* performed a theoretical study on the interaction of NH_3 and H_2O molecules on the surface of germane [46]. His study concluded that the Ge-H surface is more sensitive toward moisture than NH_3 molecules. In another study, Kannan *et al.* studied the adsorption mechanism of different pollutants on germanane, theoretically evaluating the energy involved in adsorption, charge transfer, and energetics of the band gap. They studied the adsorption behavior of CO_2, H_2O, H_2S, NH_3, and NO_2 on the Ge-H surface. DFT calculations concluded that the Ge-H sheets are susceptible to NO_2 molecules [47]. Srimathi *et al.* studied biomarkers for Liver cirrhosis such as methanol, 2-pentanone, and limonene. The DFT studies suggest that the biomarkers adsorb on the surface reversibly *via* physisorption [48]. Fig. (**5**) illustrates summary of various molecules that can be sensed using 2D germanene or its derivatives.

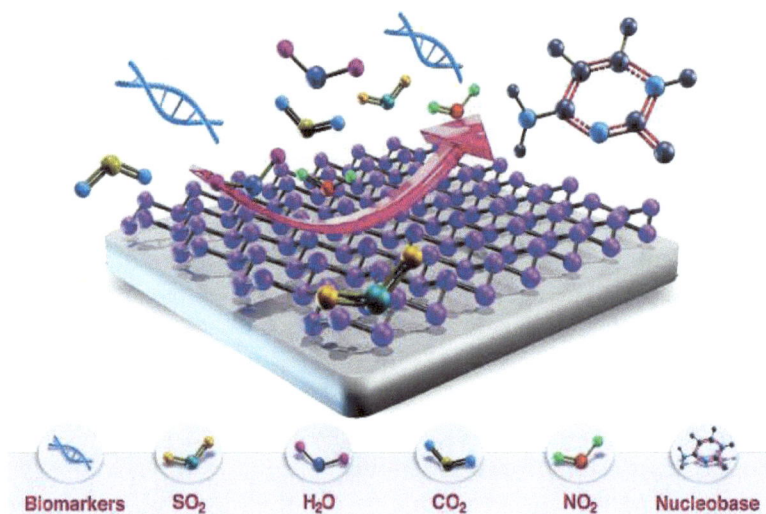

Biomarkers SO_2 H_2O CO_2 NO_2 Nucleobase

Fig. (5). Germanane in gas sensing applications.

In a similar study, Srimathi *et al.* evaluated the adsorption behaviors of different nucleobases, viz. adenine, guanine, cytosine, thymine, and uracil, on the germanane surface [49]. It is observed that the adsorption of nucleobases affects the resistance of the germanane sheet, which suggests the potential use of germanane sheets in sensing the nucleobases. The adsorption order on the Ge-H surface followed the order of cytosine > guanine > adenine > thymine > uracil. Munshi *et al.*, in another study, established theoretically that vacancy defects on the germanane layer stimulate the adsorption of H_2S and SO_2 gases. His study also confirmed that doping some aspects of the germanene sheet affects the adsorption of certain gases, such as N-doped Ge-H, which tends to adsorb CO_2 gas [50]. Moreover, the adsorption of gases affected the band gap energetics of the Ge-H sheet.

Germanene Derivatives in Biological Applications

Kouloumpis *et al.* first evaluated the antimicrobial efficacy of germanene [51]. He developed a methodology to screen the antimicrobial potential using this unique 2D material against gram-positive and gram-negative micro-organisms. His group observed the efficacy of GeH films against E. coli and *C. glutamicum,* and B. lactofermentum. In a standard method, the lethal effects of germanene were analyzed by exposing bacteria to the germanene surface for a brief period of 20 hours after exposure for 20 h. To investigate the mechanism of the lethal action of germanene toward microbes, Kouloumpis performs a TEM analysis of the C. glutamicum cell wall when microbes are placed over the germanene sheet and silicon wafers. The TEM analysis revealed that the cell wall remains undamaged when bacteria is placed over the silicon wafers. At the same time, a cellular rapture is observed when the cell wall interacts with germanene. Wang *et al.* synthesized novel nanocrystals (GeN) capable of exhibiting photothermal performance [52]. A summary of the significant studies on biological systems has been summarized in Table (**5**).

Table 5. Biological applications of germanane derivatives.

S. No.	2D Structure	Activity	Observation	DOI
1.	Germanene nanocrystals (Ge NCs)	Hydrogel-assisted photothermal therapy (PTT)	Innovative germanene-modified chitosan antimicrobial hydrogel (CS/Ge NCs0.8) exhibits antibacterial properties against E. coli and S. aureus *in vitro*.	[53]
2.	Germanene-based nanosheets	Surgical adjuvant therapeutic strategy in solid-tumor treatment	Ge-based hydrogel achieves *in situ* postoperative treatment for inhibiting tumor recurrence and wound infection.	[54]
3.	Germanene (Ge)	Biomedical applications	Ge-based strategy for surgical adjuvant treatment; drug-loaded Ge@hydrogel achieves residual tumor-eliminating and bacteria-killing effects.	[55]
4.	Germanene	Adsorption process of methyl mercury (MeHg)	MeHg chemisorbed on germanene; calculated formation energy is -1.61 eV.	[56]
5.	Germanene (2D-Ge)	Biomedical applications	Germanene (2D-Ge) categorized as a potential material for biomedical applications due to its outstanding properties.	[57]

(Table 5) cont.....

S. No.	2D Structure	Activity	Observation	DOI
6.	Germanene, phosphorene, antimonene	Biomedical applications of hydrogels	Beyond graphene 2D material/hydrogel composites with 2D materials like phosphorene, antimonene, and germanene have gained attention for biomedical applications.	[58]
7.	Silicene nanosheet	Antibacterial and antifungal activities	Silicene nanosheet exhibits efficient antibacterial and antifungal activities with good biocompatibility for biomedical applications.	[59]
8.	Germanene compounds	Fluorescent markers for tracking micromachines	Chemically modified 2D germanene compounds used as stable fluorescent markers for tracking individual micromachines in biomedical applications.	[60]
9.	Germanene, silicene, borophene, *etc.*	Detection and removal of biological contaminants	Different 2D nanomaterials, including germanene, silicene, and other Xenes, explored for the detection and removal of biological contaminants in various environments.	[61]
10.	Germanene, silicene, arsenene, borophene	Biosensors and bioimaging	Monoelemental 2D materials (Xenes) like germanene, silicene, arsenene, and borophene explored for their potential in biosensors and bioimaging in biomedicine.	[62]

Drawing parallels to graphene's journey, germanene stands poised to impact diverse sectors. Its exceptional properties open doors to advancements in electronics, energy storage, sensors, and beyond similar to that of graphene and its derivatives [63 - 65]. By harnessing cost-effective synthesis techniques, germanene becomes more accessible, igniting innovation and practical applications in various industries [66].

CONCLUSION

This book chapter reviewed the recent findings on germanane and its derivatives. The structure and reactivity of germanane were discussed in detail, focusing on the reasons for its buckled structure and the existence of the mixed hybrid orbitals. The chemical synthesis of germanene, germanane, and its derivatives was discussed with a prominent and in-depth discussion on the role of surface revamping in various properties of it. The chemical synthesis of germanene was discussed with details of the mechanism and, in principle, a discussion on the science behind it. The surface revamping of germanene with hydrogen to form germanane and further restructuring of germanane with a wide range of alkyl substituents is discussed in detail. The optoelectronic properties of germanane are discussed, keeping an eye on its origin and compared with graphene for a better understanding of the readers. Various applications of germanane are reviewed for a concise and comprehensive understanding, which includes in-depth discussions on the optoelectric, catalytic, sensing, and bio applications.

The energy applications of germanane are discussed with the potential of this wonder material in revamping the current energy demand and supply framework. Discussion on the fabrication of various highly efficient energy devices is included to develop consolidated arguments on the possibilities. In addition, the relation between the structure of germanene derivatives and their reactivity as the catalytic application has been discussed along with several examples. The discussion shows the immense potential of germanane as a catalytic material for several chemical processes. The first principle studies are included for the thorough understanding of the adsorption of various chemical moieties on the germanane surface; the discussion sheds light on the scope of using germanene and its derivatives in gas sensing applications. The chapter concludes with an extensive discussion of the biological applications of germanene derivatives in biological applications. It was established that these materials could extensively be used in various biological applications ranging from antimicrobial activities to drug delivery for cancer-related applications.

REFERENCES

[1]   Zhang, H. Introduction: 2D materials chemistry. *Chem. Rev.,* **2018**, *118*(13), 6089-6090.
[http://dx.doi.org/10.1021/acs.chemrev.8b00278] [PMID: 29991267]

[2]   Geim, A.K.; Novoselov, K.S. The rise of graphene. *Nat. Mater.,* **2007**, *6*(3), 183-191.
[http://dx.doi.org/10.1038/nmat1849] [PMID: 17330084]

[3]   Takeda, K.; Shiraishi, K. Theoretical possibility of stage corrugation in Si and Ge analogs of graphite. *Phys. Rev. B Condens. Matter,* **1994**, *50*(20), 14916-14922.
[http://dx.doi.org/10.1103/PhysRevB.50.14916] [PMID: 9975837]

[4]   Van Ngoc, H.; Thi Phuong Thuy, H.; Van On, V. Potential optoelectronic applications of C and Si-doped germanene nanoribbons. *Mater Today Commun,* **2023**, *36*, 106538.
[http://dx.doi.org/10.1016/j.mtcomm.2023.106538]

[5]   Liu, N.; Bo, G.; Liu, Y.; Xu, X.; Du, Y.; Dou, S.X. Recent progress on germanene and functionalized germanene: Preparation, characterizations, applications, and challenges. *Small,* **2019**, *15*(32), 1805147.
[http://dx.doi.org/10.1002/smll.201805147] [PMID: 30756479]

[6]   Li, L.; Lu, S.; Pan, J.; Qin, Z.; Wang, Y.; Wang, Y.; Cao, G.; Du, S.; Gao, H.J. Buckled germanene formation on Pt(111). *Adv. Mater.,* **2014**, *26*(28), 4820-4824.
[http://dx.doi.org/10.1002/adma.201400909] [PMID: 24841358]

[7]   Derivaz, M.; Dentel, D.; Stephan, R.; Hanf, M.C.; Mehdaoui, A.; Sonnet, P.; Pirri, C. Continuous germanene layer on Al(111). *Nano Lett.,* **2015**, *15*(4), 2510-2516.
[http://dx.doi.org/10.1021/acs.nanolett.5b00085] [PMID: 25802988]

[8]   Sahoo, N.G.; Esteves, R.J.; Punetha, V.D.; Pestov, D.; Arachchige, I.U.; McLeskey, J.T., Jr Schottky diodes from 2D germanane. *Appl. Phys. Lett.,* **2016**, *109*(2), 023507.
[http://dx.doi.org/10.1063/1.4955463]

[9]   Wang, Y.; Cong, C.; Yang, W.; Shang, J.; Peimyoo, N.; Chen, Y.; Kang, J.; Wang, J.; Huang, W.; Yu, T. Strain-induced direct–indirect bandgap transition and phonon modulation in monolayer WS2. *Nano Res.,* **2015**, *8*(8), 2562-2572.
[http://dx.doi.org/10.1007/s12274-015-0762-6]

[10]   Yan, J.; Zhang, Y.; Goler, S.; Kim, P.; Pinczuk, A. Raman scattering and tunable electron–phonon coupling in single layer graphene. *Solid State Commun.,* **2007**, *143*(1-2), 39-43.
[http://dx.doi.org/10.1016/j.ssc.2007.04.022]

[11]   Fukaya, Y.; Matsuda, I.; Feng, B.; Mochizuki, I.; Hyodo, T.; Shamoto, S. Asymmetric structure of germanene on an Al(111) surface studied by total-reflection high-energy positron diffraction. *2D Mater.,* **2016**, *3*(3), 035019.
[http://dx.doi.org/10.1088/2053-1583/3/3/035019]

[12]   Wang, W.; Uhrberg, R.I.G. Coexistence of strongly buckled germanene phases on Al(111). *Beilstein J. Nanotechnol.,* **2017**, *8*, 1946-1951.
[http://dx.doi.org/10.3762/bjnano.8.195] [PMID: 29046842]

[13]   Dávila, M.E.; Le Lay, G. Few layer epitaxial germanene: A novel two-dimensional Dirac material. *Sci. Rep.,* **2016**, *6*(1), 20714.
[http://dx.doi.org/10.1038/srep20714] [PMID: 26860590]

[14]   Dávila, M.E.; Xian, L.; Cahangirov, S.; Rubio, A.; Le Lay, G. Germanene: A novel two-dimensional germanium allotrope akin to graphene and silicene. *New J. Phys.,* **2014**, *16*(9), 095002.
[http://dx.doi.org/10.1088/1367-2630/16/9/095002]

[15]   Zhuang, J.; Gao, N.; Li, Z.; Xu, X.; Wang, J.; Zhao, J.; Dou, S.X.; Du, Y. Cooperative electron-phonon coupling and buckled structure in germanene on Au(111). *ACS Nano,* **2017**, *11*(4), 3553-3559.
[http://dx.doi.org/10.1021/acsnano.7b00687] [PMID: 28221757]

[16]   Wang, W.; Uhrberg, R.I.G. Investigation of the atomic and electronic structures of highly ordered two-dimensional germanium on Au(111). *Phys. Rev. Mater.,* **2017**, *1*(7), 074002.
[http://dx.doi.org/10.1103/PhysRevMaterials.1.074002]

[17]   Zhuang, J.; Liu, C.; Zhou, Z.; Casillas, G.; Feng, H.; Xu, X.; Wang, J.; Hao, W.; Wang, X.; Dou, S.X.; Hu, Z.; Du, Y. Dirac signature in germanene on semiconducting substrate. *Adv. Sci.,* **2018**, *5*(7), 1800207.
[http://dx.doi.org/10.1002/advs.201800207] [PMID: 30027050]

[18]   Gou, J.; Zhong, Q.; Sheng, S.; Li, W.; Cheng, P.; Li, H.; Chen, L.; Wu, K. Strained monolayer germanene with 1 × 1 lattice on Sb(111). *2D Mater.,* **2016**, *3*(4), 045005.
[http://dx.doi.org/10.1088/2053-1583/3/4/045005]

[19]   Zhang, L.; Bampoulis, P.; Rudenko, A.N.; Yao, Q.; van Houselt, A.; Poelsema, B.; Katsnelson, M.I.; Zandvliet, H.J.W. Structural and electronic properties of germanene on MoS 2. *Phys. Rev. Lett.,* **2016**, *116*(25), 256804.
[http://dx.doi.org/10.1103/PhysRevLett.116.256804] [PMID: 27391741]

[20]   Amlaki, T.; Bokdam, M.; Kelly, P.J. Z 2 invariance of germanene on MoS 2 from first principles. *Phys. Rev. Lett.,* **2016**, *116*(25), 256805.
[http://dx.doi.org/10.1103/PhysRevLett.116.256805] [PMID: 27391742]

[21]   Vogg, G.; Brandt, M.S.; Stutzmann, M. Polygermyne-a prototype system for layered germanium polymers. *Adv. Mater.,* **2000**, *12*(17), 1278-1281.
[http://dx.doi.org/10.1002/1521-4095(200009)12:17<1278::AID-ADMA1278>3.0.CO;2-Y]

[22]   Bianco, E.; Butler, S.; Jiang, S.; Restrepo, O.D.; Windl, W.; Goldberger, J.E. Stability and exfoliation of germanane: A germanium graphane analogue. *ACS Nano,* **2013**, *7*(5), 4414-4421.
[http://dx.doi.org/10.1021/nn4009406] [PMID: 23506286]

[23]   Cultrara, N.D.; Wang, Y.; Arguilla, M.Q.; Scudder, M.R.; Jiang, S.; Windl, W.; Bobev, S.; Goldberger, J.E. Synthesis of 1T, 2H, and 6R germanane polytypes. *Chem. Mater.,* **2018**, *30*(4), 1335-

1343.
[http://dx.doi.org/10.1021/acs.chemmater.7b04990]

[24]   Luo, X.; Zurek, E. Crystal structures and electronic properties of single-layer, few-layer, and multilayer GeH. *J. Phys. Chem. C,* **2016**, *120*(1), 793-800.
[http://dx.doi.org/10.1021/acs.jpcc.5b11770]

[25]   Tobash, P.H.; Bobev, S. Synthesis, structure and electronic structure of a new polymorph of CaGe2. *J. Solid State Chem.,* **2007**, *180*(5), 1575-1581.
[http://dx.doi.org/10.1016/j.jssc.2007.03.003]

[26]   Pinchuk, I.V.; Odenthal, P.M.; Ahmed, A.S.; Amamou, W.; Goldberger, J.E.; Kawakami, R.K. Epitaxial co-deposition growth of $CaGe_2$ films by molecular beam epitaxy for large area germanane. *J. Mater. Res.,* **2014**, *29*(3), 410-416.
[http://dx.doi.org/10.1557/jmr.2014.2]

[27]   Jiang, S.; Butler, S.; Bianco, E.; Restrepo, O.D.; Windl, W.; Goldberger, J.E. Improving the stability and optical properties of germanane *via* one-step covalent methyl-termination. *Nat. Commun.,* **2014**, *5*(1), 3389.
[http://dx.doi.org/10.1038/ncomms4389] [PMID: 24566761]

[28]   Jiang, S.; Arguilla, M.Q.; Cultrara, N.D.; Goldberger, J.E. Improved topotactic reactions for maximizing organic coverage of methyl germanane. *Chem. Mater.,* **2016**, *28*(13), 4735-4740.
[http://dx.doi.org/10.1021/acs.chemmater.6b01757]

[29]   Liu, Z. Methyl-terminated germanane $GeCH_3$ synthesized by solvothermal method with improved photocatalytic properties. *Appl. Surf. Sci.,* **2019**, *468*, 881-888.
[http://dx.doi.org/10.1016/j.apsusc.2018.10.228]

[30]   Sciacca, D.; Berthe, M.; Ryan, B.J.; Peric, N.; Deresmes, D.; Biadala, L.; Boyaval, C.; Addad, A.; Lancry, O.; Makarem, R.; Legendre, S.; Hocrelle, D.; Panthani, M.G.; Prévot, G.; Lhuillier, E.; Diener, P.; Grandidier, B. Transport properties of methyl-terminated germanane microcrystallites. *Nanomaterials,* **2022**, *12*(7), 1128.
[http://dx.doi.org/10.3390/nano12071128] [PMID: 35407246]

[31]   Hartman, T.; Konečný, J.; Mazánek, V.; Šturala, J.; Sofer, Z. A decade of germananes: Four approaches to their functionalization. *Inorg. Chem.,* **2022**, *61*(31), 12425-12432.
[http://dx.doi.org/10.1021/acs.inorgchem.2c01873] [PMID: 35877186]

[32]   Jiang, S.; Krymowski, K.; Asel, T.; Arguilla, M.Q.; Cultrara, N.D.; Yanchenko, E.; Yang, X.; Brillson, L.J.; Windl, W.; Goldberger, J.E. Tailoring the electronic structure of covalently functionalized germanane *via* the interplay of ligand strain and electronegativity. *Chem. Mater.,* **2016**, *28*(21), 8071-8077.
[http://dx.doi.org/10.1021/acs.chemmater.6b04309]

[33]   Wong, K.T.; Kim, Y.G.; Soriaga, M.P.; Brunschwig, B.S.; Lewis, N.S. Synthesis and characterization of atomically flat methyl-terminated Ge(111) surfaces. *J. Am. Chem. Soc.,* **2015**, *137*(28), 9006-9014.
[http://dx.doi.org/10.1021/jacs.5b03339] [PMID: 26154680]

[34]   Khuong Dien, V.; Li, W.B.; Lin, K.I.; Thi Han, N.; Lin, M.F. Electronic and optical properties of graphene, silicene, germanene, and their semi-hydrogenated systems. *RSC Advances,* **2022**, *12*(54), 34851-34865.
[http://dx.doi.org/10.1039/D2RA06722F] [PMID: 36540216]

[35]   Ye, X.S.; Shao, Z.G.; Zhao, H.; Yang, L.; Wang, C.L. Intrinsic carrier mobility of germanene is larger than graphene's: First-principle calculations. *RSC Advances,* **2014**, *4*(41), 21216-21220.
[http://dx.doi.org/10.1039/C4RA01802H]

[36]   Livache, C.; Ryan, B.J.; Ramesh, U.; Steinmetz, V.; Gréboval, C.; Chu, A.; Brule, T.; Ithurria, S.; Prévot, G.; Barisien, T.; Ouerghi, A.; Panthani, M.G.; Lhuillier, E. Optoelectronic properties of methyl-terminated germanane. *Appl. Phys. Lett.,* **2019**, *115*(5), 052106.
[http://dx.doi.org/10.1063/1.5111011]

[37] Serino, A.C.; Ko, J.S.; Yeung, M.T.; Schwartz, J.J.; Kang, C.B.; Tolbert, S.H.; Kaner, R.B.; Dunn, B.S.; Weiss, P.S. Lithium-ion insertion properties of solution-exfoliated germanane. *ACS Nano,* **2017,** *11*(8), 7995-8001.
[http://dx.doi.org/10.1021/acsnano.7b02589] [PMID: 28763196]

[38] Wu, B.; Šturala, J.; Veselý, M.; Hartman, T.; Kovalska, E.; Bouša, D.; Luxa, J.; Azadmanjiri, J.; Sofer, Z. Functionalized germanane/SWCNT hybrid films as flexible anodes for lithium-ion batteries. *Nanoscale Adv.,* **2021,** *3*(15), 4440-4446.
[http://dx.doi.org/10.1039/D1NA00189B] [PMID: 36133472]

[39] Loaiza, L.C.; Dupré, N.; Davoisne, C.; Madec, L.; Monconduit, L.; Seznec, V. Complex lithiation mechanism of siloxene and germanane: Two promising battery electrode materials. *J. Electrochem. Soc.,* **2021,** *168*(1), 010510.
[http://dx.doi.org/10.1149/1945-7111/abd44a]

[40] Mortazavi, B.; Dianat, A.; Cuniberti, G.; Rabczuk, T. Application of silicene, germanene and stanene for Na or Li ion storage: A theoretical investigation. *Electrochim. Acta,* **2016,** *213*, 865-870.
[http://dx.doi.org/10.1016/j.electacta.2016.08.027]

[41] Liu, Z.; Dai, Y.; Zheng, Z.; Huang, B. Covalently-terminated germanane GeH and GeCH$_3$ for hydrogen generation from catalytic hydrolysis of ammonia borane under visible light irradiation. *Catal. Commun.,* **2019,** *118*, 46-50.
[http://dx.doi.org/10.1016/j.catcom.2018.09.016]

[42] Ni, C.; Chevalier, M.; Veinot, J.G.C. Metal nanoparticle-decorated germanane for selective photocatalytic aerobic oxidation of benzyl alcohol. *Nanoscale Adv.,* **2022,** *5*(1), 228-236.
[http://dx.doi.org/10.1039/D2NA00518B] [PMID: 36605808]

[43] Liu, Z.; Lou, Z.; Li, Z.; Wang, G.; Wang, Z.; Liu, Y.; Huang, B.; Xia, S.; Qin, X.; Zhang, X.; Dai, Y. GeH: a novel material as a visible-light driven photocatalyst for hydrogen evolution. *Chem. Commun.,* **2014,** *50*(75), 11046-11048.
[http://dx.doi.org/10.1039/C4CC03636K] [PMID: 25098390]

[44] Konečný, J.; Hartman, T.; Antonatos, N.; Mazánek, V.; Sofer, Z.; Šturala, J. Photomodification of benzyl germanane with group 6 metal carbonyls. *FlatChem,* **2022,** *33*, 100354.
[http://dx.doi.org/10.1016/j.flatc.2022.100354]

[45] Nagarajan, V.; Bhattacharyya, A.; Chandiramouli, R. Adsorption of ammonia molecules and humidity on germanane nanosheet-A density functional study. *J. Mol. Graph. Model.,* **2018,** *79*, 149-156.
[http://dx.doi.org/10.1016/j.jmgm.2017.11.009] [PMID: 29169059]

[46] Kannan, V.; Ganesan, V.; Vijayakumar, V. Adsorption of CO$_2$, H$_2$O, H$_2$S, NH$_3$ and NO$_2$ on germanane nanosheet-A density functional study. *Comput. Theor. Chem.,* **2022,** *1214*, 113799.
[http://dx.doi.org/10.1016/j.comptc.2022.113799]

[47] Srimathi, U.; Nagarajan, V.; Chandiramouli, R. Detection of nucleobases using 2D germanane nanosheet: A first-principles study. *Comput. Theor. Chem.,* **2018,** *1130*, 68-76.
[http://dx.doi.org/10.1016/j.comptc.2018.03.011]

[48] Srimathi, U.; Nagarajan, V.; Chandiramouli, R. Germanane nanosheet as a novel biosensor for liver cirrhosis based on adsorption of biomarker volatiles - A DFT study. *Appl. Surf. Sci.,* **2019,** *475*, 990-998.
[http://dx.doi.org/10.1016/j.apsusc.2019.01.008]

[49] Monshi, M.M.; Aghaei, S.M.; Calizo, I. Doping and defect-induced germanene: A superior media for sensing H$_2$S, SO$_2$, and CO$_2$ gas molecules. *Surf. Sci.,* **2017,** *665*, 96-102.
[http://dx.doi.org/10.1016/j.susc.2017.08.012]

[50] Kouloumpis, A.; Chatzikonstantinou, A.V.; Chalmpes, N.; Giousis, T.; Potsi, G.; Katapodis, P.; Stamatis, H.; Gournis, D.; Rudolf, P. Germanane monolayer films as antibacterial coatings. *ACS Appl. Nano Mater.,* **2021,** *4*(3), 2333-2338.

[http://dx.doi.org/10.1021/acsanm.0c03149] [PMID: 33842855]

[51] Xu, Y.; Wang, X.; Zhang, W.L.; Lv, F.; Guo, S. Recent progress in two-dimensional inorganic quantum dots. *Chem. Soc. Rev.,* **2018**, *47*(2), 586-625.
[http://dx.doi.org/10.1039/C7CS00500H] [PMID: 29115355]

[52] Fojtů, M.; Balvan, J.; Raudenská, M.; Vičar, T.; Šturala, J.; Sofer, Z.; Luxa, J.; Plutnar, J.; Masařík, M.; Pumera, M. 2D germanane derivative as a vector for overcoming doxorubicin resistance in cancer cells. *Appl. Mater. Today,* **2020**, *20*, 100697.
[http://dx.doi.org/10.1016/j.apmt.2020.100697]

[53] Wang, X.; Sun, X.; Bu, T.; Xu, K.; Li, L.; Li, M.; Li, R.; Wang, L. Germanene-modified chitosan hydrogel for treating bacterial wound infection: An ingenious hydrogel-assisted photothermal therapy strategy. *Int. J. Biol. Macromol.,* **2022**, *221*, 1558-1571.
[http://dx.doi.org/10.1016/j.ijbiomac.2022.09.128] [PMID: 36126816]

[54] Feng, C.; Ouyang, J.; Tang, Z.; Kong, N.; Liu, Y.; Fu, L.Y.; Ji, X.; Xie, T.; Farokhzad, O.C.; Tao, W. Germanene-based theranostic materials for surgical adjuvant treatment: Inhibiting tumor recurrence and wound infection. *Matter,* **2020**, *3*(1), 127-144.
[http://dx.doi.org/10.1016/j.matt.2020.04.022]

[55] Tao, W.; Kong, N.; Ji, X.; Zhang, Y.; Sharma, A.; Ouyang, J.; Qi, B.; Wang, J.; Xie, N.; Kang, C.; Zhang, H.; Farokhzad, O.C.; Kim, J.S. Emerging two-dimensional monoelemental materials (Xenes) for biomedical applications. *Chem. Soc. Rev.,* **2019**, *48*(11), 2891-2912.
[http://dx.doi.org/10.1039/C8CS00823J] [PMID: 31120049]

[56] Al Fauzan, M.R.; Satya, T.P.; Setyawan, G.; Fahrurrozi, I.; Puspasari, F.; Partini, J.; Sholihun, S. Adsorption of toxic heavy metal methylmercury (MeHg) on germanene in aqueous environment: A first-principles study. *Indon. J. Chem.,* **2021**, *21*(6), 1484.
[http://dx.doi.org/10.22146/ijc.66902]

[57] Khan, K.; Tareen, A.K.; Iqbal, M.; Wang, L.; Ma, C.; Shi, Z.; Ye, Z.; Ahmad, W.; Rehman Sagar, R.U.; Shams, S.S.; Sophia, P.J.; Ullah, Z.; Xie, Z.; Guo, Z.; Zhang, H. Navigating recent advances in monoelemental materials (Xenes)-fundamental to biomedical applications. *Prog. Solid State Chem.,* **2021**, *63*, 100326.
[http://dx.doi.org/10.1016/j.progsolidstchem.2021.100326]

[58] Garg, M.; Thakur, A. A review: Biomedical applications of phosphorene, antimonene, and germanene-based 2D material/hydrogel complexes. *J. Mater. Sci.,* **2023**, *58*(1), 34-45.
[http://dx.doi.org/10.1007/s10853-022-07954-7]

[59] Taşaltın, N.; Taşaltın, C.; Üstün-Alkan, F.; Karakuş, S. Biocompatible, antibacterial, and antifungal two-dimensional silicene nanosheets with a honeycomb hexagonal structure. *Biomass Convers. Biorefin.,* **2023**.
[http://dx.doi.org/10.1007/s13399-022-03726-0]

[60] Maric, T.; Beladi-Mousavi, S.M.; Khezri, B.; Sturala, J.; Nasir, M.Z.M.; Webster, R.D.; Sofer, Z.; Pumera, M. Functional 2D germanene fluorescent coating of microrobots for micromachines multiplexing. *Small,* **2020**, *16*(27), 1902365.
[http://dx.doi.org/10.1002/smll.201902365] [PMID: 31433114]

[61] Jayanta Sarmah, B. Trending 2D nanomaterial composites in detection and sensing of biological contaminants. In: *2D Nanomaterials for Energy and Environmental Sustainability*; Springer, **2022**.
[http://dx.doi.org/10.1007/978-981-16-8538-5_8]

[62] Duan, X.; Liu, Z.; Xie, Z.; Tareen, A.K.; Khan, K.; Zhang, B.; Zhang, H. Emerging monoelemental 2D materials (Xenes) for biosensor applications. *Nano Res.,* **2023**, *16*(5), 7030-7052.
[http://dx.doi.org/10.1007/s12274-023-5418-3]

[63] Punetha, V.D.; Ha, Y.M.; Kim, Y.O.; Jung, Y.C.; Cho, J.W. Interaction of photothermal graphene networks with polymer chains and laser-driven photo-actuation behavior of shape memory polyurethane/epoxy/epoxy-functionalized graphene oxide nanocomposites. *Polymer,* **2019**, *181*,

121791.
[http://dx.doi.org/10.1016/j.polymer.2019.121791]

[64]   Punetha, V.D.; Ha, Y.M.; Kim, Y.O.; Jung, Y.C.; Cho, J.W. Rapid remote actuation in shape memory hyperbranched polyurethane composites using cross-linked photothermal reduced graphene oxide networks. *Sens. Actuators B Chem.,* **2020**, *321*, 128468.
[http://dx.doi.org/10.1016/j.snb.2020.128468]

[65]   Punetha, V.D.; Rana, S.; Yoo, H.J.; Chaurasia, A.; McLeskey, J.T., Jr; Ramasamy, M.S.; Sahoo, N.G.; Cho, J.W. Functionalization of carbon nanomaterials for advanced polymer nanocomposites: A comparison study between CNT and graphene. *Prog. Polym. Sci.,* **2017**, *67*, 1-47.
[http://dx.doi.org/10.1016/j.progpolymsci.2016.12.010]

[66]   Sahoo, N.G.; Sandeep, M. A process of manufacturing. Indian Patent 352780, 2016.

CHAPTER 8

Silicene - A Novel 2D Material with Potential for Nanoelectronics and Photonics

Rakshit Pathak[*, 1], **Shalini Bhatt**[1] and **Rajesh Kumar**[2]

1 Centre of Excellence for Research, P.P. Savani University, Surat-394125, Gujarat, India

2 Department of Chemistry, S.S.J. University, Campus Almora-263601, Uttarakhand, India

Abstract: Due to its distinct physicochemical properties, silicene, a silicon allotrope with a 2-D honeycomb assembly, has attracted considerable interest from the entire research community. The mixed sp^2/sp^3 hybridization of silicon atoms increases surface chemical activity and enables a range of mechanical and electronic characteristics. A new topology of silicon-based nanoparticles known as 2D silicene has recently been developed. It has a distinctive planar structure with a considerable surface, unusual physiochemical characteristics, and favorable biological effects. In theoretical observation, it exhibits remarkable characteristics and has many advantages over graphene as a 2D material, which makes it a more exciting component and a matter of deep study. So, the present chapter provides a complete overview of this 2D material covering its wide applications in different sectors. The chapter mainly provides insights into the synthesis approach and its characteristics, including its mechanical, electrical, and spintronic attributes. Then, to shed light on the various phases of silicene seen on the metal surfaces on its electrical structures, we describe the experimental characterization of silicene. The chapter also covers the most current uses of silicene outlined in the context of nanoelectronics.

Keywords: Band Structure, Density Functional Theory, Scanning Tunnel Microscopy, Spin-Orbit Coupling, Xenes, 2D materials.

INTRODUCTION

A revolution in studying and producing different 2D materials may be attributed to Novoselov and their teammates in 2004 when they discovered graphene. Graphene as a material grasps the wide attention of the entire scientific world owing to its unique and remarkable electrical and mechanical characteristics as a subject of fundamental study and future applications [1 - 3]. Graphene is considered the most recent ten years' worth of materials science research and

---

* **Corresponding author Rakshit Pathak:** Centre of Excellence for Research, P.P. Savani University, Surat-394125, Gujarat, India; E-mail: rakshit.pathak@ppsu.ac.in

Vinay Deep Punetha (Ed.)

consists of single layers of sp2- hybridized carbon atoms in the form of a crystalline structure. Moreover, it is being used catalytically in the energy devices and biomedical sectors [4 - 8]. The honeycomb design and electronic structure (Band Gap and Dirac points) of graphene make it a material having excellent properties such as charge-carrying capabilities, electron flow, and the Quantum Hall Effect (QHE) [9]. The high mobility of graphene's charge makes it a viable candidate for future nanoelectronic devices, such as in energy storage, circuits, transistors, bio-detection, *etc* [10]. It suggests that the capabilities of novel 2D materials in such applications and recent developments in the modification of the several physicochemical characteristics of graphene have opened the door to the research of other 2D materials [11, 12]. Various 2D materials have been investigated and have been categorized as a subclass of 2D materials and further named Xenes (Table 1).

Table 1. 2D material (Xenes) of different elements of the group (III-VI) [13].

Atomic Symbol	Xenes
Si	Silicene
Ge	Germanene
Sn	Stanene
Pb	Plumbene
P	Phosphorene
As	Arsenene
Sb	Antimonene
Bi	Bismuthene
Se	Selenene
Te	Tellurene

Considering their remarkable qualities, including very thin folded structure, extremely high surface-to-volume ratio, exceptional mechanical strength, and flexibility, there are a wide range of features found in this material for developing devices and applications in sensing, photonics, energy storage, *etc* [14, 15]. One of the 2D materials that mimic graphene in various ways is silicene. The structure of silicene in the form of a hexagonal lattice first came to notice after the publication of a study in 1994. The study of Takeda *et al.* [16] on the hexagonal Si and Ge rings concerning their carbon equivalent is the first basic theory where silicene and germanene first emerged. The proposed study discusses the aromatic structure of Si and Ge. Even before the emergence of graphene in 2004, silicene finished its premiere as a 2D honeycomb in a diffident manner and maybe with an uncertain goal [17]. Due to its exceptional solid-state characteristics, such as the quantum spin Hall (QSH), high spin-orbit coupling (SOC), and modulated bandgap, silicene has recently attracted more significant attention as a critical material in material chemistry. Due to such exceptional properties, silicene production and use have transformed from hypothetical assumptions to practical interpretations in recent years.

Although the electrical structures of silicon and carbon atoms are identical, synthesizing silicene is extremely difficult. For carbon, sp^2 hybridization is more stable than sp^3 hybridization, whereas silicon is the opposite. As a result, it is energetically unfavorable for silicon atoms to spontaneously create silicene, which restricts the use of traditional chemical or physical approaches to silicene synthesis. Many theoretical computations have projected intriguing and rich physics of silicene despite the difficulty in production. For example, the Dirac fermion state was demonstrated in a linear dispersion band structure close to the Fermi level by Density Functional Theory (DFT) calculations [18, 19].

Similarly, the quantum hall effects (QHEs) were also predicted in silicene [20, 21]. Silicene's electronic structure may change due to substrate defects on which it may synthesize or *via* metal interactions. These theoretical efforts have opened the road for experimental research on silicene's unusual features. Furthermore, silicon, the material used to make silicon chips, is the source of silicene.

Silicene has the potential to become a future substance for various applications. On the energy front, silicene is anticipated to increase the storage of solar cells, making solar electricity more accessible and practical. Silicene may improve DNA sequencing, X-ray imaging, and medical sensors that track blood flow and heart rhythms [22]. It will also speed up and produce more biocompatible and bio-disposable health monitoring gadgets. It can produce flexible displays, paper-thin cell phones, and more effective batteries. So, the present chapter is an effort to discuss the present scenario of silicene as a 2D material in terms of its chemistry, synthesis approaches, derivatives, and future applications associated with the nanoelectronics of this miraculous material. The chapter starts with the outline of the initial status of silicene based on different theoretical assumptions and calculations. It also covers the silicene basic hybridization properties and their impact on the formation of silicene. The chapter also provides insights into silicon synthesis based on the deposition method where Ag (111) was used as a substrate material. After covering the basic theoretical and experimental observation, the chapter is focused on the different applications of silicene 2D material, especially in nanoelectronics and energy storage devices.

STRUCTURE ASSOCIATED CHEMISTRY OF SILICENE

Si atoms are arranged in a monolayer with a hexagonal configuration to form a 2D silicene. It is forecast to be buckled in its low energy state, which consists of top and bottom Si atoms (Fig. **1**). The longer Si-Si bond length than the C-C prevents the Si from becoming "pure" sp^2 hybridized and forming π-bonds, silicene buckles. The Si atoms buckle, allowing for more orbital overlap, leads to mixed sp^2-sp^3 hybridization [23, 24]. It is just the most simplified structure of silicene to understand the arrangement of silicon in the hexagonal lattice or the freestanding.

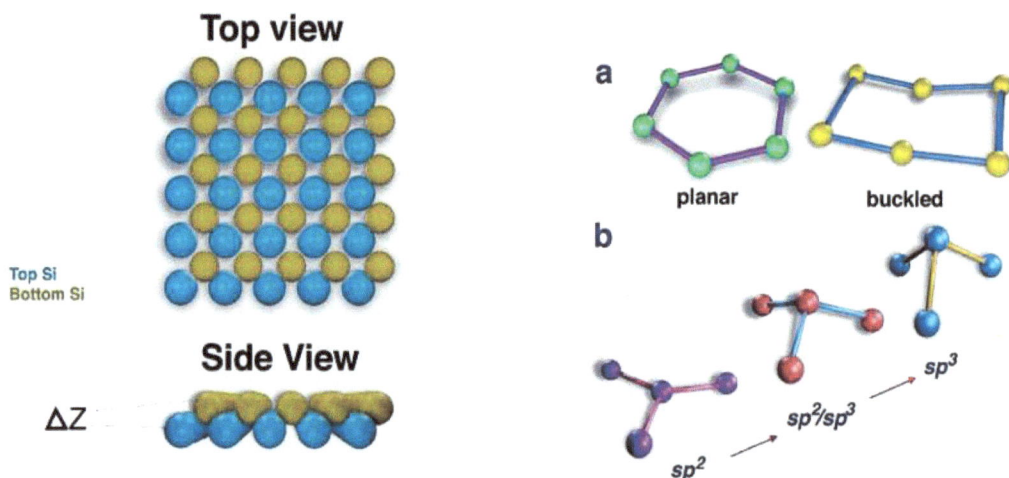

Fig. (1). Top and side views of freestanding silicene; silicene structure and bonding (a) Planar and buckled hexagonal ring, (b) Mixed sp2- sp3 hybridization.

Compared to the flat state symmetry (D_{6h}), unique to carbon, these aromatic phases were projected to be more even in a consistently wavy structure (D_{3d} symmetry). This assumption was further supported by the total energy calculation [25]. Many exclusive features are generated due to the symmetry and this unique structure of silicene, which is called buckling.

The mirror symmetry breakdown that occurs in silicene due to buckling eliminates the planar high-symmetry-related instability. s-p orbital mixing helps the system become stable by preventing the pseudo-Jahn-Teller distortion that results from the interaction between occupied and unoccupied molecular orbitals that causes buckling [26, 27]. Contrarily, it is significant that low buckling maintains the hexagonal symmetry, meaning that the two atoms in the unit cell are equivalent in the honeycomb lattice, and electrons can move between the closest neighbors, maintaining the existence of Dirac fermions [28].

Due to the longer Si-Si bond length than C-C, which prevents Si atoms from becoming "pure" sp^2 hybridized and forming π-bonds, silicene buckles more quickly than other materials. The Si atoms buckle, allowing for more orbital overlap, which leads to mixed sp^2-sp^3 hybridization. In freestanding silicene, the basic structure and parameters are as follows, as shown in Table **2**. All these values are calculated using DFT calculation [29, 30].

Table 2. Different parameters of the freestanding silicene structure.

Parameters	Unit	Values
Hybridization	-	Mixed sp^2-sp^3
Si-Si bond length	nm	2.28
Lattice parameters	-	$\alpha = \beta = 90; \gamma = 120$
Lattice Constant	Å	a = b = 3.87 Å
Buckling parameter (δ) (nm)	nm	0.045
Energy Gap (Eg)	meV	1.9

Silicene's Band Structure

Silicene's Dirac cones have linear dispersion and cross at the Fermi energy and the Brillouin zone. The two degenerate bands at a particular site arise from the A and B sub-lattices of the hexagonal structure, as illustrated in Fig. (**2**) [31, 32].

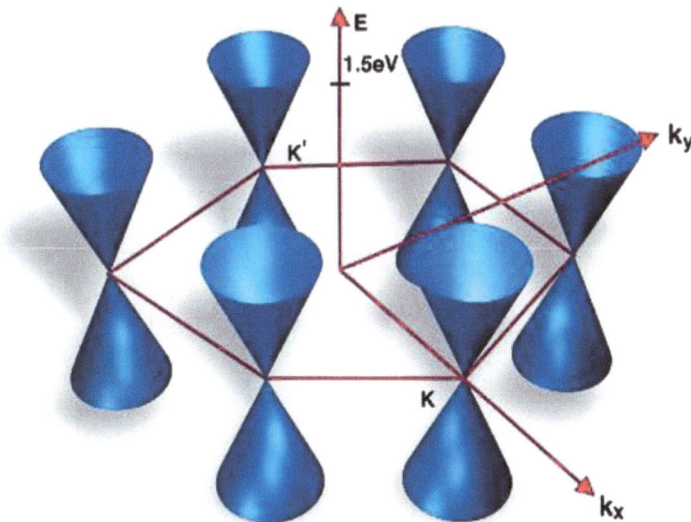

Fig. (2). Silicene hexagonal structure and the dirac cones with brillouin zone.

In the graphene structures, the p_z orbit on each sub-lattice, isolated from the s and p_x, p_y orbitals by the reflection symmetry of the plane, are the source of the Dirac bands. The optical features of freestanding silicene have been the subject of more theoretical research. Much like graphene, Sommerfeld's constant may be used to compute the low-frequency absorbance of silicene and other Xenes using equation 01 [33].

$$\alpha = {e^2}/{\hbar c} \qquad (1)$$

Although Dirac cones were not realized, the initial electronic band structure of silicene was computed using DFT. Despite early claims of observation, whether the Dirac cones are present for silicene on Ag is still in dispute, even though it has been detected for graphene [34]. Moreover, spin-orbit coupling (SOC) of bands in mxenes and graphene often eliminates degeneracies. However, for silicene, the end effect has been demonstrated to open a Significant tiny gap of 1.55 meV (whereas it is insignificant for graphene), suggesting that it may be more effective than graphene at exhibiting the QHE [35]. In another proposed model, the effective SOC in buckling, the silicene, and type structures (Xenes) create a topological gap at the Dirac points that is compatible with a conductive edge at geometrical boundaries, such as those of a silicene ribbon. In addition, the model also suggests that the stronger the spin-orbit coupling, the higher the gap opening at the Dirac points. Because of this, it is anticipated that topological features would be more stable in Xenes with increasing atomic mass [15]. Owing to this, the newly created model of 2D silicene on a substrate has been extended to other xenes with a method for synthesis and identification almost identical to that of silicene. Much work has lately been dedicated to constructing non-interacting substrates that serve as a substrate for silicene, or "template engineering." Among different substrates, the most common substrate is silver phase Ag (111). The phase structure of silicene is also varied in silicene based on the substrate.

Silicene Structure Phases on Ag (111) substrate

In silicene, the sp^2 and sp^3 hybridization occurred in combination. Owing to this, it has a low buckled structure. Silicene phases have been reported to develop on a variety of substrates, including different lattice structures of silver (Ag), gold (Au), and ZrB_2, *etc.* Due to the mixed hybridization, different phases of silicene occurred in the substrate [2, 36]. The most common substrate used to synthesize silicene is Ag (111), which is grown with different phases. The most common phases are as follows:

3 X 3 silicene is the most commonly synthesized by keeping the temperature around 420K at the time of Si deposition on Ag (111). Through the comparison of scanning tunnel microscope (STM) observation, three groups detailed the silicene model in 2012 [37 - 39]. They determined the two different phases of the silicene hexagonal structure and named them as H and T phases. Both H and T phases coexist and demonstrate similar stability. In contrast to phase H, phase T often forms with lower Si. As a result, phase T is a precursor to H, which is thought to be the stable phase of 3 X 3 silicene.

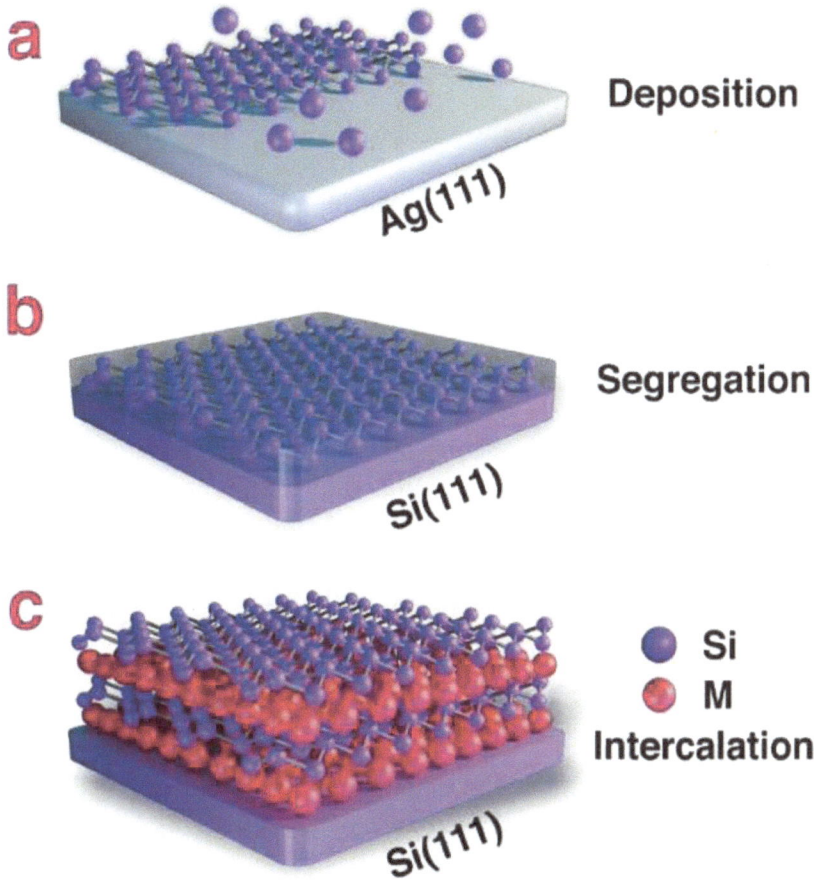

Fig. (3). Silicene synthesis *via* (a,b) epitaxial deposition and segregation, (c) intercalation.

At temp. 480K, a new silicene phase known as $\sqrt{7}$ X $\sqrt{7}$ is formed. This phase often appears as a damaged moire pattern. The hexagon at the positive and the wrecked area between them make up this moiré design, as shown in Fig. (3). The location variation of Si atoms from those of Ag (111) causes the brilliant areas of the moire pattern. Hexagonal rings will become stable and complete due to the strong connection between silicene and the substrate made of Ag (111). The contact splits these hexagonal rings into other regions of the moire pattern when there is a more significant positional divergence between the silicon atoms and the substrate. Raising the temperature to 500 K makes a silicene phase known as √3 X √3 silicene the most stable [25, 40]. The silicene phase that has been studied the most is this one. In addition to these structures, other buckling patterns and structures were also studied. However, thorough experimental and theoretical studies for these phases are absent. There are more divergent opinions than common ones, even for the three stages briefly described in this section. Still, a detailed structure study is much needed for the silicene structure.

SYNTHESIS OF SILICENE

Silicene is one of the most significant types of xenes, which have the band structure of the Dirac type. Different methods exist in the literature for the synthesis of silicene using different deposition methods and substrates [41, 42]. To develop the silicene layers and comprehend how the substrate and the phase transition between low- and high-buckling impact the electrical structure, silicene is most profoundly synthesized by depositing silicon in monolayers onto the surface of Ag (111). Different substrates are also used for depositing the silicon on their surface to synthesize the epitaxial silicene. The deposition method is a type of bottom-up approach. It is named epitaxial growth on the substrate, so the silicene structure synthesized using this technique is also termed epitaxial silicene [43, 44]. This method provides potential chip scale production targeting the technological applicability of synthesized silicene despite the higher cost. A combination of experimental techniques like STM and DFT computations is often required to physically identify silicene [45]. Three broad methods have been used so far to create silicene, as shown in Fig. (**3**). A brief description of these three methods is given below.

Epitaxial Silicene by a Deposition Method

The deposition process begins with condensing thermally dispersed silicon onto a substrate. The Ag (111) inhibits the synthesis and device mixing of Silicene, even though the number of substrates on which it may be produced is fast increasing [46]. In this regard, it is crucial to remember that while Ag (110) was first demonstrated to support silicon nanoribbons, Ag (111) has since proved a very versatile template for the synthesis of silicene *via* the deposition method.

In 2012, silicene was synthesized on Ag (111) substrate using the deposition method under high vacuum conditions [38]. After the successful synthesis, it was further synthesized using different substrates such as Ir, MoS_2, *etc* [47, 48]. Simultaneously with a 4X4 Ag (111) supercell that displays the well-known "flower pattern" in STM, silicene on silver (111) creates the classic, highly organized, and recognizable 3X3 phase [35, 49]. Due to high-resolution synchrotron radiation angle-resolved photoelectron spectroscopy (SR-ARPES) and DFT observations, the "magic" epitaxial association upon the growth of silver on Si (111) substrates, the "flower pattern" of silicene was revealed. Most often, the formation of variously rebuilt domains results in the large-scale deposition of epitaxial silicene.

Epitaxial Silicene by Segregation Method

Despite theoretical expectations, silicene synthesis still needs to be solved. Most attempts to produce silicene on crystalline surfaces used molecular beam epitaxy (MBE)/ deposition, as explained in the above section. However, a different attempt was also made to synthesize epitaxial silicene based on the segregation method. In synthesizing epitaxial silicene by segregation, the substrate works as a pool through a buffer. The most significant result was obtained in the study of Fleurence and their coworker in 2012, where they used ZrB_2 as a substrate. In this synthesis approach, epitaxial silicene was made *via* surface segregation on $ZrBr_2$ thin sheets grown on Si layers. Direct-electronic band gap at the spot is caused by a specific silicene buckling caused by the epitaxial connection with the particular surface. These findings show that epitaxial strain alters the buckling and the electrical characteristics of silicene. The total surface system's electronic band structure is determined by interaction with the substrate; however, unlike metal-supported silicene, ZrB_2-supported silicene has a gapped electronic structure [50].

Intercalated Silicene

In addition to the physical methods often used in silicene synthesis, chemical processes can also be used to create silicon nanosheets. The flexibility of the process enables the synthesis of silicon nanoparticles/dots, nanowires, nanotubes, *etc.* Silicon nanosheets organically embedded in calcium disilicide ($CaSi_2$) serve as a prime example of the synthesis of silicene in an intercalated approach [51, 52]. It was the primary method for this type of synthesis route. The goal is to identify silicene-like pre-formed structures that can be created *via* silicide topo-chemical de-intercalation. $CaSi_2$ is a Zintl silicide in which planar monolayers of Ca^{2+} or intercalated multilayer silicene are used to separate 2D silicon sheets formed of Si physical rings from one another [53, 54]. In another study related to intercalated silicene synthesis, Tokmachev (2016) synthesized multilayer silicene using the intercalation compound $SrSi_2$. It is feasible to create a polymorph of $SrSi_2$ that has the intercalated multilayer silicene structure and the simplest stacking of silicene sheets. Different spectroscopic experiments proved the quality of synthesized silicene synthesis carried out on Si (001) and Si (111) surfaces resulting in epitaxial layers of $SrSi_2$ whose orientation is regulated by the substrate. The transport characteristics of the films reflect the structural $SrSi_2/Si$ connection [55].

Besides that, other silicon-based nanostructures have also been studied for a unique and straightforward method of consistent manufacture. Any type of nano-scaled silicon acceptable to the various optoelectronic sectors is a potential way to replace cubic silicon in scalable devices and meet the technical requirements. While practical studies successfully demonstrated hexagonal silicon, the so-called lonsdaleite phase is also known as Si_{24} [56]. Theoretical research in this context was focused on exploring the allotropic phases of silicon other than silicene. Low-dimensional silicon nanosheets, which have unusual characteristics, are the subject of another burgeoning study. Chemical processing is quickly becoming a viable alternative to epitaxial techniques for producing silicon nanosheets with 4 nm thickness on a wide scale and at a reasonable cost [46].

SILICENE'S FUNCTIONALIZATION

The chemical functionalization and pile-up stacking are the key refrains that make the silicene valuable and applicable by generating the band gap on the Dirac points. The silicene single layer is intended to serve as a precursor for a new artificial material with increased functionality in both strategies. Silicene zero band gap electronic properties make it difficult to use them in optoelectronic contrivances, which are the most basic need of the present hour [57]. Silicene's band gap can be manipulated *via* the functionalization process and can be made into non-zero, which makes it suitable for various electronic applications. The main methods for functionalizing silicene are hydrogenation, halogenation, and oxidization [58].

Hydrogenation

Non zero band gap can easily be achieved by hydrogenation, which transforms sp^2 Si atoms into sp^3. The functionalized Si-H is referred to as silicone. First-principle calculations reveal that the chair and boat conformation are the two types of conformations where the produced compound is dynamically stable. The functionalized silicones showed different physicochemical characteristics due to the modulated band structure. For instance, three hydrogenated double-layer silicene sheets with well-ordered structures, namely Si_8H_4, $Si_{16}H_{12}$, and $Si_{12}H_{10}$, show direct (or quasi-direct) band gaps in the RGB color spectrum at high hydrogen concentrations and might be employed as white LED light emitters [59].

Half-silicone is also a type of hydrogen-functionalized silicene in which only the top surface of the substrate is functionalized with the hydrogen, resulting in an asymmetric out-of-plane structural structure. Reviewing the steps of silicene's production, one could be amazed if stacking single layers of silicene might be used to manufacture graphite-like silicon. A novel thermodynamically stable layered phase of silicon known as silicate, characterized by silicene sheets stacking with a dumbbell pattern and high directionality in the electrical and structural characteristics, was projected to form from such a crystal [46]. With its immense promise for future nanoelectronics and spintronics, hydrogenation offers a unique technique to customize the electrical and magnetic characteristics of silicene. The 2D semiconductor silicone provides a platform for further doping and functionalization because of its large band gap. Doping can also boost carrier mobility in silicone by lowering the holes and electron mass [42, 58].

Halogenation

Similar to hydrogenation, halogen chemisorption (X= F, Cl, Br, I) elements may also cause a non-zero band gap [35, 60]. Like silicone, the chair structure is much more stable than the boat for X-silicene. Band gaps for F, Cl, Br, and I in all X-silicene are 1.47, 1.979, 1.950, and 1.194 eV, respectively [32]. The rivalry between the bond strengths causes the fluctuating pattern in Eg. DFT calculations, and the many-body Green's function approaches have been used to conduct an in-depth study on the F-silicene with chair configuration among X-silicene [23]. Besides this study, Wang *et al.* studied half-fluorinated silicene *via* DFT calculations. They observed that all three conformers are dynamically stable, as shown by phonon dispersion charts, and similar to half-hydrogenated silicene, the zigzag structure is the most stable among them [61]. Zhao and their coworkers also studied the adsorption of a magnetic super halogen molecule $MnCl_3$ on silicene in addition to halogen element atoms [62]. A super halogen molecule, by definition, has an electron affinity greater than Chlorine. For instance, when deposited one at a time on the silicene sheet, the $MnCl_3$ retains its 4μB magnetic moment. It is interesting to note that the silicene sheet with two molecules of $MnCl_3$ on the opposite side is antiferromagnetic, the one with two MnCl3 molecules on the same side is ferromagnetic with a total spin moment of 8 μB. Silicene is half-metallic when only one $MnCl_3$ is adsorbed onto it, but when two $MnCl_3$ are adsorbed onto it, it develops as a semiconductor with a band gap around 0.63 eV.

Oxidization

The applicability of silicene is badly affected when the functionalized silicene is exposed to the oxidation process and the chemical stability of silicene declines. Using first-principles calculations (FPC), the stability of freestanding silicene in an oxygen atmosphere has also been studied by Liu and their coworkers [63]. The study observed that O_2 was discovered to be readily adsorbed on silicene and to spontaneously split into two O atoms without encountering any energy barriers. Strong Si-O bonds created by the fragmented O atoms reduce the energy of each O_2 by 4.046–5.355 eV. Additionally, oxidization offers a versatile method for tuning the band gap of silicene. The silicene bandgap may presumably be changed most easily by oxidation. Silicone can be used in place of hexagonal boron nitride (hex-BN) as a precursor step for silicene oxides in addition to the present attempt to create 2D honeycomb silica-type structures [64, 65]. Only at high O_2 supply, up to 1000 L, which is generally 104 greater than that on Si (111) 7X7 surface, does the oxidation process begin, demonstrating the silicene nanoribbons' exceptional resistance to oxidation. By using Ar+ sputtering to produce deliberate flaws, the oxidation absorption may be increased by roughly two orders of magnitude. Encasing the silicene layers in a thin Al or Al_2O_3 a few nm thick can stop oxidation [58]. Stimulatingly, oxygen intercalation enabled the creation of silicene that resembles a freestanding object. Due to the lower contact with the substrate, this layer displays the characteristics of a honeycomb lattice and forms Dirac fermions, much like the intercalation method.

Metal Functionalization

Owing to its highly reactive surface, silicene is easily modified in terms of its electronic characteristics by the adsorption of metallic moieties, such as transition metals (TM), main group elements, and alkali and alkaline earth metals. The preferred adsorption site for a single alkali atom on silicene is invariably the hollow site. After relaxing, the alkali metal atom maintains a stable position above the hollow site [66]. Alkali atom adsorption does not cause any noticeable deformation or stress on the lattice of silica [67, 68]. Alkaline earth metals are less suited to the hollow site than alkali metals. The adsorption of a transition metal atom may also cause the silicene Dirac point to exhibit a large band gap. Based on the transition metal type, the gap's size varies from 0.03 to 0.66 eV. Contrary to alkali metal adsorption, it exclusively results in n-type in silicene because of lower functions, resulting in all types of doping, depending on the work functions [69].

Additionally, different metals favor various adsorption sites; for example, the hollow site is the best adsorption site for a single alkali metal atom. Similar to this, various alkali metals, transition metals, and non-metals favor various binding sites. One of the key parameters for device use is the band gap of silicene, which is affected by different binding sites and substrate metals. Table **3** shows various substrate atoms and the band gap of silicene that corresponds to them, along with their bending height and binding energy.

Table 3. Different metals adsorbed silicene and the associated band gap.

Atom	Buckling Length (nm)	Adsorption Energy (E_{ads})	Type of Bandgap	Adsorption Site	Refs.
Li	17	2.4	MT	HL	[70]
Na	21	1.85	MT	HL	[70]
K	27	2.11	MT	HL	[70]
Ti	7.0	4.89	HMT	BR	[70]
Cr	5	3.2	HMT	HL	[70]
Co	7.2	5.61	MT	HL	[71]
Ni	2.7	4.78	HMT	HL	[71]
Mo	3.7	5.46	MT	BR	[70]
Pd	6	4.2	HMT	HL	[71]
Ag	13	1.9	MT	HL	[72]
Pt	4.3	5.87	HMT	HL	[71]
Au	10.1	2.32	MT	HL	[71]
Al	17.1	2.87	MT	VL	[73]
Ga	17	2.4	—	VL	[71]
In	20.3	2.14	—	VL	[71]
Tl	20.6	2.17	MT	HL	[74]
Sn	16.7	2.94	MT	VL	[71]
Pb	16.1	2.64	MT	HL	[74]
B	7.1	5.85	MT	VL	[73]
N	14.2	5.54	MT	BR	[73]
P	11.8	5.28	MT	Top	[73]

Abreviations: MT: Metallic; HMT: Halfmetallic; HL: Hallow; VL: Valley; BR: Bridge

CHARACTERIZATION OF SILICENE STRUCTURES

After silicene has been successfully synthesized, the main concern is whether it possesses the unusual electronic band (particularly the Dirac electronic state) that has been theorized. The two main spectroscopic methods used to describe the electronic structures of silicene up to this point are Scanning Tunnel Microscopy (STM) and Angle-Resolved Photoelectron Spectroscopy (ARPES). These are the two most common approaches to the identification of synthesized Silicene in terms of its electronic behavior. After implementing such methods, a new controversy was born around the electronics of silicene structure and the presence of the Dirac cone. In the year 2012, Vogt and their coworkers studied the Dirac cone existence in the 4X4 Silicene using ARPES spectroscopy [38], which was further evidenced by Avila *et al.* in the study of the electronic structure of 3X3 silicene on Ag (111) substrate using the Low-Energy Electron Diffraction (LEED) and ARPES examination [75]. Avila concluded that the (3X3) phase's symmetry point in multiple Brillouin zones exhibits a silicene-formed band structure with a visible gap close to the Fermi state.

Furthermore, in line with recently published photoemission studies, results demonstrated that the substantial 0.7 Å Buckling of the silicene causes the gap opening near the Fermi level. The gapped silicene band has also demonstrated the 2-D nature of the charge carriers. In 2014, Johnson and their colleagues studied the epitaxial silicene developed on Ag (111) substrate and discovered that the Si p_z states bridge the Fermi level due to substantial sp^3 hybridization between silicon atoms and the substrate d states of Ag [76]. According to the findings, epitaxial silicene (specifically Ag (111) dependent) lacks the Dirac cone electrical structure that characterizes freestanding sheets of silica and graphene. There are many different studies conducted year by year to develop a clear vision of the electronics of Si sheets based on these spectroscopic techniques. In 2016, Feng Y *et al.* observed that monolayer silicene has Dirac cones that have grown on an Ag (111) support. They demonstrate that the silicene-Ag (111) substrate contact is the source of this peculiar Dirac cone shape, proving the existence of a special kind of Dirac fermion produced by such an interaction [77]. Further, this study was supported by Feng B. in the year 2019 when they estimated the electronic structure of silicene synthesized on Ag (111) surface. In the study, they showed that six Dirac cone pairs were recently discovered near the Brillouin zone border of Ag (111) by angle-resolved photoemission spectroscopy studies. However, their origin is still unknown. They conclude that the substrate-overlayer interaction produces external periodic potentials that separate the Dirac cones of freestanding silicene [78].

Spectroscopic technique such as Raman is also a rapid and efficient method often used to describe the atomic structure and the defects (carbon-based nanomaterials) and other 2D materials [79]. For silicene characterization, Raman is also an effective spectroscopic technique used by different researchers. By comparing, the interpretation was made for the silicene grown on the Ag substrate. In 2010, the Raman spectra of Ag-based silicene were studied by Housa and their coworkers, and they found an intense peak at 516-521cm^{-1} for 4x4 silicene. The zone-center E_{2g} vibrational mode, similar to graphene in that it represents an in-plane displacement, is caused by the bond stretching of all pairs of silicon atoms arranged in six-atom rings, with the frequency being solely based on the length of the Si-Si bond. Due to the lower bond length (2.28Å), freestanding silicene is predicted to exhibit the E_{2g} mode at a very high frequency [80]. By contrasting the

Raman spectra of the encapsulated silicene sheets with the DFT-generated spectra, the spectra have been interpreted. The double degenerated E_{2g} mode obtained from DFT calculations peaked at 505 cm^{-1} and 495 cm^{-1} for the 4x4 superstructures [81]. In a most significant study, Du *et al*. reported in-situ Raman spectroscopy of epitaxial silicene on Ag(111) surfaces with different structures and coverage [65]. The Raman spectra were obtained the in-situ measurement of the 4x4 structures of silicene where E_{2g} mode comes around 530 cm^{-1} as the major peak.

SILICENE-BASED APPLICATION IN NANOELECTRONICS

The entire scientific community is hopeful that this 2D material can be potent in novel applications because it shares many graphene's features. To test its application in other industries, there are far too many presumptions and observations made. In the chapter, we compiled the most specific property of silicene, *i.e.*, the band gap and the associated electronic structure. Considering this point of view, we covered the most critical sector of the material world, *i.e.*, nanoelectronics. The use of silicene in nanoelectronics requires opening a significant gap without impairing the electronic properties of the particular band gap [82]. Silicene can open a significant band gap through covalent functionalization in the nanoribbon form (as discussed in section 4 of the chapter). However, this degrades the material's electronic properties and substantially lowers carrier mobility, as was previously observed for graphene. There are only two methods to widen a band gap in silicene while maintaining its electronic properties: intervalley interaction and breaking inversion symmetry [83, 84]. It is mainly making a silicene nanomesh by drilling holes in the more active procedure (such as metallic doping) to open the band gap deprived of disrupting the electronic properties. This is known as intervalley interaction. This is also accomplished by breaking the inversion symmetry by introducing a vertical electric field to the silicene layer. Opening a band gap in silicene by the surface adsorption of foreign atoms is one strategy based on combining these two techniques. With the band gap open, many practical silicon field transistors may be constructed.

Silicene Transistors

An efficient way to comprehend silicene's electronic characteristics and meet the needs of different electrical, sensing, or energy devices is through silicene-based transistors. Table **4** provides theoretical information based on various investigations of silicene-based field effect transistors (FETs). These silicon FETs' gate modulation differs significantly. However, there have been recent advancements in silicene transistor modeling. FETs are a valuable tool for comprehending silicene's electrical characteristics and realizing different electronic, sensor, or even energy devices. These are directly correlated with bandgap engineering and its tuning *via* different modulations. Different simulation studies of silicene-based FETs have been done previously, such as surface modified silicene based FETs, nanoribbon FETs, and dual gate FETs based on band gap and the channel length (the path that links the charge carriers between the drain and the source or the distance an electron traveled from source to drain).

Table 4. Simulated silicene transistors performances (theoretical value and observations) [56].

FET Devices	Channel Length (Å)	Band Gap (meV)
Dual-gated silicene FET	67	160
Na-adsorbed silicene FET	113.8	500
Silicene TFET	32	200-300
Silicene nanomesh FET	91	680
Silicene nanoribbon FET	44-115.1	320-400
All metallic silicene FET	100	400

Due to challenges with the synthesis portability and air stability, numerous experimental studies on silicon transistors lagged behind theoretical or modeling studies. For instance, Scalise and their colleagues studied monolayer silicene (Ag-free) transistors and revealed that the manufactured device behavior was comparable to graphene, supporting theoretical predictions on ambipolar Dirac charge transport [81]. The stability and electrical behavior of silicene-based transistors may also be affected by other factors, such as the number of layers. The lifespan of multilayer silicene devices was up to 48 hours instead of 2 minutes. Silicene also exhibits an elevated SOC than graphene, and the bandgap opening is comparatively more accessible and real [85].

Silicene for Topology-based Electronics

Silicene has many possibilities as a nanomaterial for use in low-energy devices. The Quantum Spin Hall state has become a potential candidate due to recent developments in the study of topological states of matter [57]. These include topological insulator field-effect transistors (TI-FETs) [86]. With the application of a gate voltage, the topological phase may be changed between a QSH edge state and a trivial insulating bulk state (ON/OFF state). A relatively high/Low electric field is required to achieve the topological phase transition, while SOC outcomes depend on room temperature functioning. Because of its critical field value enabling the topological phase transition, silicene, despite its modest SOC

(1.55meV) that restricts in low temperatures, is still a possible choice for gate-controllable TI-FETs [46, 87].

Silicene-Based Junctions

In addition to homogeneous silicene sheets or nanoribbons, silicene-based junctions are also gaining wide attention because they offer considerable potential for novel applications in various domains, especially in magnetic, electrical, and thermal fields. With a Fe (111)/silicene stack injector, Zhou and their coworkers theoretically examined the spin transport in a silicene channel [88]. The dominance of spin-down states above the Fermi state produces a high spin injector efficacy. They suggest the Fe (111)/silicene heterostructure as a potent device to create effective spin injection devices.

By enhancing the thermal conductivities or by creating a heterostructure with silicon, the fabricated 2D transistors are better able to dissipate heat. In recent work, silicene's band gap was created *via* hydrogenation, and the hydrogenated silicene was then investigated for its potential use in spintronics. After hydrogenation, silicene changes into a material with a broadband gap and a band gap of 3.32 eV. When using hydrogenated silicene as the scattering area and CrO_2 as semi-metallic electrodes, magnetic tunnel junction (MTJ) parameters such as magnetoresistance and spin-filtering efficiency are found to be higher than when using pristine silicene. According to the simulation results, across the whole bias range, hydrogenated silicene's magnetoresistance is greater than 85% (compared to its pure form) [89]. Another area of attention for employing silicene-based junctions is energy storage. Since silicene is an ideal material with distinguishing features like tuneable band gap and compatibility with the pervasive semiconductor sector, silicene-based junctions have a wide range of potential applications beyond the discussed applications.

CONCLUSION

The literature on 2D silicene has grown significantly since its discovery, and different assumptions and scientific observations were made based on different mathematical calculations. As a result, new research frontiers on Xenes that resemble silicene have also been sparked. There is no doubt that silicene and its derivatives can be a fascinating platform for both initial study and device applications. Silicene synthesis on Ag (111) is one of the fastest growing research in material chemistry, which showed the significance of this miraculous 2D material. However, more experimental research on these novel device concepts needs to be conducted to stimulate the ongoing efforts to understand how silicene reacts toward several energies related issues. Based on the previous work and experimental findings, the chapter concludes that silicene is a strong contender for nanoelectronic applications and devices. Additionally, we found that continued study of silicene and associated nanoparticles will advance the development of silicene-based products and the identification of novel 2D materials.

REFERENCES

[1]   Su, S.; Sun, Q.; Gu, X. Two-dimensional nanomaterials for biosensing applications. *TrAC, Trends Anal. Chem,* **2019**, *121*, 115668.
[http://dx.doi.org/10.1016/j.trac.2019.07.021]

[2]   Zhuang, J.; Xu, X.; Feng, H.; Li, Z.; Wang, X.; Du, Y. Honeycomb silicon: A review of silicene. *Sci. Bull.,* **2015**, *60*(18), 1551-1562.
[http://dx.doi.org/10.1007/s11434-015-0880-2]

[3]   Roy, A.S.; Cheruvathoor Poulose, A.; Bakandritsos, A.; Varma, R.S.; Otyepka, M. 2D graphene derivatives as heterogeneous catalysts to produce biofuels *via* esterification and trans-esterification reactions. *Appl. Mater. Today,* **2021**, *23*, 101053.
[http://dx.doi.org/10.1016/j.apmt.2021.101053]

[4]   Zhao, H.; Wang, Y.; Bao, L. Engineering nano–bio interfaces from nanomaterials to nanomedicines. *Acc. Mater. Res,* **2022**, *3*(8), 812-829.
[http://dx.doi.org/10.1021/accountsmr.2c00072]

[5]   Feng, Z.; Adolfsson, K.H.; Xu, Y. Carbon dot/polymer nanocomposites: From green synthesis to energy, environmental and biomedical applications. *Sustain. Mater. Technol,* **2021**, e00304.
[http://dx.doi.org/10.1016/j.susmat.2021.e00304]

[6]   Pathak, R.; Guleria, K.; Kumari, A.; Mehta, S.P.S. Deacidification of Camelina sativa L. seed oil by Physisorption method and characterization of produced biodiesel. *J. Appl. Nat. Sci.,* **2021**, *13*(1), 287-294.
[http://dx.doi.org/10.31018/jans.v13i1.2555]

[7]   Yaragalla, S.; Bhavitha, K.B.; Athanassiou, A. A review on graphene based materials and their antimicrobial properties. *Coatings,* **2021**, *11*(10), 1197.
[http://dx.doi.org/10.3390/coatings11101197]

[8]   Kumar, P.; Huo, P.; Zhang, R. Antibacterial properties of graphene-based nanomaterials. *Nanomaterials,* **2019**, *9*(5), 737.
[http://dx.doi.org/10.3390/nano9050737]

[9]   Shi, Z.; Khaledialidusti, R.; Malaki, M.; Zhang, H. MXene-based materials for solar cell applications. *Nanomaterials,* **2021**, *11*(12), 3170.
[http://dx.doi.org/10.3390/nano11123170] [PMID: 34947518]

[10] Punetha, V.D.; Dhali, S.; Rana, A.; Karki, N.; Tiwari, H.; Negi, P.; Basak, S.; Sahoo, N.G. Recent advancements in green synthesis of nanoparticles for improvement of bioactivities: A review. *Curr. Pharm. Biotechnol.,* **2022**, *23*(7), 904-919.
[http://dx.doi.org/10.2174/1389201022666210812115233] [PMID: 34387160]

[11] Mas-Ballesté, R.; Gómez-Navarro, C.; Gómez-Herrero, J.; Zamora, F. 2D materials: To graphene and beyond. *Nanoscale,* **2011**, *3*(1), 20-30.
[http://dx.doi.org/10.1039/C0NR00323A] [PMID: 20844797]

[12] Nguyen, D.K.; Hoang, D.Q.; Hoat, D.M. Exploring a silicene monolayer as a promising sensor platform to detect and capture NO and CO gas. *RSC Advances,* **2022**, *12*(16), 9828-9835.
[http://dx.doi.org/10.1039/D2RA00442A] [PMID: 35424916]

[13] Zhao, A.; Wang, B. Two-dimensional graphene-like Xenes as potential topological materials. *APL Mater.,* **2020**, *8*(3), 030701.
[http://dx.doi.org/10.1063/1.5135984]

[14]   Braga, D.; Gutiérrez Lezama, I.; Berger, H.; Morpurgo, A.F. Quantitative determination of the band gap of WS2 with ambipolar ionic liquid-gated transistors. *Nano Lett.,* **2012**, *12*(10), 5218-5223.
[http://dx.doi.org/10.1021/nl302389d] [PMID: 22989251]

[15]   Grazianetti, C.; Martella, C. The rise of the xenes: From the synthesis to the integration processes for electronics and photonics. *Materials,* **2021**, *14*(15), 4170.
[http://dx.doi.org/10.3390/ma14154170]

[16]   Takeda, K.; Shiraishi, K. Theoretical possibility of stage corrugation in Si and Ge analogs of graphite. *Phys. Rev. B Condens. Matter,* **1994**, *50*(20), 14916-14922.
[http://dx.doi.org/10.1103/PhysRevB.50.14916] [PMID: 9975837]

[17]   Rafi-Ul-Islam, S.M.; Siu, Z.B.; Sahin, H.; Jalil, M.B.A. Valley and spin quantum Hall conductance of silicene coupled to a ferroelectric layer. *Front. Phys.,* **2022**, *10*, 1021192.
[http://dx.doi.org/10.3389/fphy.2022.1021192]

[18]   De Padova, P.; Vogt, P.; Resta, A.; Avila, J.; Razado-Colambo, I.; Quaresima, C.; Ottaviani, C.; Olivieri, B.; Bruhn, T.; Hirahara, T.; Shirai, T.; Hasegawa, S.; Carmen Asensio, M.; Le Lay, G. Evidence of Dirac fermions in multilayer silicene. *Appl. Phys. Lett.,* **2013**, *102*(16), 163106.
[http://dx.doi.org/10.1063/1.4802782]

[19]   Podsiadły-Paszkowska, A.; Krawiec, M. Dirac fermions in silicene on Pb(111) surface. *Phys. Chem. Chem. Phys.,* **2015**, *17*(3), 2246-2251.
[http://dx.doi.org/10.1039/C4CP05104A] [PMID: 25485668]

[20]   Ezawa, M. Quantum hall effects in silicene. *J. Phys. Soc. Jpn.,* **2012**, *81*(6), 064705.
[http://dx.doi.org/10.1143/JPSJ.81.064705]

[21]   Zhan, L.; Fang, Y.; Zhang, R.; Lu, X.; Lü, T.; Cao, X.; Zhu, Z.; Wu, S. Quantum spin Hall effect in tilted penta silicene and its isoelectronic substitutions. *Phys. Chem. Chem. Phys.,* **2022**, *24*(25), 15201-15207.
[http://dx.doi.org/10.1039/D2CP01390H] [PMID: 35612307]

[22]   Ma, L.; Song, X.; Yu, Y.; Chen, Y. Two-Dimensional silicene/silicon nanosheets: An emerging silicon-composed nanostructure in biomedicine. *Adv. Mater.,* **2021**, *33*(31), 2008226.
[http://dx.doi.org/10.1002/adma.202008226] [PMID: 34050575]

[23]   Bechstedt, F.; Gori, P.; Pulci, O. Beyond graphene: Clean, hydrogenated and halogenated silicene, germanene, stanene, and plumbene. *Prog. Surf. Sci.,* **2021**, *96*(3), 100615.
[http://dx.doi.org/10.1016/j.progsurf.2021.100615]

[24]   Jose, D.; Datta, A. Structures and chemical properties of silicene: Unlike graphene. *Acc. Chem. Res.,* **2014**, *47*(2), 593-602.
[http://dx.doi.org/10.1021/ar400180e] [PMID: 24215179]

[25]   Cahangirov, S.; Topsakal, M.; Aktürk, E.; Šahin, H.; Ciraci, S. Two- and one-dimensional honeycomb structures of silicon and germanium. *Phys Rev Lett.,* **2009**, *102*(23), 236804.
[http://dx.doi.org/10.1103/PhysRevLett.102.236804]

[26]   Soto, J.R.; Molina, B.; Castro, J.J. Reexamination of the origin of the pseudo Jahn-Teller puckering instability in silicene. *Phys. Chem. Chem. Phys.,* **2015**, *17*(12), 7624-7628.
[http://dx.doi.org/10.1039/C4CP05912C] [PMID: 25715330]

[27]   Bhattacharjee, R.; Majumder, T.; Datta, A. Analysis of pseudo jahn–teller distortion based on natural bond orbital theory: Case study for silicene. *J. Comput. Chem.,* **2019**, *40*(15), 1488-1495.
[http://dx.doi.org/10.1002/jcc.25815] [PMID: 30854679]

[28]   López Jiménez, F.; Triantafyllidis, N. Buckling of rectangular and hexagonal honeycomb under combined axial compression and transverse shear. *Int. J. Solids Struct.,* **2013**, *50*(24), 3934-3946.
[http://dx.doi.org/10.1016/j.ijsolstr.2013.08.001]

[29]   Houssa, M.; Dimoulas, A.; Molle, A. Silicene: A review of recent experimental and theoretical

investigations. *J. Phys. Condens. Matter,* **2015**, *27*(25), 253002.
[http://dx.doi.org/10.1088/0953-8984/27/25/253002] [PMID: 26045468]

[30]   Balendhran, S.; Walia, S.; Nili, H.; Sriram, S.; Bhaskaran, M. Elemental analogues of graphene: Silicene, germanene, stanene, and phosphorene. *Small,* **2015**, *11*(6), 640-652.
[http://dx.doi.org/10.1002/smll.201402041] [PMID: 25380184]

[31]   Qin, R.; Zhu, W.; Zhang, Y.; Deng, X. Uniaxial strain-induced mechanical and electronic property modulation of silicene. *Nanoscale Res. Lett.,* **2014**, *9*(1), 521.
[http://dx.doi.org/10.1186/1556-276X-9-521] [PMID: 25276108]

[32]   Gao, N.; Zheng, W.T.; Jiang, Q. Density functional theory calculations for two-dimensional silicene with halogen functionalization. *Phys. Chem. Chem. Phys.,* **2012**, *14*(1), 257-261.
[http://dx.doi.org/10.1039/C1CP22719J] [PMID: 22083171]

[33]   Matthes, L.; Pulci, O.; Bechstedt, F. Optical properties of two-dimensional honeycomb crystals graphene, silicene, germanene, and tinene from first principles. *New J. Phys.,* **2014**, *16*(10), 105007.
[http://dx.doi.org/10.1088/1367-2630/16/10/105007]

[34]   Molle, A.; Goldberger, J.; Houssa, M.; Xu, Y.; Zhang, S.C.; Akinwande, D. Buckled two-dimensional Xene sheets. *Nat. Mater.,* **2017**, *16*(2), 163-169.
[http://dx.doi.org/10.1038/nmat4802] [PMID: 28092688]

[35]   Dávila, M.E.; Lew Yan Voon, L.C.; Zhao, J.; Le Lay, G. Elemental Group IV two-dimensional materials beyond graphene. *Semicond. Semimet.,* **2016**, *95*, 149-188.
[http://dx.doi.org/10.1016/bs.semsem.2016.04.003]

[36]   Salomon, E.; El Ajjouri, R.; Lay, G.L.; Angot, T. Growth and structural properties of silicene at multilayer coverage. *J. Phys. Condens. Matter,* **2014**, *26*(18), 185003.
[http://dx.doi.org/10.1088/0953-8984/26/18/185003] [PMID: 24728034]

[37]   Kalwar, B.A.; Fangzong, W.; Ahmed, I.; Saeed, M.H. Ti atom doped single vacancy silicene for hydrogen energy storage: DFT study. *J. Chin. Chem. Soc.,* **2021**, *68*(12), 2243-2253.
[http://dx.doi.org/10.1002/jccs.202100369]

[38]   Vogt, P.; De Padova, P.; Quaresima, C.; Avila, J.; Frantzeskakis, E.; Asensio, M.C.; Resta, A.; Ealet, B.; Le Lay, G. Silicene: Compelling experimental evidence for graphenelike two-dimensional silicon. *Phys. Rev. Lett.,* **2012**, *108*(15), 155501.
[http://dx.doi.org/10.1103/PhysRevLett.108.155501] [PMID: 22587265]

[39]   Liu, H.; Han, N.; Zhao, J. Band gap opening in bilayer silicene by alkali metal intercalation. *J. Phys. Condens. Matter,* **2014**, *26*(47), 475303.
[http://dx.doi.org/10.1088/0953-8984/26/47/475303] [PMID: 25351483]

[40]   Resta, A.; Leoni, T.; Barth, C.; Ranguis, A.; Becker, C.; Bruhn, T.; Vogt, P.; Le Lay, G. Atomic structures of silicene layers grown on Ag(111): Scanning tunneling microscopy and noncontact atomic force microscopy observations. *Sci. Reports,* **2013**, *3*(1), 1-7.
[http://dx.doi.org/10.1038/srep02399]

[41]   Pawlak, R.; Drechsel, C.; D'Astolfo, P.; Kisiel, M.; Meyer, E.; Cerda, J.I. Quantitative determination of atomic buckling of silicene by atomic force microscopy. *Proc. Natl. Acad. Sci.,* **2020**, *117*(1), 228-237.
[http://dx.doi.org/10.1073/pnas.1913489117] [PMID: 31871150]

[42]   Zhao, J.; Wu, K. Surface functionalization of silicene. *NanoSci. Technol.,* **2018**, 211-233.
[http://dx.doi.org/10.1007/978-3-319-99964-7_11]

[43]   Mattox, D.M. Thermal evaporation and deposition in vacuum. *Foundat. Vacuum Coat. Technol.,* **2018**, (Jan), 151-184.
[http://dx.doi.org/10.1016/B978-0-12-813084-1.00005-4]

[44]   Kim, U.; Kim, I.; Park, Y.; Lee, K.Y.; Yim, S.Y.; Park, J.G.; Ahn, H.G.; Park, S.H.; Choi, H.J. Synthesis of Si nanosheets by a chemical vapor deposition process and their blue emissions. *ACS*

Nano, **2011**, *5*(3), 2176-2181.
[http://dx.doi.org/10.1021/nn103385p] [PMID: 21322533]

[45]   Kupchak, I.; Fabbri, F.; De Crescenzi, M.; Scarselli, M.; Salvato, M.; Delise, T.; Berbezier, I.; Pulci, O.; Castrucci, P. Scanning tunneling microscopy and Raman evidence of silicene nanosheets intercalated into graphite surfaces at room temperature. *Nanoscale,* **2019**, *11*(13), 6145-6152.
[http://dx.doi.org/10.1039/C9NR00343F] [PMID: 30874280]

[46]   Molle, A.; Grazianetti, C.; Tao, L.; Taneja, D.; Alam, M.H.; Akinwande, D. Silicene, silicene derivatives, and their device applications. *Chem. Soc. Rev.,* **2018**, *47*(16), 6370-6387.
[http://dx.doi.org/10.1039/C8CS00338F] [PMID: 30065980]

[47]   Meng, L.; Wang, Y.; Zhang, L.; Du, S.; Wu, R.; Li, L.; Zhang, Y.; Li, G.; Zhou, H.; Hofer, W.A.; Gao, H.J. Buckled silicene formation on Ir(111). *Nano Lett.,* **2013**, *13*(2), 685-690.
[http://dx.doi.org/10.1021/nl304347w] [PMID: 23330602]

[48]   Chiappe, D.; Scalise, E.; Cinquanta, E.; Grazianetti, C.; van den Broek, B.; Fanciulli, M.; Houssa, M.; Molle, A. Two-dimensional Si nanosheets with local hexagonal structure on a MoS_2 surface. *Adv. Mater.,* **2014**, *26*(13), 2096-2101.
[http://dx.doi.org/10.1002/adma.201304783] [PMID: 24347540]

[49]   Bottomley, L. A.; Gadsby, E. D.; Poggi, M. A. Microscopy techniques | Atomic force and scanning tunneling microscopy. In: *Encyclopedia of Analytical Science*; Elsevier, **2005**.
[http://dx.doi.org/10.1016/B0-12-369397-7/00386-1]

[50]   Fleurence, A.; Yoshida, Y.; Lee, C.C.; Ozaki, T.; Yamada-Takamura, Y.; Hasegawa, Y. Microscopic origin of the π states in epitaxial silicene. *Appl. Phys. Lett.,* **2014**, *104*(2), 021605.
[http://dx.doi.org/10.1063/1.4862261]

[51]   Nagoya, A.; Yaokawa, R.; Ohba, N. Mechanism of monolayer to bilayer silicene transformation in $CaSi_2$ due to fluorine diffusion. *Phys. Chem. Chem. Phys.,* **2021**, *23*(15), 9315-9324.
[http://dx.doi.org/10.1039/D0CP06644C] [PMID: 33885084]

[52]   Sun, L.; Xie, J.; Huang, S.; Liu, Y.; Zhang, L.; Wu, J.; Jin, Z. Rapid CO_2 exfoliation of Zintl phase $CaSi_2$-derived ultrathin free-standing Si/SiOx/C nanosheets for high-performance lithium storage. *Sci. China Mater.,* **2022**, *65*(1), 51-58.
[http://dx.doi.org/10.1007/s40843-021-1708-6]

[53]   Noguchi, E.; Sugawara, K.; Yaokawa, R.; Hitosugi, T.; Nakano, H.; Takahashi, T. Direct observation of Dirac cone in multilayer silicene intercalation compound CaSi2. *Adv. Mater.,* **2015**, *27*(5), 856-860.
[http://dx.doi.org/10.1002/adma.201403077] [PMID: 25502913]

[54]   Yaokawa, R.; Ohsuna, T.; Morishita, T.; Hayasaka, Y.; Spencer, M.J.S.; Nakano, H. Monolayer-t--bilayer transformation of silicenes and their structural analysis. *Nat. Commun.,* **2016**, *7*(1), 10657.
[http://dx.doi.org/10.1038/ncomms10657] [PMID: 26847858]

[55]   Tokmachev, A.M.; Averyanov, D.V.; Karateev, I.A.; Parfenov, O.E.; Vasiliev, A.L.; Yakunin, S.N.; Storchak, V.G. Topotactic synthesis of the overlooked multilayer silicene intercalation compound $SrSi_2$. *Nanoscale,* **2016**, *8*(36), 16229-16235.
[http://dx.doi.org/10.1039/C6NR04573A] [PMID: 27469172]

[56]   Kim, D.Y.; Stefanoski, S.; Kurakevych, O.O.; Strobel, T.A. Synthesis of an open-framework allotrope of silicon. *Nat. Mater.,* **2015**, *14*(2), 169-173.
[http://dx.doi.org/10.1038/nmat4140] [PMID: 25401923]

[57]   Kharadi, M.A.; Malik, G.F.A.; Khanday, F.A.; Shah, K.A.; Mittal, S.; Kaushik, B.K. Review-silicene: From material to device applications. *ECS J. Solid State Sci. Technol.,* **2020**, *9*(11), 115031.
[http://dx.doi.org/10.1149/2162-8777/abd09a]

[58]   Zhao, J.; Liu, H.; Yu, Z.; Quhe, R.; Zhou, S.; Wang, Y.; Liu, C.C.; Zhong, H.; Han, N.; Lu, J.; Yao, Y.; Wu, K. Rise of silicene: A competitive 2D material. *Prog. Mater. Sci.,* **2016**, *83*, 24-151.
[http://dx.doi.org/10.1016/j.pmatsci.2016.04.001]

[59]   Lin, S.Y.; Chang, S.L.; Thuy Tran, N.T.; Yang, P.H.; Lin, M.F. H–Si bonding-induced unusual electronic properties of silicene: A method to identify hydrogen concentration. *Phys. Chem. Chem. Phys.,* **2015,** *17*(39), 26443-26450.
[http://dx.doi.org/10.1039/C5CP04841A] [PMID: 26392324]

[60]   Abdelsalam, H.; Saroka, V.A.; Ali, M.; Teleb, N.H.; Elhaes, H.; Ibrahim, M.A. Stability and electronic properties of edge functionalized silicene quantum dots: A first principles study. *Physica E,* **2019,** *108*, 339-346.
[http://dx.doi.org/10.1016/j.physe.2018.07.022]

[61]   Wang, X.; Liu, H.; Tu, S.T. First-principles study of half-fluorinated silicene sheets. *RSC Advances,* **2015,** *5*(9), 6238-6245.
[http://dx.doi.org/10.1039/C4RA12257G]

[62]   Zhao, T.; Zhang, S.; Wang, Q.; Kawazoe, Y.; Jena, P. Tuning electronic and magnetic properties of silicene with magnetic superhalogens. *Phys. Chem. Chem. Phys.,* **2014,** *16*(42), 22979-22986.
[http://dx.doi.org/10.1039/C4CP02758B] [PMID: 25144623]

[63]   Liu, G.; Lei, X.L.; Wu, M.S.; Xu, B.; Ouyang, C.Y. Is silicene stable in O $_2$? -First-principles study of O $_2$ dissociation and O $_2$ -dissociation-induced oxygen atoms adsorption on free-standing silicene. *Europhys. Lett.,* **2014,** *106*(4), 47001.
[http://dx.doi.org/10.1209/0295-5075/106/47001]

[64]   Gao, N.; Lu, G.Y.; Wen, Z.; Jiang, Q. Electronic structure of silicene: effects of the organic molecular adsorption and substrate. *J. Mater. Chem. C Mater. Opt. Electron. Devices,* **2017,** *5*(3), 627-633.
[http://dx.doi.org/10.1039/C6TC04943E]

[65]   Du, Y.; Zhuang, J.; Liu, H.; Xu, X.; Eilers, S.; Wu, K.; Cheng, P.; Zhao, J.; Pi, X.; See, K.W.; Peleckis, G.; Wang, X.; Dou, S.X. Tuning the band gap in silicene by oxidation. *ACS Nano,* **2014,** *8*(10), 10019-10025.
[http://dx.doi.org/10.1021/nn504451t] [PMID: 25248135]

[66]   Bao, A.; Li, X.; Guo, X.; Yao, H.; Chen, M. Tuning the structural, electronic, mechanical and optical properties of silicene monolayer by chemical functionalization: A first-principles study. *Vacuum,* **2022,** *203*, 111226.
[http://dx.doi.org/10.1016/j.vacuum.2022.111226]

[67]   Hussain, T.; Kaewmaraya, T.; Chakraborty, S.; Ahuja, R.; Silva, C. Functionalization of hydrogenated silicene with alkali and alkaline earth metals for efficient hydrogen storage. *Phys. Chem. Chem. Phys.,* **2013,** *15*(43), 18900-18905.
[http://dx.doi.org/10.1039/c3cp52830h] [PMID: 24091878]

[68]   Li, X.D.; Fang, Y.M.; Wu, S.Q. Adsorption of alkali, alkaline-earth, simple and 3 d transition metal, and nonmetal atoms on monolayer MoS2. *AIP Adv.,* **2015,** *5*(5)
[http://dx.doi.org/10.1063/1.4921564]

[69]   Ni, Z.; Zhong, H.; Jiang, X.; Quhe, R.; Luo, G.; Wang, Y.; Ye, M.; Yang, J.; Shi, J.; Lu, J. Tunable band gap and doping type in silicene by surface adsorption: Towards tunneling transistors. *Nanoscale,* **2014,** *6*(13), 7609-7618.
[http://dx.doi.org/10.1039/C4NR00028E] [PMID: 24896227]

[70]   Sahin, H.; Peeters, F.M. Adsorption of alkali, alkaline-earth, and 3 d transition metal atoms on silicene. *Phys. Rev. B Condens. Matter Mater. Phys.,* **2013,** *87*(8), 085423.
[http://dx.doi.org/10.1103/PhysRevB.87.085423]

[71]   Lin, X.; Ni, J. Much stronger binding of metal adatoms to silicene than to graphene: A first-principles study. *Phys. Rev. B Condens. Matter Mater. Phys.,* **2012,** *86*(7), 075440.
[http://dx.doi.org/10.1103/PhysRevB.86.075440]

[72]   Ersan, F.; Arslanalp, Ö.; Gökoğlu, G.; Aktürk, E. Effects of silver adatoms on the electronic structure of silicene. *Appl. Surf. Sci.,* **2014,** *311*, 9-13.

[http://dx.doi.org/10.1016/j.apsusc.2014.04.176]

[73]   Sivek, J.; Sahin, H.; Partoens, B.; Peeters, F.M. Adsorption and absorption of boron, nitrogen, aluminum, and phosphorus on silicene: Stability and electronic and phonon properties. *Phys. Rev. B Condens. Matter Mater. Phys.,* **2013**, *87*(8), 085444.
        [http://dx.doi.org/10.1103/PhysRevB.87.085444]

[74]   Kaloni, T.P.; Schwingenschlögl, U. Effects of heavy metal adsorption on silicene. *Phys. Status Solidi Rapid Res. Lett.,* **2014**, *8*(8), 685-687.
        [http://dx.doi.org/10.1002/pssr.201409245]

[75]   Avila, J.; De Padova, P.; Cho, S.; Colambo, I.; Lorcy, S.; Quaresima, C.; Vogt, P.; Resta, A.; Le Lay, G.; Asensio, M.C. Presence of gapped silicene-derived band in the prototypical (3 × 3) silicene phase on silver (111) surfaces. *J. Phys. Condens. Matter,* **2013**, *25*(26), 262001.
        [http://dx.doi.org/10.1088/0953-8984/25/26/262001] [PMID: 23759650]

[76]   Johnson, N.W.; Vogt, P.; Resta, A.; De Padova, P.; Perez, I.; Muir, D.; Kurmaev, E.Z.; Le Lay, G.; Moewes, A. The metallic nature of epitaxial silicene monolayers on Ag(111). *Adv. Funct. Mater.,* **2014**, *24*(33), 5253-5259.
        [http://dx.doi.org/10.1002/adfm.201400769]

[77]   Feng, Y.; Liu, D.; Feng, B.; Liu, X.; Zhao, L.; Xie, Z.; Liu, Y.; Liang, A.; Hu, C.; Hu, Y.; He, S.; Liu, G.; Zhang, J.; Chen, C.; Xu, Z.; Chen, L.; Wu, K.; Liu, Y.T.; Lin, H.; Huang, Z.Q.; Hsu, C.H.; Chuang, F.C.; Bansil, A.; Zhou, X.J. Direct evidence of interaction-induced Dirac cones in a monolayer silicene/Ag(111) system. *Proc. Natl. Acad. Sci.,* **2016**, *113*(51), 14656-14661.
        [http://dx.doi.org/10.1073/pnas.1613434114] [PMID: 27930314]

[78]   Feng, B.; Zhou, H.; Feng, Y.; Liu, H.; He, S.; Matsuda, I.; Chen, L.; Schwier, E.F.; Shimada, K.; Meng, S.; Wu, K.; Wu, K. Superstructure-induced splitting of Dirac cones in silicene. *Phys. Rev. Lett.,* **2019**, *122*(19), 196801.
        [http://dx.doi.org/10.1103/PhysRevLett.122.196801] [PMID: 31144949]

[79]   Yin, P.; Lin, Q.; Duan, Y. Applications of Raman spectroscopy in two-dimensional materials. *J. Innov. Opt. Health Sci.,* **2020**, *13*(5), 2030010.
        [http://dx.doi.org/10.1142/S1793545820300104]

[80]   Houssa, M.; Pourtois, G.; Afanas'ev, V.V.; Stesmans, A. Can silicon behave like graphene? A first-principles study. *Appl. Phys. Lett.,* **2010**, *97*(11), 112106.
        [http://dx.doi.org/10.1063/1.3489937]

[81]   Scalise, E.; Cinquanta, E.; Houssa, M.; van den Broek, B.; Chiappe, D.; Grazianetti, C.; Pourtois, G.; Ealet, B.; Molle, A.; Fanciulli, M.; Afanas'ev, V.V.; Stesmans, A. Vibrational properties of epitaxial silicene layers on (111) Ag. *Appl. Surf. Sci.,* **2014**, *291*, 113-117.
        [http://dx.doi.org/10.1016/j.apsusc.2013.08.113]

[82]   Pan, F.; Wang, Y.; Jiang, K.; Ni, Z.; Ma, J.; Zheng, J.; Quhe, R.; Shi, J.; Yang, J.; Chen, C.; Lu, J. Silicene nanomesh. *Sci. Reports,* **2015**, *5*(1), 1-8.
        [http://dx.doi.org/10.1038/srep09075]

[83]   Lee, K. W.; Lee, C. E. Quantum spin-valley Hall effect in AB-stacked bilayer silicene. *Sci. Reports,* **2019**, *9*(1), 1-9.
        [http://dx.doi.org/10.1038/s41598-019-55927-9]

[84]   Shahabi, N.; Phirouznia, A. Photogalvanic effect in silicene. *Physica E,* **2021**, *133*, 114808.
        [http://dx.doi.org/10.1016/j.physe.2021.114808]

[85]   Salimian, F.; Dideban, D. Comparative study of nanoribbon field effect transistors based on silicene and graphene. *Mater. Sci. Semicond. Process.,* **2019**, *93*, 92-98.
        [http://dx.doi.org/10.1016/j.mssp.2018.12.032]

[86]   Ezawa, M. Topological electronics and topological field effect transistor in silicene, germanene and stanene. *IEEE 15th International Conference on Nanotechnology (IEEE-NANO),,* Rome, Italy, 27-30

July, 2015.
[http://dx.doi.org/10.1109/NANO.2015.7388677]

[87]   Liu, C.C.; Feng, W.; Yao, Y. Quantum spin Hall effect in silicene and two-dimensional germanium. *Phys. Rev. Lett.,* **2011**, *107*(7), 076802.
[http://dx.doi.org/10.1103/PhysRevLett.107.076802] [PMID: 21902414]

[88]   Zhou, J.; Bournel, A.; Wang, Y.; Lin, X.; Zhang, Y.; Zhao, W. Silicene spintronics: Fe(111)/silicene system for efficient spin injection. *Appl. Phys. Lett.,* **2017**, *111*(18), 182408.
[http://dx.doi.org/10.1063/1.4999202]

[89]   Kharadi, M.A.; Malik, G.F.A.; Khanday, F.A.; Shah, K.A. Hydrogenated silicene based magnetic junction with improved tunneling magnetoresistance and spin-filtering efficiency. *Phys. Lett. A,* **2020**, *384*(32), 126826.
[http://dx.doi.org/10.1016/j.physleta.2020.126826]

CHAPTER 9

Stanene, Mxene and Transition Metal Chalcogenides

Rakshit Pathak[1,*], Mayank Punetha[1], Rajesh Kumar[2] and Anshu Tamta[2]

[1] Centre of Excellence for Research, P.P. Savani University, Surat-394125, Gujrat, India

[2] Department of Chemistry, S.S.J. University, Campus Almora-263601, Uttarakhand, India

Abstract: In recent years, there has been a notable surge of interest in Stanene, MXene, and Transition Metal Chalcogenides. The chapter offers a comprehensive exploration of these cutting-edge 2D materials and their multifaceted applications. Stanene, with its remarkable quantum effects and physicochemical properties, holds promise for the future of nanoelectronics and optoelectronics. Similarly, MXene and Transition Metal Chalcogenides exhibit exceptional characteristics that make them indispensable in various fields, from theranostics to sensor nano-systems and spintronics. Practical applications often hinge on the successful manipulation of molecules through quantum dynamics, but limited synthesis methods for 2D materials pose challenges in this regard. The chapter delves into the structures, synthesis techniques, and applications associated with these materials, providing a comprehensive overview of their potential and current advancements. While a substantial portion of research on these materials has remained theoretical, the chapter underscores the pressing need for increased experimental endeavours. It serves as an invaluable resource for researchers, scientists, and professionals interested in harnessing the unique properties of Stanene, MXene, and Transition Metal Chalcogenides across a spectrum of innovative applications.

Keywords: Chalcogenides, Doping, Graphene, MXenes, Nanomaterials, Stanene, 2D material.

STANENES

Introduction

In the wake of the ground-breaking finding of graphene, several scientific groups have commended the advancement of this wonder material in science and technology-related innovations. Graphene has many exceptional qualities, including extraordinarily high carrier mobilities, optical transparency, huge

---

* **Corresponding author Rakshit Pathak:** Centre of Excellence for Research, P.P. Savani University, Surat-394125, Gujrat, India; E-mail: rakshit.pathak@ppsu.ac.in

Vinay Deep Punetha (Ed.)

surface area, thermal conductivity, and the capacity to be accommodated in scalable manufacturing processes. Graphene, the first discovered 2D material, is applicable in several applications such as bioenergy-related applications, biology, and advanced, multipurpose wearable devices [1 - 6]. However, when used in nano-electronic logic gates, it encounters unfavorable fundamental issues due to its disappearing electronic bandgap. Following significant developments in the material science field, specific issues have come to light, including interoperability with the semiconductor sector and oxidative environment sensitivity [7]. As a consequence, researchers are looking for additional native or synthetic 2D analogs that can overcome the challenges associated with graphene. Beyond graphene, numerous articles have detailed a broad category of stacked 2D materials, each with distinctive properties determined by technology. These layered atomic crystals are grouped based on the component's chemical bonding states and shapes. The majority of these are transition metals (TM) to chalcogenide groups (*e.g.*, MoS_2, $MoSe_2$, WS_2, TiS_2), metal-organic frameworks (MOFs), h-BNs and mono elemental 2D nanomaterials [8, 9].

The novel 2D nanomaterials, synthesized from 3D van der Waals phases or wet-chemical or vapor phases, are encouraged by developing new intriguing physical and chemical characteristics (bottom-up approaches). Exfoliation from 3D structures makes up most of the earlier synthetic strategies. These processes include physical exfoliation, liquid phase exfoliation, and ion intercalation [10]. These procedures for creating 2D sheets are frequently used because they are straightforward and reliable enough for mass production. However, they might be difficult to employ when creating nanosheets with consistent thicknesses and lateral dimensions. Nonetheless, the scalability of these technologies has been widely utilized, for instance, in energy storage devices, sensors, electronics, and photonic devices, as well as in the biomedicine sector [11].

Recently, elemental 2D systems from groups 13 to 16 have become experimentally feasible [12]. These mono-elemental heterostructures have developed strong physical and chemical characteristics due to the epitaxial development of their atomic layers. These atomic layers are created by interfacial solid bonds on the substrate surface and in-plane bonding of homogenous atomic units. Because of their enormous atomic radii, they create stable buckling structures instead of graphene. Moreover, the manufactured structures of the atomic layers in these 2D systems exhibit quantum confinement and a great affinity for the applied electric field bias, strain, doping, adsorption, and functionalization [13].

In contrast to the lower graphene bond length, these puckered 2D materials favor sp^2-sp^3 mixed hybridized bonds due to their comparatively long Si–Si and Ge–Ge bond lengths. Significant experiments on Si and Ge puckered structured 2D materials have been attained as a promising candidate for applications in field-effect transistors (FETs) and electronics and other devices as predicted by a plethora of research on these mechanically 2D structures [14, 15]. The primary method of the next-gen epitaxial tin (Sn) atomic layer ("stanene") on a Bi_2Te_3 substrate occurred in 2015, and the realization of its dissipation less conductivity at room temperature in 2018 encouraged further research into its fundamental properties in comparison to those of its rival elemental 2D analog [16]. The current chapter briefly introduces stanene, including information on its structure, synthesis, and various uses related to its different quantum characteristics. The chapter offers a comprehensive overview of the synthesis process, emphasizing the various methods. This study focused on the fabrication of stanene using an epitaxial method, which is currently the maximum suitable, standard, and viable technique. This chapter is an additional knowledge resource, particularly for budding scholars in the area.

Structure of Stanene

Each 2D material must possess structural stability and stable surface qualities to be used in various fields. Tin (Sn), as a bulk material, occurs predominantly in two forms: α-Sn (grey tin) with a less-bulk diamond structure and β-Sn (White), a bendable Td structure under normal circumstances. The Monoatomic Sn layer is known as stanene with a hexagonal lattice structure as graphene. According to the generalized gradient approximation (GGA), it has an optimized lattice constant of 4.67 [17]. It has a wavy assembly with its neighboring atom preferring an out-o--plane alignment and a mix of sp^2/ sp^3 mixed orbital hybridization states (Fig. **1**).

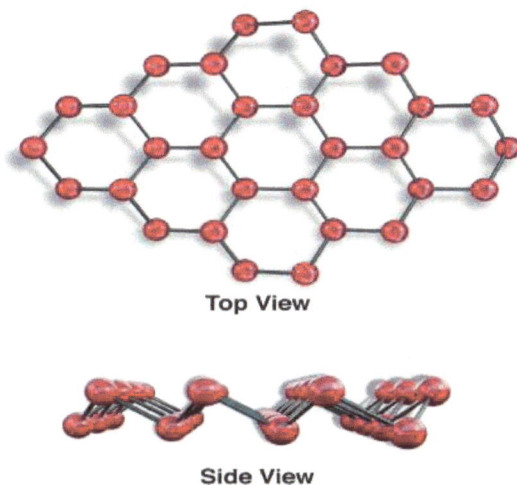

Top View

Side View

Fig. (1). Structure of stanene with top and side views.

Stanene cannot form a flawless flat honeycomb structure due to the straight Sn–Sn bond. Instead, it stabilizes as a buckled pattern, producing smaller bonds by moving nearby sublattices in the out-of-plane direction. In its quasi-fre--standing forms, stanene's puckered structure is divided into two categories: high buckled (HB) and low buckled (LB). The HB structure is a hexagonal bilayer alignment with a 9-fold atomic coordination with a relatively substantial differential in its out-of-planar height (z) between nearby sublattices [18]. According to reports, the hexagonal stanene structures with the dumbbell (DB) geometric shape are the most stable due to the cohesive energy of associated atoms and the +ve frequencies in the Brillouin zone [19].

Synthesis of Stanene

The two primary methods for producing 2D materials are top-down and bottom-up synthesis [20]. The epitaxial growth is extensively used *via* physical vapor deposition, chemical vapor deposition (CVD), interfacial breakdown, and surface partition [21, 22]. Such a bottom-up approach requires careful selection of the growth substrate to be compatible with the planned layered products. The growing substrate must be meticulously selected for the bottom-up method to produce the desired layered 2D material. Bottom-up nanosheets as produced have the drawbacks of poor output, uneven thickness, and damage-related challenges.

On the other hand, the top-down process is not limited to layered precursors and needs pure crystalline structures and regulated chemical conditions over the resulting layer states. Numerous methods, such as solution-based topotactic transformation, can satisfy these requirements [23]. The standard synthesis approach for the formation of stanene is described below.

Bottom-up Synthesis

Via Physical Vapour Deposition (PVD) Method

To ensure a regulated material deposition on a specific substrate, PVD employs heated atomic sources [24, 25]. Under ultrahigh vacuum environments, high-transparency atomic sources are often used in PVD. The products generated in a UHV atmosphere may be analyzed using surface-sensitive methods like scanning tunneling microscopy (STM) to retain the samples' original condition.

Via Chemical Vapour Deposition

To produce layered materials, CVD involves controlled reactions of gas, liquid, or solid precursors [26, 27]. Active catalytic substrate materials are the most concrete for conventional CVD; preparation can be done under atm pressure to ultrahigh vacuum conditions. The development path diverges depending on the substrate: one with low precursor insolubility catalyzes the formation of atomically thick films. Other substrates with adequate precursor dissolution may cause the precursor to segregate to the surface after cooling. Solid-state sources that are integrated into the supporting substrate are used for surface breakdown or segregation. A surface coating provided by the precursor that later shortens into a 2D material grows on the substrate, which is heated adequately to activate diffusion techniques [28].

Epitaxial Bottom-up Growth of Stanene

Due to sp^3 hybridization being physically more advantageous than sp^2, the experimental production of stanene is quite challenging. As a consequence, nature does not contain a stratified form of Sn that is comparable to graphite. Creating a material that would facilitate the development of a Sn monolayer is the answer to this problem. The selection of an appropriate material presents the most significant difficulty in this strategy. A substrate needs to have a few qualities to be a strong contender. The heterostructure of stanene and substrate must first be steady. To reduce strain, a minor lattice mismatch and hexagonal symmetry of the substrate's outermost layer are used to ensure stability. A representative technique of epitaxial synthesis of Stanene is depicted in Fig. (**2**).

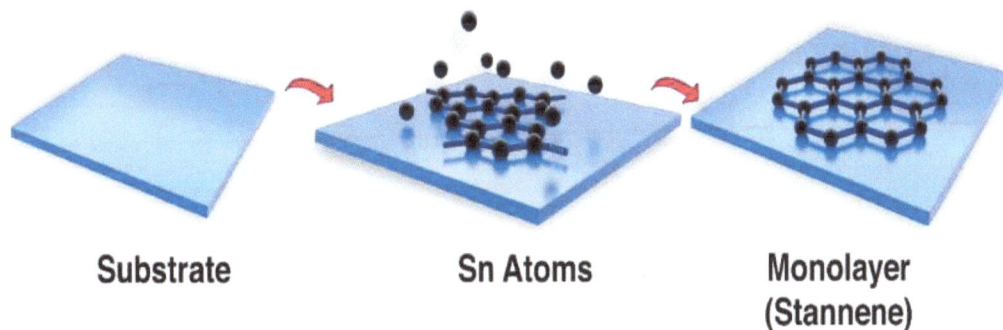

Substrate   **Sn Atoms**   **Monolayer (Stannene)**

Fig. (2). Epitaxial synthesis of stanene in a substrate.

The potential viability of various substrates to sustain stanene and maintain, or even improve, its properties have been investigated by several researchers by using different types of substrates (Metal, nonmetal, or transition complex) [23, 29]. We mention that the Sn growth on different surfaces has long been researched. The attention focused on the quantum characteristics of Sn due to the emergence of graphene and the forecast of the striking characteristics in 2D materials. In this context, Zhu and their co-workers did a primary or the very first epitaxial synthesis of Stanene in 2015. On bismuth telluride, epitaxial growth of the Sn layer was performed. In another study, the pure Al(111) surface was expected to sustain buckled, metallic stanene [30]. However, a square-like structure can be seen in the high-resolution scanning tunneling microscope (STM) picture of Sn monolayers on Al (111). Gou *et al.* [23] also conducted an intriguing experiment using a related substrate and described a formed stanene layer on an Sb (111) substrate. The band gap at the K points is increased by the substantial tension that the stanene layer experiences to 0.2 eV, which is as large as the estimates for the freestanding material. Besides these explained substrates, different substrates have been investigated for the synthesis of epitaxial Stanene; the brief of these substrates is given in Table **1** with the substrate's electronic properties.

Table 1. Epitaxial synthesis of stanene in various substrates.

Substrate	Substrate Electronic Properties	Buckling Length (nm)	Bond Length (Å)	Refs.
Bi_2Te_3	Insulator	44	3.5	[31]
Freestanding	Without substrate	47	3.3	[23]
Sb (111)	SMT	43	—	[32]
Cu (111)	MT	51	1.8	[28]
hBN/Ir (111)	Insulator/MT	72		[29]
InSb (111)	SC	45.8	2.85	[33]
Bi (111)	MT	45.4	4	[34]
MoS_2	SC	ND	2.9	[21]
Au (111)	MT	5.1–5.7	2.42.4	[21]
Ag (111)	Metal	4.98	2.6	[35]

Abreviations: MT: Metal; SMT: Semimetal; SC: Semiconductor; ND: Not Determined

Characterization of Stanene

Stanene has an extraordinary ability to display topological effects at ambient temperature due to the spin-orbit gap (SOG) and create a significant band gap. The existence of Sn atoms in a hexagonal lattice is confirmed by Saxena and his two co-workers, who studied atomic-scale morphological and elemental characterization using High-resolution transmission electron microscopy (HRTEM) fitted with Selected Area Electron diffraction (SAED) and Energy Dispersive X-Ray (EDAX) detectors. The location of the Raman peak and 'd' spacing determined by SAED are in good arrangement with the results of first-principles calculations (FPC). Atomic force microscopy (AFM) measurements indicate an interspacing of about 0.33 nm. The lack of any oxidized phases is suggested by the fact that no oxygen traces were found in the EDAX spectrum, and the same is confirmed by the Raman spectra of stanene [32]. Only a few studies have been reported on the characterization of stanene, as this 2D material is more studied in theoretical calculation and prediction. However, with the embedded-atom approach, Khan *et al.* (2017) executed an equilibrium molecular dynamics simulation to describe the thermal and mechanical characteristics of STNRs. The new zigzag and armchair stanene nanoribbons' room temperature thermal conductivities were calculated to be 0.95, 0.89, and 0.026 W/m-K, respectively. The thermal conductivity of flawed stanene nanoribbon is reduced by 30 to 50 percent at a defect concentration of 1.5%. However, a 70-90% loss was seen at a vacancy concentration of 5% for various defects [36].

Properties of Stanene

Sn primarily occurs in two forms with a less dense diamond cubic structure (Grey Sn) and -Sn (white), which has a bendable Td structure. Structural parameters such as lattice constant and buckling length of stanene are highly dependent on its primary structure (Table **2**). Density functional theory (DFT) observation concluded that the chair-like structure remains steadier than the flat morphology in stanene.

Table 2. Structure parameter of stanene.

Structure	Lattice Constant (in nm)	Buckling Length (in nm)	Refs.
low-buckled (LB) stanene	46.7	8.5	[37]
dumbbell (DB) Stanene	90.5	34.1	[37]

The low-buckled (LB) stanene structure is seen in Fig. (**3a**), and it has two atomic planes for the Sn atoms in the unit cell. An LB hexagonal lattice is created by each Sn atom forming a covalent link with three neighboring Sn atoms. Because of the buckle, stanene can have a wide range of electrical characteristics depending on external influences. The LB stanene's and orbitals overlap due to the buckling, but the π- bond is not very strong. More sp^3-like hybridizations can be created by including more Sn atoms, resulting in a highly stable dumbbell unit (DB). Exothermic and spontaneous, with no obstacles to overcome, is how the formation process occurs [19]. The DB stanene structure is seen in Fig. (**3b**).

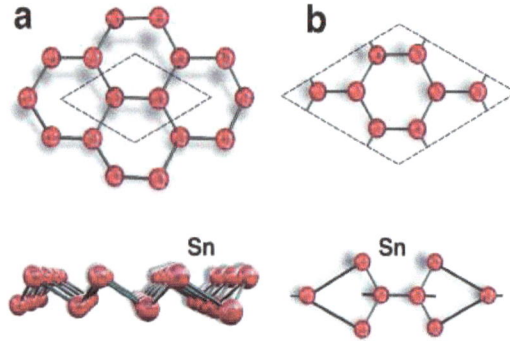

Fig. (3). a: LB stanene structure. b. DB stanene structure (Dashed line shows unit cell).

Stanene is well-versed in several properties, which make it one of the significant 2D materials in nanoelectronics and energy-related applications. Some of its properties are discussed below.

Mechanical Strength

To retain its intrinsic distinctive features, confrontation with distortion in the geometric assembly of dense Sn is a crucial problem. To assess the mechanical strengths, computational analyses of different loadings laterally of the zigzag and armchair boundaries of stable stanene were analyzed by several researchers [38]. These investigations have shown that stanene stability is anisotropic, with variations in Young's value and tensile strength about the edge. According to calculations, stanene has a Young's modulus of around 26.684 N/m. According to the stress-strain response, the tensile strength has values of 4 and 3.6 N/m in zigzag and armchair positions [13]. Table **3** compares the mechanical strength of this material to other hypothetical 2D nanomaterials.

Table 3. Mechanical strength comparison of various 2D nanomaterials.

2D-NM	Young Modulus (in N/m)	Tensile Strength (N/m)	Refs.
Stanene	26.7	3.63	[39]
Graphene	335	0.15	[40]
Silicene	32	7.2	[41]
Germanene	44	5	[41]
Hexagonal- Boron Nitride	267	0.21	[42]
Phosphorene	22	2.1	[43]

Electronic and Optical Properties

There may be a revolution in nanoelectronics due to the emergence of novel quantum states enabling spin-polarized transport in diverse 2D-NMs. The characteristics of next-generation electrical applications will likely be influenced by bandgaps and Spin-Orbit Coupling (SOC) [44]. Condensed quantum physics is very interested in the solid electronic properties of graphene [45]. These properties comprise a linear band dispersion in the position of a Dirac point. These characteristics of graphene motivate research on stanene sheet growth on diverse substrates to observe its nano-electronic applications. Due to interactions between the hybridized outer electrons and the substrate, strain from lattice divergence, and the applied electric field, these heterostructures show modulation of the topological bandgap. Stanene nanoribbons' electrical characteristics are further influenced by the SOC of topological behavior after functionalization [46]. According to the first principal calculations, the intrinsic electron and hole mobility of the buckling 2D stanene is 103 cm^2/V-Sec at ambient temperature, which is exceptionally high compared to the monolayer of phosphorene [47]. As a result, due to its unusual quantum characteristics in the QSH state, stanene emerges as a very intriguing new 2D material for several energies related and electronic applications. The different electronic properties of several 2D materials are given in Table **4**.

Table 4. Comparison of electronic bandgap of several 2D nanomaterials.

2D-NM	Electronic Bandgap (in eV)		Refs.
	Without Spin-Orbit Coupling	**With Spin-Orbit Coupling**	
Stanene	0	0.1-0.5	[31]
Graphene	0	0.15	[40]
Silicene	0	0.00	[41]
Germanene	0	0.024	[42]
Hexagonal- Boron Nitride	5-6.5	-	[48]
Phosphorene	1.5	0.02	[49]

A wide range of optical responses is made possible by dielectric transmission and SOC splitting because the bandgap of the Sn layer may be controlled by generating strain at the hetero-layer junctions or geometric modifications [50]. According to first-principles calculations (FPC), the quantum confinement has no visible impact. It offers widespread optical features since the form and peak locations are autonomous of the tapering of the nanoribbon structure [51].

Thermal Properties

Both pathways (electron or photon transport) affect the thermal characteristics of 2D nanostructured materials. With the use of FPC and Boltzmann transport equation, the value of K_{ph} of the phonon to the Sn monolayer has been investigated by several researchers. Due to slow phonon group velocities and brief lifetimes, the optical mode's contributions to thermal conduction are often negligible. The buckled stanene structure's linear phonon dispersion relationship prohibits the complete decoupling of the in-plane and out-of-plane phonon modes. Due to these properties and theoretical projections, stanene may have low K values as low as 11.9 W/mK at 300 K [52]. Table **5** lists the thermal conductivity of several 2D nanomaterials.

Table 5. Comparison of carrier mobility and thermal conductivity of several 2D nanomaterials.

2D-NM	Carrier Mobility (cm²/V-Sec)	Thermal Conductivity (W/m-K)	Refs.
Stanene	2500-3000	1.25-12	[53]
Graphene	20 X 10⁴	~5.3 X 10³	[54]
Silicene	100-200	3.3- 29	[54]
Germanene	6.2 X 10⁶	3-10	[13]
Hexagonal- Boron Nitride	~2300	~280	[13]
Phosphorene	~1000	110	[47]

Applications of Stanene

Stanene is the most exciting of the new 2D NM that has been created since the creation of graphene in 2004. Tin atoms comprise the typical hexagonal configuration found in 2D materials in this substance. Like other 2D elements, their existence, as well as many of their properties, were predicted by theoretical calculations earlier. The different properties of stanene, discussed in the above sections, are associated with several electronics and energy-related applications of stanene. Some applications are briefly described in this section.

Stanene as a Topological Insulator (TI)

Topological insulators (TIs) are insulating materials inside the device but can transport a superconducting current at the surface or outside of the fabricated device. The electrons which are at the edged position can move without any resistance. It creates Sn layers very prominent towards the Nanoelectronics wiring. The topological insulator's edge-moving electrons are protected from quantum processes that may cause their spin to flip. Chiral currents, in which the electron's spin is inaccessible for transport, can be carried by 2D topological insulators. These materials might be beneficial for people who want to use the spin of electrons to store and transport information as conventional digital electronics [55, 56]. A schematic topological insulator sheet is shown in Fig. (4), made up of a 2D nanomaterial.

Fig. (4). A type of 2D-TI sheet.

Conducting channels without backscattering at the edges and an insulating space in the defined 2D topological insulators are also called quantum spin hall insulators. These dissipation-less conducting channels offer a significant deal for low electronic devices since they can drastically minimize power dissipation. Graphene is the first anticipated 2D-TI. At low temperatures, the energy gap caused by SOC changes graphene from a semimetal to an insulator, and helical edge states have been seen at the borders. The SOC in graphene, however, is very weak, providing an insulating gap as small as 103 meV [57]. After that, too many 2D nanomaterials are studied to fabricate topological insulators. In the case of stanene, the first reported stanene-based TI was given by Zhang *et al.* in 2013. The only factors initially considered were the freestanding state and chemical functionalization for fabricating Sn-based TI. Bare Sn with less than 0.1 eV after the spin-orbit coupling effect involves an insulating gap that unlocks with an inverted band shape. The SOC effect is significantly boosted in the bare case because of the significant influence of first-order coupling, which is prohibited in graphene. In the modest situation, the band inversion of Sn layers is similar to graphene [58].

Stanene as Superconductor

The superconducting equivalent of TI replaces the gap (due to which the insulation occurred) with a superconducting gap in topological superconductors

(TSC). Majorana modes or self-conjugate excitations, which are building blocks for fault-tolerant quantum computers, exist at their limits. Time-reversal equilibrium is fragmented by the most prevalent kind of TSC, which contains 0D Majorana states in the edges [59]. These TSCs may be created by adding strong SOC to the semiconductor to induce superconductivity. Chiral TSC is another type of superconductivity in the case of 2D nanomaterials. Chiral TSC similarly breaks time-reversal symmetry, but the boundary state shifts from a localized 0D Majorana state to a chiral edge mode, as seen in several 2D-based heterostructures [60]. Fig. (**5**) gives a summarized account of various energy applications of stanane.

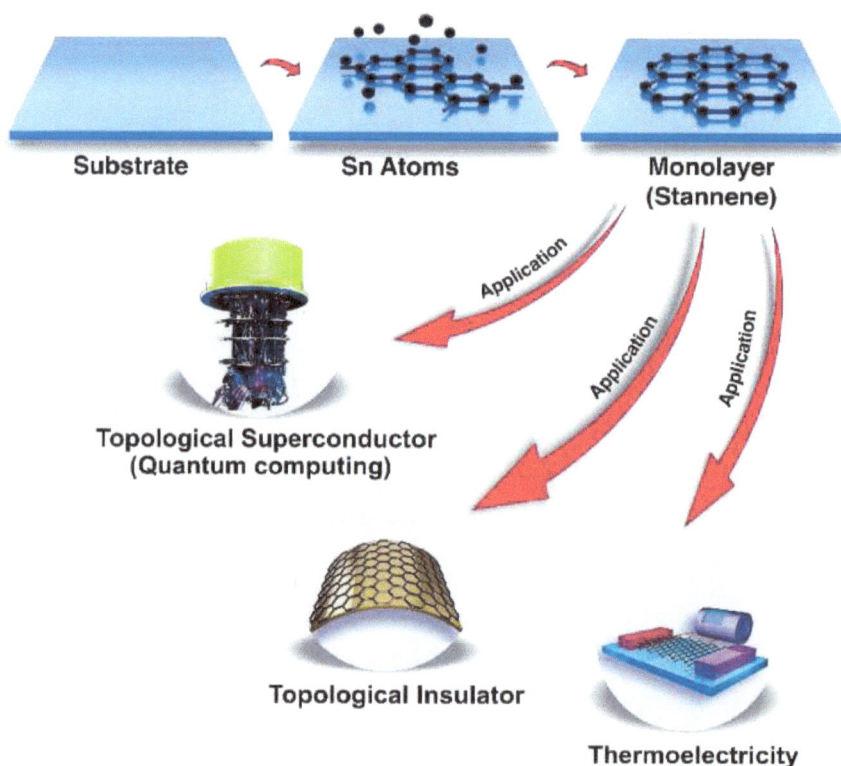

Fig. (5). Various energy applications of 2D stanene and its derivatives.

In 2014, Zhang *et al.* [56] proposed stanene as a time reversal invariant TSC. 3.7 K is the necessary temperature for -Sn to become a superconductor, according to initial tests; however, bulk -Sn exhibits no superconductivity [61]. Nonetheless, transport testing has revealed that few-layer stanene is superconducting, further raising the idea that stanene is a TSC. Based on the relative strength of the intra-orbital interaction and the SOC effect, the superconducting symmetry may be singlet or triplet. Unfortunately, this superconducting stanene characteristic has yet to be demonstrated by experimentation. It is primarily due to the discovery that critical temp. for Sn, a single sheet, is shallow, necessitating experiments at shallow temperatures [58]. This problem may be resolved by making stanene thicker. Transport experiments show that film thickness and substrate-induced doping affect the threshold temperature of stanene superconductivity [61]. The critical temperature can therefore be raised to a level that is simpler to detect with the use of an appropriate substrate and thickness. Stanene produced on Bi (111) has recently shown superconductivity, as shown by STM/STS. Monolayer stanene's tiny superconducting gap, even at 350 mK, prevents further research into Majorana states. Stanene's thickness causes the superconducting gap to grow steadily. With a superconducting gap of 0.33 meV, four-layer stanene can offer a sizable superconducting environment that can be used to examine Majorana edge modes and triplet pairing in the Stanene-based TSCs [58, 62].

MXENES

Structure

MXenes take their origin from MAX phases, which are a fascinating class of materials that have garnered significant interest due to their exceptional properties and diverse applications. These ternary compounds possess a unique layered structure, combining characteristics of both ceramics and metals. The acronym "MAX" refers to the composition of these materials, with "M" representing a transition metal, "A" denoting a group A element (typically early transition metals or metalloids), and "X" representing carbon and/or nitrogen [63 - 65]. The layered structure consists of alternating MX and A layers, creating an intriguing combination of metallic bonding within the layers and covalent bonding between them. One of the key features of MAX phases is their remarkable mechanical and thermal properties, making them excellent candidates for use in high-temperature environments and extreme conditions. They exhibit good thermal and electrical conductivity, as well as high resistance to thermal shock and deformation. Additionally, MAX phases are known for their damage tolerance and can absorb substantial amounts of energy before fracturing, making them potential materials for protective coatings and structural applications. The versatility of MAX phases extends to their tunable properties, which can be modified by adjusting the composition, layer stacking, and surface termination [66]. This property tunability has led to various applications in areas such as aerospace, nuclear energy, electronic devices, and corrosion-resistant coatings. Researchers are actively exploring the potential of MAX phases in advanced manufacturing techniques, such as additive manufacturing, to create complex and customized components. Furthermore, the MAX phases' layered structure has spurred investigations into their exfoliation and delamination, leading to the emergence of two-dimensional

(2D) materials known as MXenes. MXenes are obtained by selectively etching the A layers from the MAX phase precursor, resulting in ultrathin 2D sheets with intriguing electronic, optical, and electrochemical properties. These MXene materials are finding applications in energy storage (such as supercapacitors and batteries) and catalysts for various chemical reactions.

Classification of MXenes

The classification scheme delineates their elemental constituents and bonding arrangements. "2-1 MXenes" encompass MXenes derived from an early transition metal (M) combined with carbon (C) and sometimes nitrogen (N), with examples including Ti_2C, V_2C, Nb_2C, Mo_2C, Mo_2N, and Ti_2N, indicating the ratio of metal atoms to carbon (or nitrogen) atoms in the compound. "3-2 MXenes" include MXenes formed by combining an early transition metal (M) with three carbon atoms (C) and two nitrogen atoms (N), with instances like Ti_3C_2, TiCN, Zr_3C_2, and Hf_3C2. "4-3 MXenes" consist of early transition metals (M) bound to four carbon atoms (C) and three nitrogen atoms (N), exemplified by Ti_4N_3, Nb_4C_3, Ta_4C_3, V_4C3, and $(Mo, V)_4C_3$. A singular representative, "5-4 MXenes," results from combining an early transition metal (M) with five carbon atoms (C) and four nitrogen atoms (N), while "Double Transition Metal MXenes" feature two distinct early transition metals often accompanied by carbon and/or nitrogen, with examples including Mo_2TiC_2, Cr_2TiC_2, and Mo_2ScC_2. The subgroup "2-1-2 MXenes" further specifies double transition metal MXenes with a composition involving two carbon atoms (C) and one nitrogen atom (N) positioned between the two metals, such as Mo_2TiC_2, Cr_2TiC_2, and Mo_2ScC_2. Finally, "2-2-3 MXenes" represent double transition metal compounds with two carbon atoms (C) and three nitrogen atoms (N) interposed between the two metals, with an exemplar being $Mo_2Ti_2C_3$ [63 - 67].

Applications of MXenes

MXenes, a fascinating class of two-dimensional (2D) materials, have garnered significant attention due to their versatile properties and potential applications. These materials, derived from MAX phases through chemical exfoliation, exhibit a wide range of properties that make them promising candidates for various technological advancements. The rich diversity of MXenes and their intriguing characteristics have led to the exploration of numerous applications, offering substantial contributions to different fields. One notable area where MXenes have shown remarkable potential is in catalysis. MXenes such as Ti_2CO_2, W_2CO_2, and $TiVCO_2$ have been identified as catalysts for the hydrogen evolution reaction (HER), facilitating efficient and sustainable hydrogen generation. These catalysts hold promise for addressing energy needs and promoting clean energy production, contributing to a more sustainable future [63]. Moreover, MXenes have extended the possibilities of 2D materials by expanding the family of structures and chemistries. Materials like Mo_2TiC_2Tx, $Mo_2Ti_2C_3Tx$, and Cr_2TiC_2Tx exhibit unique properties, revealing novel structures and opening avenues for chemical diversity in 2D materials [64]. These discoveries broaden the understanding of materials science and offer prospects for innovative technologies. The electronic properties of MXenes have also attracted significant interest. Compounds like $Ti_2C(OH)_2$, $Zr_2C(OH)_2$, $Zr_2N(OH)_2$, $Hf_2C(OH)_2$, $Hf_2N(OH)_2$, $Nb_2C(OH)_2$, and $Ta_2C(OH)_2$ exhibit intriguing electronic band structures, shedding light on their potential as components for nanoelectronic applications [65]. These MXenes pave the way for the development of nanoscale electronic devices and hold promise for future advances in electronics. In energy storage, MXenes have demonstrated their utility. MXenes such as P and Si MXenes have shown enhanced capacities for metal-ion batteries, indicating their potential to contribute to the advancement of energy storage technologies [66]. Additionally, MXenes have been explored for their reactivity and selectivity towards gas molecules, offering potential applications in catalysis, gas separation, and sensing [67]. Medical applications have also been explored, with MXenes showing promise in the field of theranostics. The development of a novel superparamagnetic MXene-based theranostic nanoplatform, such as Ta_4C_3-IONP-SPs, holds the potential for efficient breast cancer theranostics, combining diagnostic imaging and therapy [68]. Furthermore, MXenes have demonstrated exceptional properties for electromagnetic (EM) absorbing and shielding composites, making them suitable candidates for EM absorbing and shielding applications [69]. Their tunable properties enable the design of materials with desired EM characteristics, paving the way for innovative solutions in communication and technology.

The above discussion shows that MXenes exhibit a diverse range of properties that hold promise for various applications across multiple fields. From catalysis and energy storage to electronics and medical applications, MXenes continue to captivate researchers with their unique properties and potential contributions to technological advancements. As research in MXenes advances, their transformative impact on various industries is becoming increasingly evident, opening up new horizons for innovative materials and technologies. A summarized account of MXenes and their application has been mentioned in the following Table **6**.

Table 6. Various MXenes, their properties and applications.

S. No.	Name of MXene	Property	Application	Refs.
1	Ti_2CO_2	Catalyst for HER	Efficient hydrogen evolution	[63]
2	W_2CO_2	Catalyst for HER	Sustainable hydrogen generation	[63]
3	$TiVCO_2$	Catalyst for HER	Enhanced hydrogen evolution	[63]
4	Mo_2TiC_2Tx	Material properties	Chemical diversity in 2D materials	[64]
5	Mo_2Ti2C_3Tx	Material properties	Novel structures and chemistries	[64]
6	Cr_2TiC_2Tx	Material properties	Expanding family of 2D materials	[64]
7	$Ti_2C(OH)_2$	Electronic properties	Unique band structures in MXenes	[65]
8	$Zr_2C(OH)_2$	Electronic properties	Tuning electronic behavior	[65]
9	$Zr_2N(OH)_2$	Electronic properties	Exploring MXenes' electronic structures	[65]
10	$Hf_2C(OH)_2$	Electronic properties	Nearly free electron states in MXenes	[65]
11	$Hf_2N(OH)_2$	Electronic properties	Unique electronic band structures	[65]
12	$Nb_2C(OH)_2$	Electronic properties	Electronic transport in MXenes	[65]
13	$Ta_2C(OH)_2$	Electronic properties	Potential nanoelectronic applications	[65]
14	P, Si MXenes	Material properties	Enhanced metal-ion battery capacities	[66]
1	M_2C (M = Ti, V, Nb, Mo)	Reactivity and selectivity towards gas molecules	Catalysis, gas separation, sensing	[67]
2	Ta_4C_3-IONP-SPs	Superparamagnetic theranostic nanoplatforms	Breast cancer theranostics, imaging, therapy	[68]
3	MXene Defects	Structural and magnetoelectronic behavior of defects	Modulation of MXene properties, device applications	[70]
4	Ti_2CF_2, Ti_2CO_2	Conductive anchoring materials for Li-S batteries	Lithium-sulfur battery improvement	[71]
5	S-doped Ti_3C_2Tx	Enhanced sodium storage performance	Sodium-ion battery anode material	[72]
1	Ti_3C_2 MXenes	Tunable electromagnetic (EM) absorbing and shielding composites	EM absorbing, shielding	[73]

(Table 6) cont.....

S. No.	Name of MXene	Property	Application	Refs.
2	MXene Structures	Formation mechanisms and properties of different MXene structures	Materials optimization	[74]
3	$Mo_2M_2C_3O_2$ MXenes	Quantum spin Hall (QSH) phase prediction	Lower-power electronics, spintronics	[75]
4	Few-layer MXenes	Enhanced electrochemical properties for sodium-ion batteries	Energy storage	[76]
5	Ti_3C_2 MXenes	Influence of basal plane functionalization on HER activity	Electrocatalysis, energy storage	[77]
6	Sc_2CT_2 (T = F, OH) MXenes	Electrical and thermal properties study	Next-gen electronic devices	[78]
1	MAX to MXene Exfoliation	Exfoliation of MAX phases into 2D MXenes	Novel 2D systems synthesis	[79]
2	MXene Colloidal Solutions	Stability and degradation of MXene colloidal solutions	Colloidal solutions stability	[80]
3	2D Transition-Metal Carbides	CO_2 conversion catalysts	CO_2 reduction, catalysts	[81]
4	Nb_4C_3-based MXenes	Solid solutions synthesis and energy storage	Supercapacitors, energy storage	[82]
5	MXene Quantum Capacitance	Quantum capacitance and work function investigation	Supercapacitors, electronic devices	[83]
6	2D Carbon Wrapped TiN	One-step synthesis of 2D carbon-wrapped TiN	Material synthesis	[84]

TRANSITION-METAL CHALCOGENIDES (TMDS)

Introduction: Structure and Classification

Transition-metal dichalcogenides (TMDs), also referred to as transition-metal dichalcogenide monolayers (TMDCs), constitute a remarkable class of atomically thin semiconductors within the MX_2 composition, where M signifies a transition-metal atom, such as Mo or W, and X represents a chalcogen atom like S, Se, or Te. These materials manifest an intriguing structural arrangement, with a single layer of M atoms nestled between two layers of X atoms, giving rise to their unique properties. Part of the expansive category of two-dimensional (2D) materials, TMDs are characterized by their extraordinary thinness, thus earning the moniker "2D materials." For instance, the thickness of a monolayer of MoS_2 measures a mere 6.5 angstroms (Å). The distinctiveness of TMDs stems from the interaction of large atoms within their two-dimensional architecture, a property that sets them apart from first-row transition-metal dichalcogenides. This distinction becomes particularly pronounced in compounds like WTe_2, which exhibits exceptional phenomena such as anomalous giant magnetoresistance and superconductivity, underscoring the captivating attributes of these materials. As researchers delve into the fascinating world of TMDs, their remarkable electronic, optical, catalytic, and magnetic properties come to the fore, rendering them a subject of extensive investigation for both fundamental studies and technological applications [85 - 87].

Semimetal few-layered $PtSe_2$ and monolayer WS_2 were synthesized through thermally assisted selenization for a metal-semiconductor-metal (MSM) photodetector. Fabrication involved photolithography and direct laser patterning, with varied scanning steps to achieve diverse channel widths. The PtSe2/WS2 heterojunction displayed barrier height modulation with increasing laser power due to the photogating effect, thereby widening material options for 2D electrodes and optoelectronic device fabrication [85]. Defect formation energies and charge transition levels in bulk and 2D MX_2 (M = Mo or W; X = S, Se, or Te) TMDs were determined through a blend of experimental deep-level transient spectroscopy (DLTS) and computational methods. The dimensionality reduction from bulk to 2D significantly influenced defect properties, shedding light on differences in optical properties among 2D TMDs synthesized using diverse methods, with potential implications for enhancing TMD-based devices [86]. Ruthenium dichalcogenide crystals (RuX_2, X = S, Se, Te) were synthesized as uniform nanoparticles on carbon nanotubes through a scaled production strategy. The RuX_2 catalysts exhibited remarkable hydrogen evolution reaction (HER) catalytic behavior, outperforming commercial Pt/C counterparts. This advancement holds promise for efficient proton and anion exchange membrane electrolyzers, propelling the exploration of high-activity transition metal dichalcogenides for water electrolysis [87]. Layered iron dichalcogenides (FeX2, X = S, Se, Te) were predicted as promising anode materials for alkali metal-ion batteries (AIBs) through first-principles calculations. The high surface-to-bulk ratio and low ion transport resistance of layered FeX2 materials exhibited exceptional performance, including high ionic conductivity, excellent rate capability, and cycling performance, making them potential high-performance anode materials for AIBs [88]. A rapid method employing 1-dodecanol encapsulation and gold-tape-assisted exfoliation produced macroscopic-scale TMD monolayers of uniform, high optical quality, expanding research beyond micron-sized devices. The encapsulation isolated the TMD from the substrate and passivated chalcogen vacancies. Integration with photonic crystal cavities enabled polariton array creation, exemplifying the utility of encapsulated monolayers in photonics and low-dimensional systems [89]. Alloying emerged as a strategic solution to the inertness of the basal plane in two-dimensional transition metal dichalcogenides (2D TMDCs). A machine learning approach predicted the hydrogen evolution reaction (HER) activity and the stability of 2D cation-mixed TMDC alloys, highlighting alloying's potential in reducing the Gibbs free energy of hydrogen adsorption on the basal plane. This study paves the way for rational design and discovery of catalytically active TMDC alloys [90]. Nanocomposites of molybdenum dichalcogenides ($MoCH_2$/Gr) with hybrid and non-hybrid structures exhibited promising electrocatalytic performances for the hydrogen evolution reaction (HER). Hydrogen binding energetics, computed with density functional theory, elucidated catalytic sites. Materials synthesized through a microwave-assisted approach demonstrated high activity and durability for HER, highlighting the potential of tellurium (Te)-rich nanocomposites and indicating the catalytic advantage of hybrid structures [91]. The spontaneous linear ordered unzipping of bi-elemental 2D MX2 transition metal chalcogenides, exemplified by 1T' MoS2, was achieved through self-linearized oxygenation at chalcogenides. The process yielded stable dispersions of 1D nanoribbon structures in water, with potential applications in electrocatalysis for hydrogen evolution reactions, showcasing a reliable route for synthesizing nanoribbons with controlled material chemistry [92]. A molecular precursor strategy using [$MIV(SC_2H_4N(Me)C_2H_4S)_2$] complexes enabled reliable synthesis of layered 2D transition metal dichalcogenides (TMDCs). The chelating ligand formed monomeric complexes, producing stable precursors for 2D TMDCs like TiS_2. The approach's versatility and effectiveness offer a promising synthetic route for scalable and reproducible fabrication of van der Waals 2D heterostructures [93]. Electrodeposition directly onto electrodes enabled the production of transition metal dichalcogenides (TMDs) and TMD bilayer electrodes. The strategy yielded enhanced proton reduction activity in sequentially deposited bilayer TMD structures, suggesting superior electron transfer kinetics and offering a promising avenue for designing high-performance electrocatalysts for the hydrogen evolution reaction [94]. Colloidal synthesis proved valuable for producing tunable transition-metal dichalcogenide (TMD) nanostructures with diverse compositions, structures, and thicknesses. This synthetic approach revealed unique catalytic properties, such as the selective hydrogenation of substituted nitroarenes, facilitated by high defect densities in the colloidal TMD nanostructures. This versatile method allows systematic studies of structure-property relationships and chemical reactivity in 2D materials [95]. Uniform confinement of monolayer and interlayer-expanded few-layer MoS_2 within an N-doped carbon matrix resulted in hollow hierarchical spheres (H-MoS_2@NC) with advantageous sodium storage properties. The flexible MX-H-MoS_2NC electrode demonstrated improved sodium storage performance compared to PVDF-bonded H-MoS_2@NC and bare MoS_2, highlighting a promising avenue for scalable and efficient sodium-ion battery anodes [96]. A novel vapor transport and liquid sulfur synthesis route produced large single crystals of transition metal dichalcogenides (TMDs). The investigation of the physical property of the synthesized TMDs, such as CoS_2, ReS_2, NbS_2, and TaS_2, demonstrated high quality and potential suitability for emerging electronic devices, offering an efficient method for obtaining high-quality TMD single crystals [97]. Single-layer technetium-based TMDs (TcX_2, where X = S, Se, or Te) were systematically studied for structural stability, electronic properties, and thermoelectric performance. Monolayer 1Tdp-TcX_2 exhibited enlarged indirect band gaps and remarkably high thermoelectric Figs of merit (ZT) at elevated temperatures, showcasing their potential for optoelectronic and thermoelectric applications [98].

Applications of TMDCs

Transition metal dichalcogenides (TMDCs) have garnered immense attention in recent years due to their diverse and intriguing properties, leading to a wide range of potential applications in various fields. The reduction of band gap through partial oxidation has been demonstrated, allowing for the modulation of electronic properties and potential use in catalysis or energy storage applications [99]. Furthermore, the discovery of helical edge states isolated within the band gap presents promising prospects for practical applications in quantum spin Hall (QSH) edge state devices [100]. Spectroscopic identification of mid-gap states and the influence of many-body interactions on energy gaps have implications for designing novel electronic and optoelectronic devices [101]. Anisotropic properties of certain TMDCs have led to the development of highly efficient energy conversion materials, such as those exhibiting up to 12.59% solar-t--hydrogen (STH) efficiency and slow electron-hole recombination [102]. Tunable band gaps and the confinement of specific electrons offer opportunities for tailoring electronic and optical properties for diverse applications [103]. Strain engineering has been found to significantly impact the band gap of TMDCs, providing a means for sensor development or electronic tuning [104]. Notably, distinct trends in defect tolerance have been identified among different TMDC groups, suggesting potential applications in semiconducting devices and catalysis [105]. Moreover, changes in nanosheet dimensions have been harnessed for systematic variations in optical properties, enabling tailored light-matter interactions [106]. The application potential of TMDCs extends to electrocatalysis, where redox peaks observed in cyclic voltammetry highlight the influence of material conductivity and active sites on hydrogen evolution reaction (HER) catalysis [107]. Layer-dependent shifts in indirect band gaps and van Hove singularities introduce the possibility of tunable properties for next-generation electronic and optoelectronic devices [108]. Investigations into specific TMDCs, such as (2H-WTe_2), have revealed the sophisticated relation between spin-orbit coupling and interlayer coupling, with potential implications for quantum electronic applications [109]. Distinct lattice thermal conductivity behavior observed in TaSe2 compared to NbSe$_2$ offers insights into the thermal transport properties of TMDCs, relevant for thermal management and thermoelectric applications [110]. Among the TMDC family, novel families of heterostructured transition-metal dichalcogenides (TMDCs) with incommensurate spatial arrangements have been prepared, demonstrating semiconducting behavior with an indirect band gap around 1 eV and opening avenues for tunable physical properties [111]. Additionally, the generation of spinless and spinful k · p parameter sets for circular and linear dichroism under strain offers opportunities for advanced optical and optoelectronic applications [112]. The design of PN-junctions using TMDCs has led to photovoltaic efficiencies exceeding 14%, displaying their potential in solar energy conversion [113]. Investigations into energy transfer mechanisms in layered TMDCs and heterostructures reveal insights into exciton recombination dynamics and potential pathways for controlling energy conversion processes [114]. Surface doping of Rb atoms has emerged as a universal method to tune band gaps in 2D TMDCs, offering a wide range of energy levels and promising applications in electronic and optoelectronic devices [115]. Furthermore, the determination of exciton binding energies and band gaps in monolayer TMDCs through optical reflectivity/absorption spectra provides crucial information for designing novel semiconductor devices [116]. The diverse and tunable properties of transition metal dichalcogenides (TMDCs) have sparked a wave of research and innovation across various scientific and technological domains. The presented numerical results highlight the potential of TMDCs in applications ranging from catalysis and energy conversion to quantum electronics and optoelectronics. As our understanding of these materials continues to deepen, TMDCs will likely play an increasingly important role in shaping the future of advanced materials and devices. A summary of recent advancements has been given in Table **7**.

Table 7. Various chalcogenides, their key properties and applications

S. No.	Chalcogenide	Key Property	Application	Refs.
1.	Monolayer and Few-Layer Dichalcogenides	Controlled Oxidation	Reduction of band gap for partial oxidation, complete oxidation induces metallic behavior.	[99]
2.	Mo and W in 1T′ Crystalline Phase	Quantum Spin Hall Edge States	Helical edge states isolated within the band gap, guidance for practical QSH edge state applications.	[100]
3.	WS_2, MoS_2, WSe_2, $MoSe_2$	Atomic Defect States	Spectroscopic location of mid-gap states, many-body interactions enlarge energy gaps.	[101]

(Table 7) cont.....

S. No.	Chalcogenide	Key Property	Application	Refs.
4.	PdSe$_2$	Solar-to-Hydrogen Efficiency	Anisotropic properties, high STH efficiency up to 12.59%, slow electron-hole recombination (~1.9 ns).	[102]
5.	FeTe$_2$, NiTe$_2$	Half-Metallic Heterostructures	Tunable band gaps, confinement of spin-down electrons.	[103]
6.	MX$_2$ (M = Mo, W, X = S, Se)	Strain-Modulated Band Gaps	Evident band gap increases under strain, sensitivity to strain.	[104]
7.	Transition Metal Dichalcogenides (TMDs)	Defect Tolerance	Group IV TMDs (*e.g.*, WS$_2$, MoS$_2$) are defect-tolerant, and group VI and X TMDs (*e.g.*, WSe$_2$, MoSe$_2$) form deep gap states upon vacancy creation.	[105]
8.	WS$_2$, MoS$_2$, WSe$_2$, MoSe$_2$	Size-Dependent Optical Properties	Changes in nanosheet dimensions lead to systematic variations in optical extinction.	[106]
9.	GaS, GaSe, GaTe, InSe	Inherent Electrochemistry and Electrocatalysis	Redox peaks observed in cyclic voltammetry, materials' conductivity, and active sites influence hydrogen evolution reaction (HER) catalysis.	[107]
10.	PtX$_2$ (X = S, Se, Te)	Electronic Properties with Film Thickness	Layer-dependent shifts in indirect band gaps and van Hove singularities, potential for tunable properties.	[108]
11.	2H-WSe$_2$, 2H-WTe$_2$	Indirect-to-Direct Band Gap Transition	(2H-WTe$_2$), competition between spin-orbit coupling and interlayer coupling influences transition.	[95]
12.	NbSe$_2$, TaSe$_2$	Thermal Transport Properties	Abnormal lattice thermal conductivity behavior in TaSe$_2$ compared to NbSe$_2$	[110]
13.	Heterostructured TMDCs	Incommensurate Heterostructuring	Semiconducting behavior with an indirect band gap of around 1 eV.	[111]
14.	MoS$_2$, MoSe$_2$, MoTe$_2$, WS$_2$, WSe$_2$, WTe$_2$, MoSSe, MoSeTe, WSSe, WSeTe	Valleytronics and Optical Properties	Spinless and spinful k · p parameter sets generated for circular and linear dichroism under strain.	[112]
15.	TMDs (MoS$_2$, MoSe$_2$, *etc.*)	Photovoltaic Efficiency	PN-junctions of MoSe$_2$ on h-BN exhibit photovoltaic efficiencies exceeding 14%	[113]
16.	Layered TMDCs (MoS$_2$, WS2, WSe$_2$) and CdSe/ZnS QDs	Exciton Recombination Dynamics and Energy Transfer	Energy transfer investigated in layered TMDCs and CdSe/ZnS QD heterostructures.	[114]
17.	Bulk 2H Transition Metal Dichalcogenides	Universally Tunable Band Gap	Surface doping of Rb atoms leads to universally tunable band gap in 2H TMDCs, ranging from 0.8 eV to 2.0 eV,	[115]

(Table 7) cont.....

S. No.	Chalcogenide	Key Property	Application	Refs.
18.	Monolayer WS$_2$, WSe$_2$	Exciton Binding Energies	Determined exciton binding energies and band gaps for monolayer WS2 andWSe$_2$ using optical reflectivity/absorption spectra.	[116]

Promising Horizons: Unraveling the Potential of 2D Materials

Two-dimensional (2D) materials exhibit exceptional and distinct properties stemming from their reduced dimensionality. Such materials, which primarily consist of monolayers, have garnered considerable attention in the scientific community due to their unprecedented electronic, optical, mechanical, and thermal attributes, setting them apart from their bulk counterparts. The most notable feature of 2D materials is their remarkable electronic behavior. The quantum confinement effect engendered by their two-dimensional structure leads to discrete energy levels, dramatically altering their electronic band structures. This confinement begets an assortment of novel phenomena, such as anisotropic electrical transport, sizable bandgap tunability, and the emergence of Dirac or Weyl fermions, thereby enabling tailor-made electronic applications with enhanced efficiency and speed. Furthermore, 2D materials display extraordinary mechanical properties attributed to their reduced dimensionality. Their inherent flexibility, engendered by their single-atom thickness, allows facile bending and stretching, rendering them ideal candidates for flexible electronics and stretchable sensors. Simultaneously, their inherent fragility necessitates a meticulous balance between mechanical integrity and ductility, thereby posing intriguing challenges and opportunities in their manipulation and integration into diverse nanoscale devices.

Moreover, these materials play a key role in shaping the optics with a spectrum of unique properties. The ultrathin nature of these materials facilitates strong light-matter interactions, enabling substantial light absorption and emission in nanoscale volumes. The emergence of excitons, bound electron-hole pairs, and their subsequent recombination dynamics lead to striking photoluminescence phenomena. Moreover, their extraordinary nonlinear optical responses, such as second-harmonic generation and Kerr effect, offer prospects for ultrafast optical modulation and all-optical signal processing, paving the way for advanced photonic applications. Thermal transport in 2D materials is also distinctive due to their constrained lattice structures and reduced phonon mean free paths. Consequently, these materials exhibit exceptionally high intrinsic thermal conductivities, promising efficient heat dissipation in nanoscale devices. Conversely, their low out-of-plane thermal conductivities suggest opportunities for designing efficient thermal barriers and thermoelectric materials for waste heat recovery.

Beyond their intrinsic properties, 2D materials possess unparalleled surface-t--volume ratios, conferring remarkable surface chemistries and interactions. Functionalization of these surfaces enables precise control over chemical reactivity, adsorption, and catalysis, heralding potential advancements in heterogeneous catalysis and gas sensing. The close association of these distinct properties culminates in complex heterostructures and van der Waals assemblies, offering an avenue for engineering multifunctional devices with unprecedented capabilities. Stacking disparate 2D materials creates diverse heterostructures, where interlayer interactions can be finely tuned to yield tailored electronic and optical properties. This strategy has yielded a plethora of emergent phenomena, including twist-induced superconductivity and topologically protected states, ushering in a new era of quantum materials engineering. These properties collectively establish 2D materials as a burgeoning field of research with transformative potential across a plethora of scientific and technological domains.

CONCLUSION

This chapter has elucidated substantial advancements within the realm of 2D materials, specifically focusing on stanene, MXene, and transition metal chalcogenides. It has meticulously examined their structural intricacies, encompassing their chemical and physical attributes, as well as contemplated their prospective applications, with the overarching potential to revolutionize diverse domains within the purview of science and technology. Stanene, possessing striking attributes as a topological insulator, stands poised to inaugurate a novel era distinguished by highly efficient electronic apparatus and quantum computational paradigms. The exceptional properties exhibited by stanene render it an object of profound scientific intrigue, meriting dedicated investigation and development endeavors. Conversely, MXenes have arisen as multifaceted two-dimensional materials, distinguished by their superlative conductivity, mechanical robustness, and chemical resilience. Their envisaged roles in energy storage, pliable electronics, and electromagnetic shielding applications hold significant promise, and continual scientific exploration is anticipated to unveil further pioneering utilities. Transition metal chalcogenides have manifested an expansive spectrum of material properties, ranging from superconductivity to semiconducting behavior, thereby affording them versatile candidacy for electronic and optoelectronic pursuits. The malleable characteristics intrinsic to these materials present opportunities for tailored material design to align with specific prerequisites.

REFERENCES

[1]   Punetha, V.D.; Dhali, S.; Rana, A.; Karki, N.; Tiwari, H.; Negi, P.; Basak, S.; Sahoo, N.G. Recent advancements in green synthesis of nanoparticles for improvement of bioactivities: A review. *Curr. Pharm. Biotechnol.,* **2022**, *23*(7), 904-919.
[http://dx.doi.org/10.2174/1389201022666210812115233] [PMID: 34387160]

[2]   Su, S.; Sun, Q.; Gu, X. Two-dimensional nanomaterials for biosensing applications. *TrAC, Trends Anal. Chem,* **2019**, *121*, 115668.
[http://dx.doi.org/10.1016/j.trac.2019.07.021]

[3]   Zhao, H.; Wang, Y.; Bao, L. Engineering nano–bio interfaces from nanomaterials to nanomedicines. *Acc. Mater. Res.,* **2022**, *3*(8), 812-829.
[http://dx.doi.org/10.1021/accountsmr.2c00072]

[4]   Jahangir, A.; Singh, P.K.; Singh, R.C. Ion Conducting Polyethylene Oxide-Based Polymer Electrolyte Dopped with Ionic Liquid-Trifluoromethanesulfonic Chloride. *Macromol. Symp.,* **2024**, *413*(1)
[http://dx.doi.org/10.1002/masy.202300139]

[5]   Dhapola, P.S.; Karakoti, M.; Kumar, S. Environment-friendly approach for synthesis of promising porous carbon: empowering supercapacitors for a sustainable future. *Mater. Adv,* **2024**, *5*(6), 2430-2440.
[http://dx.doi.org/10.1039/d3ma00984j]

[6]   Bhatt, S.; Pathak, R.; Punetha, V.D.; Punetha, M. Recent advances and mechanism of antimicrobial efficacy of graphene-based materials: A review. *J. Mater. Sci.,* **2023**, *58*(19), 7839-7867.
[http://dx.doi.org/10.1007/s10853-023-08534-z] [PMID: 37200572]

[7]   Zhou, Y.; Zhuge, X.; An, P.; Du, S. First-principles investigations on MXene-blue phosphorene and MXene-MoS $_2$ transistors. *Nanotechnology,* **2020**, *31*(39), 395203.
[http://dx.doi.org/10.1088/1361-6528/ab95b4] [PMID: 32442982]

[8]   Kolobov, A.V.; Tominaga, J. Chemistry of chalcogenides and transition metals. *Springer Series Mater. Sci.,* **2016**, *239*, 7-27.
[http://dx.doi.org/10.1007/978-3-319-31450-1_2]

[9]   Cho, A.J.; Kwon, J.Y. Hexagonal boron nitride for surface passivation of two-dimensional van der waals heterojunction solar cells. *ACS Appl. Mater. Interfaces,* **2019**, *11*(43), 39765-39771.
[http://dx.doi.org/10.1021/acsami.9b11219] [PMID: 31577117]

[10]   Wang, G.E.; Luo, S.; Di, T.; Fu, Z.; Xu, G. Layered Organic Metal Chalcogenides (OMCs): From bulk to two-dimensional materials. *Angew. Chem. Int. Ed.,* **2022**, *61*(27), e202203151.
[http://dx.doi.org/10.1002/anie.202203151] [PMID: 35441775]

[11]   Erdem, Ö.; Derin, E.; Zeibi Shirejini, S.; Sagdic, K.; Yilmaz, E.G.; Yildiz, S.; Akceoglu, G.A.; Inci, F. Carbon-based nanomaterials and sensing tools for wearable health monitoring devices. *Adv. Mater. Technol.,* **2022**, *7*(3), 2100572.
[http://dx.doi.org/10.1002/admt.202100572]

[12]   Mannix, A. J.; Kiraly, B.; Hersam, M. C.; Guisinger, N. P. Synthesis and chemistry of elemental 2D materials. *Nat. Rev. Chem.,* **2017**, *1*(2), 1-14.
[http://dx.doi.org/10.1038/s41570-016-0014]

[13]   Sahoo, S.K.; Wei, K.H. A perspective on recent advances in 2D stanene nanosheets. *Adv. Mater. Interfaces,* **2019**, *6*(18), 1900752.
[http://dx.doi.org/10.1002/admi.201900752]

[14]   Tian, Y.; Chen, Y.; Liu, Y.; Li, H.; Dai, Z. Elemental two-dimensional materials for Li/Na-Ion battery anode applications. *Chem. Rec.,* **2022**, *22*(10), e202200123.
[http://dx.doi.org/10.1002/tcr.202200123] [PMID: 35758546]

[15]   Venkateshalu, S.; Subashini, G.; Bhardwaj, P.; Jacob, G.; Sellappan, R.; Raghavan, V.; Jain, S.; Pandiaraj, S.; Natarajan, V.; Al Alwan, B.A.M.; Al Mesfer, M.K.M.; Alodhayb, A.; Khalid, M.;

Grace, A.N. Phosphorene, antimonene, silicene and siloxene based novel 2D electrode materials for supercapacitors-A brief review. *J. Energy Storage,* **2022**, *48*, 104027.
[http://dx.doi.org/10.1016/j.est.2022.104027]

[16]   Lozovoy, K.A.; Izhnin, I.I.; Kokhanenko, A.P.; Dirko, V.V.; Vinarskiy, V.P.; Voitsekhovskii, A.V.; Fitsych, O.I.; Akimenko, N.Y. Single-element 2D materials beyond graphene: Methods of epitaxial synthesis. *Nanomaterials,* **2022**, *12*(13), 2221.
[http://dx.doi.org/10.3390/nano12132221] [PMID: 35808055]

[17]   Nakamura, Y.; Zhao, T.; Xi, J.; Shi, W.; Wang, D.; Shuai, Z. Intrinsic charge transport in stanene: Roles of bucklings and electron-phonon couplings. *Adv. Electron. Mater.,* **2017**, *3*(11), 1700143.
[http://dx.doi.org/10.1002/aelm.201700143]

[18]   Rivero, P.; Yan, J.A.; García-Suárez, V.M.; Ferrer, J.; Barraza-Lopez, S. Stability and properties of high-buckled two-dimensional tin and lead. *Phys. Rev. B Condens. Matter Mater. Phys.,* **2014**, *90*(24), 241408.
[http://dx.doi.org/10.1103/PhysRevB.90.241408]

[19]   Mahmud, S.; Haque. Md, M.; Alam. Md, K. Strain Tunable Quantum Spin Hall Insulating Phase in Halogen-Decorated Double-Layer Stanene. *Phys. Status Solidi RRL.,* **2022**, *16*(8)
[http://dx.doi.org/10.1002/pssr.202200106]

[20]   Ferdous, N.; Islam, M.S.; Park, J.; Hashimoto, A. Tunable electronic properties in stanene and two dimensional silicon-carbide heterobilayer: A first principles investigation. *AIP Adv.,* **2019**, *9*(2), 025120.
[http://dx.doi.org/10.1063/1.5066029]

[21]   Chen, K.C.; Lee, L.M.; Chen, H.A.; Sun, H.; Wu, C.L.; Chen, H.A.; Lin, K.B.; Tseng, Y.C.; Kaun, C.C.; Pao, C.W.; Lin, S.Y. Multi-layer elemental 2D materials: Antimonene, germanene and stanene grown directly on molybdenum disulfides. *Semicond. Sci. Technol.,* **2019**, *34*(10), 105020.
[http://dx.doi.org/10.1088/1361-6641/ab3c8a]

[22]   Eltinge, S.; Ismail-Beigi, S. Epitaxial binding and strain effects of monolayer stanene on the Al_2O_3 (0001) surface. *Phys. Rev. Mater.,* **2022**, *6*(1), 014007.
[http://dx.doi.org/10.1103/PhysRevMaterials.6.014007]

[23]   Gou, J.; Kong, L.; Li, H.; Zhong, Q.; Li, W.; Cheng, P.; Chen, L.; Wu, K. Strain-induced band engineering in monolayer stanene on Sb(111). *Phys. Rev. Mater.,* **2017**, *1*(5), 054004.
[http://dx.doi.org/10.1103/PhysRevMaterials.1.054004]

[24]   Matusalem, F.; Koda, D. S.; Bechstedt, F.; Marques, M.; Teles, L. K. Deposition of topological silicene, germanene and stanene on graphene-covered SiC substrates. *Scientific Reports,* **2017**, *7*(1), 1-7.
[http://dx.doi.org/10.1038/s41598-017-15610-3]

[25]   Matusalem, F.; Bechstedt, F.; Marques, M.; Teles, L.K. Quantum spin Hall phase in stanene-derived overlayers on passivated SiC substrates. *Phys. Rev. B,* **2016**, *94*(24), 241403.
[http://dx.doi.org/10.1103/PhysRevB.94.241403]

[26]   Lee, J.H.; Lee, E.K.; Joo, W.J.; Jang, Y.; Kim, B.S.; Lim, J.Y.; Choi, S.H.; Ahn, S.J.; Ahn, J.R.; Park, M.H.; Yang, C.W.; Choi, B.L.; Hwang, S.W.; Whang, D. Wafer-scale growth of single-crystal monolayer graphene on reusable hydrogen-terminated germanium. *Science,* **2014**, *344*(6181), 286-289.
[http://dx.doi.org/10.1126/science.1252268] [PMID: 24700471]

[27]   Edwards, R.S.; Coleman, K.S. Graphene film growth on polycrystalline metals. *Acc. Chem. Res.,* **2013**, *46*(1), 23-30.
[http://dx.doi.org/10.1021/ar3001266] [PMID: 22891883]

[28]   Deng, J.; Xia, B.; Ma, X.; Chen, H.; Shan, H.; Zhai, X.; Li, B.; Zhao, A.; Xu, Y.; Duan, W.; Zhang, S.C.; Wang, B.; Hou, J.G. Epitaxial growth of ultraflat stanene with topological band inversion. *Nat. Mater.,* **2018**, *17*(12), 1081-1086.
[http://dx.doi.org/10.1038/s41563-018-0203-5] [PMID: 30397308]

[29]   Dong, X.; Zhang, L.; Yoon, M.; Zhang, P. The role of substrate on stabilizing new phases of two-dimensional tin. *2d Mater.,* **2021**, *8*(4), 045003.
       [http://dx.doi.org/10.1088/2053-1583/ac1255]

[30]   Guo, Y.; Pan, F.; Ye, M.; Wang, Y.; Pan, Y.; Zhang, X.; Li, J.; Zhang, H.; Lu, J. Interfacial properties of stanene–metal contacts. *2d Mater.,* **2016**, *3*(3), 035020.
       [http://dx.doi.org/10.1088/2053-1583/3/3/035020]

[31]   Zhu, F.; Chen, W.; Xu, Y.; Gao, C.; Guan, D.; Liu, C.; Qian, D.; Zhang, S.C.; Jia, J. Epitaxial growth of two-dimensional stanene. *Nat. Mater.,* **2015**, *14*(10), 1020-1025.
       [http://dx.doi.org/10.1038/nmat4384] [PMID: 26237127]

[32]   Saxena, S.; Chaudhary, R. P.; Shukla, S. Stanene: Atomically thick freestanding layer of 2D hexagonal tin. *Sci. Reports.,* **2016**, *6*(1), 1-4.
       [http://dx.doi.org/10.1038/srep31073]

[33]   Xu, C.Z.; Chan, Y.H.; Chen, P.; Wang, X.; Flötotto, D.; Hlevyack, J.A.; Bian, G.; Mo, S.K.; Chou, M.Y.; Chiang, T.C. Gapped electronic structure of epitaxial stanene on InSb(111). *Phys. Rev. B,* **2018**, *97*(3), 035122.
       [http://dx.doi.org/10.1103/PhysRevB.97.035122]

[34]   Zang, Y.; Zhu, K.; Li, L.; He, K. Molecular beam epitaxy growth of few-layer stanene. *Quant. Front.,* **2022**, *1*(1), 1-7.
       [http://dx.doi.org/10.1007/s44214-022-00012-y]

[35]   Yuhara, J.; le Lay, G. Beyond silicene: Synthesis of germanene, stanene and plumbene. *Jpn. J. Appl. Phys.,* **2020**, *59*, SN0801.
       [http://dx.doi.org/10.35848/1347-4065/ab8410]

[36]   Khan, M.H.; Liu, H.K.; Sun, X.; Yamauchi, Y.; Bando, Y.; Golberg, D.; Huang, Z. Few-atomi-layered hexagonal boron nitride: CVD growth, characterization, and applications. *Mater. Today,* **2017**, *20*(10), 611-628.
       [http://dx.doi.org/10.1016/j.mattod.2017.04.027]

[37]   Lyu, J.K.; Zhang, S.F.; Zhang, C.W.; Wang, P.J. Stanene: A promising material for new electronic and spintronic applications. *Ann. Phys.,* **2019**, *531*(10), 1900017.
       [http://dx.doi.org/10.1002/andp.201900017]

[38]   Shodja, H.M.; Ojaghnezhad, F.; Etehadieh, A.; Tabatabaei, M. Elastic moduli tensors, ideal strength, and morphology of stanene based on an enhanced continuum model and first principles. *Mech. Mater.,* **2017**, *110*, 1-15.
       [http://dx.doi.org/10.1016/j.mechmat.2017.04.001]

[39]   Shi, Z.; Singh, C.V. The ideal strength of two-dimensional stanene may reach or exceed the Griffith strength estimate. *Nanoscale,* **2017**, *9*(21), 7055-7062.
       [http://dx.doi.org/10.1039/C7NR00010C] [PMID: 28287225]

[40]   Kudin, K.N.; Scuseria, G.E.; Yakobson, B.I. C 2 F, BN, and C nanoshell elasticity from *ab initio* computations. *Phys. Rev. B Condens. Matter,* **2001**, *64*(23), 235406.
       [http://dx.doi.org/10.1103/PhysRevB.64.235406]

[41]   Mortazavi, B.; Rahaman, O.; Makaremi, M.; Dianat, A.; Cuniberti, G.; Rabczuk, T. First-principles investigation of mechanical properties of silicene, germanene and stanene. *Physica E,* **2017**, *87*, 228-232.
       [http://dx.doi.org/10.1016/j.physe.2016.10.047]

[42]   Şahin, H.; Cahangirov, S.; Topsakal, M.; Bekaroglu, E.; Akturk, E.; Senger, R.T.; Ciraci, S. Monolayer honeycomb structures of group-IV elements and III-V binary compounds: First-principles calculations. *Phys. Rev. B Condens. Matter Mater. Phys.,* **2009**, *80*(15), 155453.
       [http://dx.doi.org/10.1103/PhysRevB.80.155453]

[43]   Nguyen, D.T.; Le, M.Q.; Nguyen, V.T.; Bui, T.L. Effects of various defects on the mechanical

properties of black phosphorene. *Superlattices Microstruct.,* **2017**, *112*, 186-199.
[http://dx.doi.org/10.1016/j.spmi.2017.09.021]

[44]   Gong, Q.; Zhang, G. Spin-orbit coupling electronic structures of organic-group functionalized Sb and Bi topological monolayers. *Nanomaterials,* **2022**, *12*(12), 2041.
[http://dx.doi.org/10.3390/nano12122041] [PMID: 35745380]

[45]   Ares, P.; Novoselov, K.S. Recent advances in graphene and other 2D materials. *Nano Mater. Sci.,* **2022**, *4*(1), 3-9.
[http://dx.doi.org/10.1016/j.nanoms.2021.05.002]

[46]   van den Broek, B.; Houssa, M.; Lu, A.; Pourtois, G.; Afanas'ev, V.; Stesmans, A. Silicene nanoribbons on transition metal dichalcogenide substrates: Effects on electronic structure and ballistic transport. *Nano Res.,* **2016**, *9*(11), 3394-3406.
[http://dx.doi.org/10.1007/s12274-016-1217-4]

[47]   Li, L.; Yu, Y.; Ye, G.J.; Ge, Q.; Ou, X.; Wu, H.; Feng, D.; Chen, X.H.; Zhang, Y. Black phosphorus field-effect transistors. *Nat. Nanotechnol.,* **2014**, *9*(5), 372-377.
[http://dx.doi.org/10.1038/nnano.2014.35] [PMID: 24584274]

[48]   Kaloni, T.P.; Joshi, R.P.; Adhikari, N.P.; Schwingenschlögl, U. Band gap tunning in BN-doped graphene systems with high carrier mobility. *Appl. Phys. Lett.,* **2014**, *104*(7), 073116.
[http://dx.doi.org/10.1063/1.4866383]

[49]   Qiao, J.; Kong, X.; Hu, Z.X.; Yang, F.; Ji, W. High-mobility transport anisotropy and linear dichroism in few-layer black phosphorus. *Nat. Commun.,* **2014**, *5*(1), 4475.
[http://dx.doi.org/10.1038/ncomms5475] [PMID: 25042376]

[50]   Chaves, A. J.; Ribeiro, R. M.; Frederico, T.; Peres, N. M. R. Excitonic effects in the optical properties of 2D materials:an equation of motion approach. *2d Mater.,* **2017**, *4*(2), 025086.
[http://dx.doi.org/10.1088/2053-1583/aa6b72]

[51]   Wu, H.; Li, F. First-principles calculation on geometric, electronic and optical properties of fully fluorinated stanene: A large-gap quantum spin hall insulator. *Chin. Phys. Lett.,* **2016**, *33*(6), 067101.
[http://dx.doi.org/10.1088/0256-307X/33/6/067101]

[52]   Kuang, Y.D.; Lindsay, L.; Shi, S.Q.; Zheng, G.P. Tensile strains give rise to strong size effects for thermal conductivities of silicene, germanene and stanene. *Nanoscale,* **2016**, *8*(6), 3760-3767.
[http://dx.doi.org/10.1039/C5NR08231E] [PMID: 26815838]

[53]   Tang, P.; Chen, P.; Cao, W.; Huang, H.; Cahangirov, S.; Xian, L.; Xu, Y.; Zhang, S.C.; Duan, W.; Rubio, A. Stable two-dimensional dumbbell stanene: A quantum spin Hall insulator. *Phys. Rev. B Condens. Matter Mater. Phys.,* **2014**, *90*(12), 121408.
[http://dx.doi.org/10.1103/PhysRevB.90.121408]

[54]   Zhang, X.; Xie, H.; Hu, M.; Bao, H.; Yue, S.; Qin, G.; Su, G. Thermal conductivity of silicene calculated using an optimized Stillinger-Weber potential. *Phys. Rev. B Condens. Matter Mater. Phys.,* **2014**, *89*(5), 054310.
[http://dx.doi.org/10.1103/PhysRevB.89.054310]

[55]   Qi, X.L.; Zhang, S.C. Topological insulators and superconductors. *Rev. Mod. Phys.,* **2011**, *83*(4), 1057-1110.
[http://dx.doi.org/10.1103/RevModPhys.83.1057]

[56]   Wang, J.; Xu, Y.; Zhang, S.C. Two-dimensional time-reversal-invariant topological superconductivity in a doped quantum spin-Hall insulator. *Phys. Rev. B Condens. Matter Mater. Phys.,* **2014**, *90*(5), 054503.
[http://dx.doi.org/10.1103/PhysRevB.90.054503]

[57]   Yao, Y.; Ye, F.; Qi, X.L.; Zhang, S.C.; Fang, Z. Spin-orbit gap of graphene: First-principles calculations. *Phys. Rev. B Condens. Matter Mater. Phys.,* **2007**, *75*(4), 041401.
[http://dx.doi.org/10.1103/PhysRevB.75.041401]

[58]   Zhao, C.X.; Jia, J.F. Stanene: A good platform for topological insulator and topological superconductor. *Front. Phys.,* **2020**, *15*(5), 53201.
[http://dx.doi.org/10.1007/s11467-020-0965-5]

[59]   Sarma, S. Majorana zero modes and topological quantum computation. *npj Quantum Inf.,* **2015**, *1*(1), 1-13.
[http://dx.doi.org/10.1038/npjqi.2015.1]

[60]   Zhao, A.; Wang, B. Two-dimensional graphene-like Xenes as potential topological materials. *APL Mater.,* **2020**, *8*(3), 030701.
[http://dx.doi.org/10.1063/1.5135984]

[61]   Liao, M.; Zang, Y.; Guan, Z.; Li, H.; Gong, Y.; Zhu, K.; Hu, X.P.; Zhang, D.; Xu, Y.; Wang, Y.Y.; He, K. Superconductivity in few-layer stanene. *Nat. Phy.,* **2018**, *14*(4), 344-348.
[http://dx.doi.org/10.1038/s41567-017-0031-6]

[62]   Tian, Z.; Chen, K.; Sun, S.; Zhang, J.; Cui, W.; Xie, Z.; Liu, G. Crystalline boron nitride nanosheets by sonication-assisted hydrothermal exfoliation. *J. Adv. Ceram.,* **2019**, *8*(1), 72-78.
[http://dx.doi.org/10.1007/s40145-018-0293-1]

[63]   Ling, C.; Shi, L.; Ouyang, Y.; Wang, J. Searching for highly active catalysts for hydrogen evolution reaction based on O-terminated mxenes through a simple descriptor. *Chem. Mater.,* **2016**, *28*(24), 9026-9032.
[http://dx.doi.org/10.1021/acs.chemmater.6b03972]

[64]   Anasori, B.; Xie, Y.; Beidaghi, M.; Lu, J.; Hosler, B.C.; Hultman, L.; Kent, P.R.C.; Gogotsi, Y.; Barsoum, M.W. Two-dimensional, ordered, double transition metals carbides (MXenes). *ACS Nano,* **2015**, *9*(10), 9507-9516.
[http://dx.doi.org/10.1021/acsnano.5b03591] [PMID: 26208121]

[65]   Khazaei, M.; Ranjbar, A.; Ghorbani-Asl, M.; Arai, M.; Sasaki, T.; Liang, Y.; Yunoki, S. Nearly free electron states in MXenes. *Phys. Rev. B,* **2016**, *93*(20), 205125.
[http://dx.doi.org/10.1103/PhysRevB.93.205125]

[66]   Zhu, J.; Schwingenschlögl, U. P and Si functionalized MXenes for metal-ion battery applications. *2d Mater.,* **2017**, *4*(2), 025073.
[http://dx.doi.org/10.1088/2053-1583/aa69fe]

[67]   Junkaew, A.; Arróyave, R. Enhancement of the selectivity of MXenes (M_2C, M = Ti, V, Nb, Mo) *via* oxygen-functionalization: promising materials for gas-sensing and -separation. *Phys. Chem. Chem. Phys.,* **2018**, *20*(9), 6073-6082.
[http://dx.doi.org/10.1039/C7CP08622A] [PMID: 29457806]

[68]   Liu, Z.; Lin, H.; Zhao, M.; Dai, C.; Zhang, S.; Peng, W.; Chen, Y. 2D superparamagnetic tantalum carbide composite MXenes for efficient breast-cancer theranostics. *Theranostics,* **2018**, *8*(6), 1648-1664.
[http://dx.doi.org/10.7150/thno.23369] [PMID: 29556347]

[69]   Han, M.; Yin, X.; Wu, H.; Hou, Z.; Song, C.; Li, X.; Zhang, L.; Cheng, L. Ti_3C_2 MXenes with Modified Surface for High-Performance Electromagnetic Absorption and Shielding in the X-Band. *ACS Appl. Mater. Interfaces,* **2016**, *8*(32), 21011-21019.
[http://dx.doi.org/10.1021/acsami.6b06455] [PMID: 27454148]

[70]   Bandyopadhyay, A.; Ghosh, D.; Pati, S.K. Effects of point defects on the magnetoelectronic structures of MXenes from first principles. *Phys. Chem. Chem. Phys.,* **2018**, *20*(6), 4012-4019.
[http://dx.doi.org/10.1039/C7CP07165E] [PMID: 29350724]

[71]   Sim, E.S.; Yi, G.S.; Je, M.; Lee, Y.; Chung, Y.C. Understanding the anchoring behavior of titanium carbide-based MXenes depending on the functional group in Li S batteries: A density functional theory study. *J. Power Sources,* **2017**, *342*, 64-69.
[http://dx.doi.org/10.1016/j.jpowsour.2016.12.042]

[72]   Li, J.; Yan, D.; Hou, S.; Li, Y.; Lu, T.; Yao, Y.; Pan, L. Improved sodium-ion storage performance of Ti₃C₂Tₓ MXenes by sulfur doping. *J. Mater. Chem. A Mater. Energy Sustain.,* **2018**, *6*(3), 1234-1243.
[http://dx.doi.org/10.1039/C7TA08261D]

[73]   Wang, Z.; Zhang, Z.; Zhang, Y. MXenes-Au NPs modified electrochemical biosensor for multiple exosome surface proteins analysis. *Talanta,* **2023**, *265*.
[http://dx.doi.org/10.1016/j.talanta.2023.124848]

[74]   Wen, J.; Zhang, X.; Gao, H. Structural formation and charge storage mechanisms for intercalated two-dimensional carbides MXenes. *Phys. Chem. Chem. Phys.,* **2017**, *19*(14), 9509-9518.
[http://dx.doi.org/10.1039/C7CP00670E] [PMID: 28338131]

[75]   Si, C.; You, J.; Shi, W.; Zhou, J.; Sun, Z. Quantum spin Hall phase in Mo₂M₂C₃O₂ (M = Ti, Zr, Hf) MXenes. *J. Mater. Chem. C Mater. Opt. Electron. Devices,* **2016**, *4*(48), 11524-11529.
[http://dx.doi.org/10.1039/C6TC04560J]

[76]   Wu, Y.; Nie, P.; Wang, J.; Dou, H.; Zhang, X. Few-layer mxenes delaminated *via* high-energy mechanical milling for enhanced sodium-ion batteries performance. *ACS Appl. Mater. Interfaces,* **2017**, *9*(45), 39610-39617.
[http://dx.doi.org/10.1021/acsami.7b12155] [PMID: 29039906]

[77]   Handoko, A.D.; Fredrickson, K.D.; Anasori, B.; Convey, K.W.; Johnson, L.R.; Gogotsi, Y.; Vojvodic, A.; Seh, Z.W. Tuning the basal plane functionalization of two-dimensional metal carbides (MXenes) to control hydrogen evolution activity. *ACS Appl. Energy Mater.,* **2018**, *1*(1), 173-180.
[http://dx.doi.org/10.1021/acsaem.7b00054]

[78]   Zha, X.H.; Zhou, J.; Zhou, Y.; Huang, Q.; He, J.; Francisco, J.S.; Luo, K.; Du, S. Promising electron mobility and high thermal conductivity in Sc₂CT₂ (T = F, OH) MXenes. *Nanoscale,* **2016**, *8*(11), 6110-6117.
[http://dx.doi.org/10.1039/C5NR08639F] [PMID: 26932122]

[79]   Khazaei, M.; Ranjbar, A.; Esfarjani, K.; Bogdanovski, D.; Dronskowski, R.; Yunoki, S. Insights into exfoliation possibility of MAX phases to MXenes. *Phys. Chem. Chem. Phys.,* **2018**, *20*(13), 8579-8592.
[http://dx.doi.org/10.1039/C7CP08645H] [PMID: 29557432]

[80]   Zhang, C.J.; Pinilla, S.; McEvoy, N.; Cullen, C.P.; Anasori, B.; Long, E.; Park, S-H.; Seral-Ascaso, A.; Shmeliov, A.; Krishnan, D.; Morant, C.; Liu, X.; Duesberg, G.S.; Gogotsi, Y.; Nicolosi, V. Oxidation stability of colloidal two-dimensional titanium carbides (MXenes). *Chem. Mater.,* **2017**, *29*(11), 4848-4856.
[http://dx.doi.org/10.1021/acs.chemmater.7b00745]

[81]   Li, N.; Chen, X.; Ong, W.J.; MacFarlane, D.R.; Zhao, X.; Cheetham, A.K.; Sun, C. Understanding of electrochemical mechanisms for CO₂ capture and conversion into hydrocarbon fuels in transition-metal carbides (MXenes). *ACS Nano,* **2017**, *11*(11), 10825-10833.
[http://dx.doi.org/10.1021/acsnano.7b03738] [PMID: 28892617]

[82]   Yang, J.; Naguib, M.; Ghidiu, M.; Pan, L-M.; Gu, J.; Nanda, J.; Halim, J.; Gogotsi, Y.; Barsoum, M.W. Two-dimensional Nb-Based M₄C₃ solid solutions (MXenes). *J. Am. Ceram. Soc.,* **2016**, *99*(2), 660-666.
[http://dx.doi.org/10.1111/jace.13922]

[83]   Xin, Y.; Yu, Y.X. Possibility of bare and functionalized niobium carbide MXenes for electrode materials of supercapacitors and field emitters. *Mater. Des.,* **2017**, *130*, 512-520.
[http://dx.doi.org/10.1016/j.matdes.2017.05.052]

[84]   Yuan, W.; Cheng, L.; Wu, H.; Zhang, Y.; Lv, S.; Guo, X. One-step synthesis of 2D-layered carbon wrapped transition metal nitrides from transition metal carbides (MXenes) for supercapacitors with ultrahigh cycling stability. *Chem. Commun.,* **2018**, *54*(22), 2755-2758.
[http://dx.doi.org/10.1039/C7CC09017J] [PMID: 29479591]

[85]   Hou, L.; Xu, W.; Zhang, Q.; Shautsova, V.; Chen, J.; Shu, Y.; Li, X.; Bhaskaran, H.; Warner, J.H. Ultrathin lateral 2d photodetectors using transition-metal dichalcogenides PtSe$_2$-WS$_2$-PtSe$_2$ by direct laser patterning. *ACS Appl. Electron. Mater.,* **2022**, *4*(3), 1029-1038.
[http://dx.doi.org/10.1021/acsaelm.1c01194]

[86]   Kim, J. Y. Experimental and theoretical studies of native deep-level defects in transition metal dichalcogenides. *NPJ 2D Mater. Appl.,* **2022**, *6*(1), 75.
[http://dx.doi.org/10.1038/s41699-022-00350-4]

[87]   Zhang, Z.; Jiang, C.; Li, P.; Yao, K.; Zhao, Z.; Fan, J.; Li, H.; Wang, H. Benchmarking phases of ruthenium dichalcogenides for electrocatalysis of hydrogen evolution: Theoretical and experimental insights. *Small,* **2021**, *17*(13), 2007333.
[http://dx.doi.org/10.1002/smll.202007333] [PMID: 33590693]

[88]   Wang, Y.; Xie, Q.; Zhang, J.; Zheng, J.; Nai, J.; Liu, T.; Liu, Y.; Tao, X. Layered iron dichalcogenides with high ion mobility and capacity as promising anode materials for alkali metal-ion batteries: A first-principles study. *Comput. Mater. Sci.,* **2022**, *211*, 111523.
[http://dx.doi.org/10.1016/j.commatsci.2022.111523]

[89]   Li, Q.; Alfrey, A.; Hu, J.; Lydick, N.; Paik, E.; Liu, B.; Sun, H.; Lu, Y.; Wang, R.; Forrest, S.; Deng, H. Macroscopic transition metal dichalcogenides monolayers with uniformly high optical quality. *Nat. Commun.,* **2023**, *14*(1), 1837.
[http://dx.doi.org/10.1038/s41467-023-37500-1] [PMID: 37005420]

[90]   Chen, Y.; Zhao, Y.; Ou, P.; Song, J. Basal plane activation of two-dimensional transition metal dichalcogenides *via* alloying for the hydrogen evolution reaction: First-principles calculations and machine learning prediction. *J. Mater. Chem. A Mater. Energy Sustain.,* **2023**, *11*(18), 9964-9975.
[http://dx.doi.org/10.1039/D3TA01361H]

[91]   Ali, A.; Sarwar, S.; Pollard, D.R.; Wei, Z.; Wang, R.; Zhang, X.; Adamczyk, A.J. Systematic mapping of electrocatalytic descriptors for hybrid and non-hybrid molybdenum dichalcogenides with graphene support for cathodic hydrogen generation. *J. Phys. Chem. C,* **2022**, *126*(40), 17011-17024.
[http://dx.doi.org/10.1021/acs.jpcc.2c02116]

[92]   Padmajan Sasikala, S.; Singh, Y.; Bing, L.; Yun, T.; Koo, S.H.; Jung, Y.; Kim, S.O. Longitudinal unzipping of 2D transition metal dichalcogenides. *Nat. Commun.,* **2020**, *11*(1), 5032.
[http://dx.doi.org/10.1038/s41467-020-18810-0] [PMID: 33024113]

[93]   Brune, V.; Hegemann, C.; Wilhelm, M.; Ates, N.; Mathur, S. Molecular precursors to group IV dichalcogenides MS$_2$ (M=Ti, Zr, Hf). *Z. Anorg. Allg. Chem.,* **2022**, *648*(23), e202200049.
[http://dx.doi.org/10.1002/zaac.202200049]

[94]   Strange, L.E.; Garg, S.; Kung, P.; Ashaduzzaman, M.; Szulczewski, G.; Pan, S. Electrodeposited transition metal dichalcogenides for use in hydrogen evolution electrocatalysts. *J. Electrochem. Soc.,* **2022**, *169*(2), 026510.
[http://dx.doi.org/10.1149/1945-7111/ac4f25]

[95]   Sun, Y.; Terrones, M.; Schaak, R.E. Colloidal nanostructures of transition-metal dichalcogenides. *Acc. Chem. Res.,* **2021**, *54*(6), 1517-1527.
[http://dx.doi.org/10.1021/acs.accounts.1c00006] [PMID: 33662209]

[96]   Wu, Y.; Zhong, W.; Yang, Q.; Hao, C.; Li, Q.; Xu, M.; Bao, S. Flexible MXene-Ti3C2Tx bond few-layers transition metal dichalcogenides MoS2/C spheres for fast and stable sodium storage. *Chem. Eng. J.,* **2022**, *427*, 130960.
[http://dx.doi.org/10.1016/j.cej.2021.130960]

[97]   Chareev, D.A.; Khan, M.E.H.; Karmakar, D.; Nekrasov, A.N.; Nickolsky, M.S.; Eriksson, O.; Delin, A.; Vasiliev, A.N.; Abdel-Hafiez, M. Stable sulfuric vapor transport and liquid sulfur growth on transition metal dichalcogenides. *Cryst. Growth Des.,* **2023**, *23*(4), 2287-2294.
[http://dx.doi.org/10.1021/acs.cgd.2c01318] [PMID: 37038405]

[98] Purwitasari, W.; Villaos, R.A.B.; Verzola, I.M.R.; Sufyan, A.; Huang, Z-Q.; Hsu, C-H.; Chuang, F-C. High thermoelectric performance in 2D technetium dichalcogenides TcX$_2$ (X = S, Se, or Te). *ACS Appl. Energy Mater.,* **2022**, *5*(7), 8650-8657.
[http://dx.doi.org/10.1021/acsaem.2c01170]

[99] Das, S.R.; Wakabayashi, K.; Yamamoto, M.; Tsukagoshi, K.; Dutta, S. Layer-by-layer oxidation induced electronic properties in transition-metal dichalcogenides. *J. Phys. Chem. C,* **2018**, *122*(29), 17001-17007.
[http://dx.doi.org/10.1021/acs.jpcc.8b05857]

[100] Pulkin, A.; Yazyev, O.V. Controlling the quantum spin hall edge states in two-dimensional transition metal dichalcogenides. *J. Phys. Chem. Lett.,* **2020**, *11*(17), 6964-6969.
[http://dx.doi.org/10.1021/acs.jpclett.0c00859] [PMID: 32787191]

[101] Jeong, T.Y.; Kim, H.; Choi, S.J.; Watanabe, K.; Taniguchi, T.; Yee, K.J.; Kim, Y.S.; Jung, S. Spectroscopic studies of atomic defects and bandgap renormalization in semiconducting monolayer transition metal dichalcogenides. *Nat. Commun.,* **2019**, *10*(1), 3825.
[http://dx.doi.org/10.1038/s41467-019-11751-3] [PMID: 31444331]

[102] Long, C.; Liang, Y.; Jin, H.; Huang, B.; Dai, Y. PdSe$_2$: Flexible two-dimensional transition metal dichalcogenides monolayer for water splitting photocatalyst with extremely low recombination rate. *ACS Appl. Energy Mater.,* **2019**, *2*(1), 513-520.
[http://dx.doi.org/10.1021/acsaem.8b01521]

[103] Aras, M.; Kılıç, Ç.; Ciraci, S. Magnetic heterostructures of transition metal dichalcogenides: Antiparallel magnetic moments and half-metallic state. *J. Phys. Chem. C,* **2020**, *124*(42), 23352-23360.
[http://dx.doi.org/10.1021/acs.jpcc.0c06917]

[104] Zhang, Z.; Wang, J.; Song, C.; Mao, H.; Zhao, Q. Tuning band gaps of transition metal dichalcogenides WX$_2$ (X = S, Se) nanoribbons by external strain. *J. Nanosci. Nanotechnol.,* **2016**, *16*(8), 8090-8095.
[http://dx.doi.org/10.1166/jnn.2016.12798]

[105] Pandey, M.; Rasmussen, F.A.; Kuhar, K.; Olsen, T.; Jacobsen, K.W.; Thygesen, K.S. Defect-tolerant monolayer transition metal dichalcogenides. *Nano Lett.,* **2016**, *16*(4), 2234-2239.
[http://dx.doi.org/10.1021/acs.nanolett.5b04513] [PMID: 27027786]

[106] Synnatschke, K.; Cieslik, P.A.; Harvey, A.; Castellanos-Gomez, A.; Tian, T.; Shih, C-J.; Chernikov, A.; Santos, E.J.G.; Coleman, J.N.; Backes, C. Length- and thickness-dependent optical response of liquid-exfoliated transition metal dichalcogenides. *Chem. Mater.,* **2019**, *31*(24), 10049-10062.
[http://dx.doi.org/10.1021/acs.chemmater.9b02905]

[107] Luxa, J.; Wang, Y.; Sofer, Z.; Pumera, M. Layered post-transition-metal dichalcogenides (X-M-M-X) and their properties. *Chemistry,* **2016**, *22*(52), 18810-18816.
[http://dx.doi.org/10.1002/chem.201604168] [PMID: 27865023]

[108] Villaos, R. A. B. Thickness dependent electronic properties of Pt dichalcogenides. *NPJ 2D Mater. Appl.,* **2019**, *3*(1), 2.
[http://dx.doi.org/10.1038/s41699-018-0085-z]

[109] Sun, Y.; Wang, D.; Shuai, Z. Indirect-to-direct band gap crossover in few-layer transition metal dichalcogenides: A theoretical prediction. *J. Phys. Chem. C,* **2016**, *120*(38), 21866-21870.
[http://dx.doi.org/10.1021/acs.jpcc.6b08748]

[110] Wang, H.; Qin, G.; Li, G.; Wang, Q.; Hu, M. Unconventional thermal transport enhancement with large atom mass: A comparative study of 2D transition dichalcogenides. *2d Mater.,* **2017**, *5*(1), 015022.
[http://dx.doi.org/10.1088/2053-1583/aa9822]

[111] Hlova, I.Z.; Singh, P.; Malynych, S.Z.; Gamernyk, R.V.; Dolotko, O.; Pecharsky, V.K.; Johnson,

D.D.; Arroyave, R.; Pathak, A.K.; Balema, V.P. Incommensurate transition-metal dichalcogenides *via* mechanochemical reshuffling of binary precursors. *Nanoscale Adv.,* **2021**, *3*(14), 4065-4071.
[http://dx.doi.org/10.1039/D1NA00064K] [PMID: 36132842]

[112] Korkmaz, Y.A.; Bulutay, C.; Sevik, C. k · p Parametrization and linear and circular dichroism in strained Monolayer (Janus) transition metal dichalcogenides from first-principles. *J. Phys. Chem. C,* **2021**, *125*(13), 7439-7450.
[http://dx.doi.org/10.1021/acs.jpcc.1c00714]

[113] Memaran, S.; Pradhan, N.R.; Lu, Z.; Rhodes, D.; Ludwig, J.; Zhou, Q.; Ogunsolu, O.; Ajayan, P.M.; Smirnov, D.; Fernández-Domínguez, A.I.; García-Vidal, F.J.; Balicas, L. Pronounced photovoltaic response from multilayered transition-metal dichalcogenides PN-junctions. *Nano Lett.,* **2015**, *15*(11), 7532-7538.
[http://dx.doi.org/10.1021/acs.nanolett.5b03265] [PMID: 26513598]

[114] Liu, H.; Wang, T.; Wang, C.; Liu, D.; Luo, J. Exciton radiative recombination dynamics and nonradiative energy transfer in two-dimensional transition-metal dichalcogenides. *J. Phys. Chem. C,* **2019**, *123*(15), 10087-10093.
[http://dx.doi.org/10.1021/acs.jpcc.8b12179]

[115] Kang, M.; Kim, B.; Ryu, S.H.; Jung, S.W.; Kim, J.; Moreschini, L.; Jozwiak, C.; Rotenberg, E.; Bostwick, A.; Kim, K.S. Universal mechanism of band-gap engineering in transition-metal dichalcogenides. *Nano Lett.,* **2017**, *17*(3), 1610-1615.
[http://dx.doi.org/10.1021/acs.nanolett.6b04775] [PMID: 28118710]

[116] Hanbicki, A.T.; Currie, M.; Kioseoglou, G.; Friedman, A.L.; Jonker, B.T. Measurement of high exciton binding energy in the monolayer transition-metal dichalcogenides WS2 and WSe2. *Solid State Commun.,* **2015**, *203*, 16-20.
[http://dx.doi.org/10.1016/j.ssc.2014.11.005]

SUBJECT INDEX

A

Acids 23, 24, 25, 26, 59, 62, 103, 108, 130, 134, 152
  boric 130
  folic 25, 26
  fulvic 24
  hydriodic 108
  nucleic 23, 59
  sulfuric 103, 134
Actin-dependent macro-pinocytosis 70
Activity, smooth electrochemical 153
Acute toxicity 62
Adenosine triphosphate 71
Adsorption 7, 23, 24, 38, 56, 68, 70, 177, 178, 179, 181, 197, 198, 212
  electrochemical 177
  electrostatic 7
  metal ion 24
  organic molecule 24
  protein 68
  skin 56
  technique 23
Advanced manufacturing techniques 223
Aerogels 1, 7, 8, 9, 10, 108
  graphene-based 1, 7
  hybrid 8
Alkali metal atom 198
Allotropic form 165
Amino acids 67, 125
Ammonia borane process 176
Angle-resolved photoelectron spectroscopy (ARPES) 195, 199
Annealing, thermal 128
Anodic electrodes, traditional 130
Antibacterial agent 27, 28
Antigen-presenting cells (APCs) 60
Antimicrobial 28, 156, 157, 174, 179
  agents 174
  coating 157
  efficacy 179
  inhibition 157

resistance 28
Antioxidant enzymes 61
Apoptosis 57, 69, 70, 71
  mitochondrial-induced 71
Applications 26, 27, 35, 118, 119, 134, 135, 136, 157, 178, 181, 204, 218, 219, 220, 225
  biosensing 26
  cancer-related 181
  electrical 219
  energy-related 118, 119, 134, 136, 218, 220
  hybrid materials 27
  nanoelectronic 204, 225
  of boron nitride in bioengineering 157
  sensing 178, 181
  solar cell 135
  transportation 35
ARPES spectroscopy 200
Atherosclerosis 61
Atomic force microscopy (AFM) 216

B

Bacterial respiration 28
Behavior 88, 103, 133, 202
  electrical 202
  electrochemical 88, 133
  pseudo-capacitive 133
Bio-corrosion prevention 157
Bioenergy-related reactions 117
Biomass waste 88
Biomedical applications 118, 146, 155, 179, 180
Biosensor 25, 101
  application 25
  transducers 101
Boltzmann transport equation 220
Borazine concentration 128
Boron, heating 147
Bovine serum albumin (BSA) 68
Brain malignancies 62
Breast cancer theranostics 226

www.ingramcontent.com/pod-product-compliance
Lightning Source LLC
Chambersburg PA
CBHW050822220326
41598CB00006B/287